Methods in Molecular Biology

For further volumes:
http://www.springer.com/series/7651

For over 35 years, biological scientists have come to rely on the research protocols and methodologies in the critically acclaimed *Methods in Molecular Biology* series. The series was the first to introduce the step-by-step protocols approach that has become the standard in all biomedical protocol publishing. Each protocol is provided in readily-reproducible step-by-step fashion, opening with an introductory overview, a list of the materials and reagents needed to complete the experiment, and followed by a detailed procedure that is supported with a helpful notes section offering tips and tricks of the trade as well as troubleshooting advice. These hallmark features were introduced by series editor Dr. John Walker and constitute the key ingredient in each and every volume of the *Methods in Molecular Biology* series. Tested and trusted, comprehensive and reliable, all protocols from the series are indexed in PubMed.

Viruses as Therapeutics

Methods and Protocols

Edited by

Alexandra R. Lucas

The Biodesign Institute, Arizona State University, Tempe, AZ, USA

Humana Press

Editor
Alexandra R. Lucas
The Biodesign Institute
Arizona State University
Tempe, AZ, USA

ISSN 1064-3745 ISSN 1940-6029 (electronic)
Methods in Molecular Biology
ISBN 978-1-0716-1011-4 ISBN 978-1-0716-1012-1 (eBook)
https://doi.org/10.1007/978-1-0716-1012-1

This Humana imprint is published by the registered company Springer Science+Business Media, LLC, part of Springer Nature.
The registered company address is: 1 New York Plaza, New York, NY 10004, U.S.A.

Preface

Viruses as Therapeutics: Fighting Fire with Fire

New world
New lead
New light
Invisible power
Life and death
Evolution's weapon
Disease and cure
Hidden nature
Stealth in life
Amazing and rare
The virus
Alexandra R. Lucas, MD

Viruses are incredibly profuse and also diverse "life-forms," affecting all the kingdoms of life. Viruses do cause disease, but some viruses also regulate other pathogens, as with bacteriophage that infects bacteria. Viruses help to maintain the health of the entire ecosystem. Viruses have long been used as therapeutics; in the initial work using viruses as therapeutics, viruses were developed as vaccines both via natural exposure to infections and through man-made vaccines, vaccines that have been developed now for several hundred years. Viruses are also used as therapeutics providing gene expression vehicles as engineered delivery systems, and now more recently viruses have been exploited as cancer-seeking oncolytic agents. My research group has worked with virus-derived immune modulators for over 25 years, exploring their potential as a new class of therapeutics to treat immune-based disorders.

While we have studied mechanisms of viruses in infectious diseases, as well as now using viruses as highly effective sources for new therapeutics, we do not understand the exact origins of viruses. The origins of viruses are considered unknown, and have been attributed to sources as diverse as meteorites and cosmic visitors from other worlds (panspermia), as well as via abiosis or even as cellular fragments, such as transposons [1–7]. It is not known whether unicellular organisms predate viruses, or *vice versa*, or whether there is true development in parallel. Evidence for the existence of viruses has been found from the Precambrian era or earlier, and they have probably been coexistent with the very origins of life as we now know it. We do know that viruses have coevolved with all kingdoms in life including archaea, bacteria, and eukaryotes. Viruses differ extensively from all other known life-forms, sharing some genes in common with cellular life but also having many genetic sequences that differ completely from all known prokaryotic or eukaryotic genetic sequences. There exists the potential that the coevolution of eukaryotes and viruses began with common genetic elements, or conversely that viruses originated independently and have subsequently donated genes to mammalian cell organelles or have functioned as genetic vectors for transfer from cell to cell (*e.g.*, representing genetic fragments from bacteria and eukaryotes, plant and animal).

The first virus discovered was a plant virus, tobacco mosaic virus (TMV), and was found in the contagious "slimy fluid" left after filtering out bacteria (*contagium vivum fluidum*). However they are analyzed, viruses do not fit neatly within the general genetic pathways discovered that are conserved within archaea, bacteria, and eukaryotes, whether plant or animal. Viruses are obligate intracellular parasites and are a class apart, lacking nuclei and requiring host cells in which to proliferate. It could be argued that only virus-infected cells are "alive," whereas extracellular virus particles are nonliving inert vehicles, but the distinctions are probably semantic.

Viruses have been calculated as constituting up to at least 10^{31} particles in the biosphere, far exceeding the genetic signatures of all other organisms. Viruses have an extensive, far-reaching presence, and are found to be closely associated with all archaea, bacteria, plants, and animals. While often identified as causing disease states and widespread pandemics, *i.e.*, as pathogens that cause disease and loss of life with loss of crops, algae, insects, and animals, it has now become evident that viruses are also closely entwined with all life on earth. Indeed, in addition to causing death and disease, viruses play beneficial roles coexisting at all levels of life, ranging from the deep ocean to all plants and animals on the earth, and are central to maintaining a normal ecology and in mammals maintaining a healthy homeostasis, helping to maintain the microbiome.

Relevant to this methods book, viruses have coevolved very closely with man and mammals [1–7]. As noted, viruses are often feared as hidden killers, unseen foes that are difficult to identify, track, and cure. As with the recent coronavirus SARS-CoV2 pandemic (COVID 19), and also prior influenza, SARS, MERS, or even Ebola epidemics, viruses are often feared, in part because they are perceived as an unseen menace and treatments have remained limited. We now know that many viruses maintain a normal balance in the human body, for example in the microbiome of the gut, skin, lungs, pharynx, or even eye. Viruses are purported to function as regulators of many kingdoms of life, ranging from regulation of invasive plant species to phages that may modify the mammalian bacterial microbiome. Associations between differing viruses and their targets (archaea, bacteria, plants, fungi, and mammals) may be due to ancient mutually beneficial interactions. Thus, while viruses are often feared as hidden killers, in truth, they have many central roles in maintaining a healthy ecosystem. In addition to maintaining a healthy ecosystem, viruses have now been identified as a rich source for new therapeutics.

Here, we discuss methods for the development and study of viruses as therapeutic agents. We begin with the uses of viruses as methods for environmental control of overgrowth and invasion of exogenous or foreign species, *e.g.*, virus to control invasive, imported European rabbits in Australia (Chapter 1). We discuss two methods for developing viruses as vaccines in plants and salmonella (Chapters 2 and 3) and as live virus for cancer oncolysis (Chapter 4), or as viral vectors designed to express genes with therapeutic potential (Chapter 5). In this coevolution of viruses with mammals, viruses, which are considered cell-dependent symbionts or parasites, have independently developed unique and potent mechanisms designed to subvert or evade the host immune responses intended to target and remove viral infections. Thus, methods to study virus-derived immune-modulating proteins that function as highly effective blocking agents for key steps in activated, infected host immune pathways are presented here [8] (Chapter 6). We discuss unique and highly potent immune-modulating proteins (Chapter 6), as well as active peptide metabolites from these proteins [8, 9] (Chapter 7).

As therapeutics, whether as vaccines or genetic carriers or factories that develop new immune-modulating reagents, viruses have proven extraordinarily potent, often targeting pathways we have only begun to recognize and study in science. Thus, we are now actively exploring methods to develop viral oncolytics as agents to kill invasive cancers, as well as virus-derived immune-modulating proteins as a new class of biologic therapeutics.

In this methods book, we present current techniques and approaches used to develop new treatments derived from viruses, ranging from virus-derived vaccines to vectors and proteins. Newer techniques and methods that have been developed to investigate the molecular targets and mechanism of action of both virus infections and virus-derived therapeutics are described from crystallography (Chapter 8), microscopy (Chapter 9), metabolomics (Chapter 10), and analysis of necroptosis (Chapter 11). Further, a range of models and methods designed to measure therapeutic efficacy of virus-derived biologics and newer virus-derived therapeutic proteins are provided [8, 9] (Chapters 11–16).

In conclusion, viruses have long been thought of as hidden killers, pathogens, or invisible foes causing disease and death, but their role in nature is now better understood. Viruses are now known to also provide a beneficial force for balancing the ecosystem and all levels of life as well as the mammalian microbiota. Virus-derived immune modulators are in development as new therapeutics and new directions for treating severe and even incurable diseases including other severe viral infections, and as a new source for cancer-targeting oncolytics [9]. We have in fact demonstrated that a gamma herpesviral infection, MHV68, is suppressed by a poxvirus-derived myxoma virus protein, Serp-1. MHV68 infection and its treatment are now proven to be dependent upon an intact gut bacterial microbiota [9].

Thus, with the use of viral agents as therapeutics, we now are "fighting fire with fire."

Tempe, AZ, USA ***Alexandra R. Lucas***

References

1. Godoy-Vitorino F. (2019) Human microbial ecology and the rising new medicine. Ann Transl Med 7:342. https://doi.org/10.21037/atm.2019.06.56
2. Lefeuvre P, Martin DP, Santiago F. Elena SF, et al (2019) Evolution and ecology of plant viruses. Nat Rev Microbiol 17: 633
3. Banjara S, Suraweera CD, Hinds MG et al. (2020) The Bcl-2 family: ancient origins, conserved structures, and divergent mechanisms. Biomolecules 10: 128. https://doi.org/10.3390/biom10010128
4. Mustafin RN (2018) Hypothesis on the origin of viruses from transposons. Mol Genet Microbiol Virol 33: 223–232
5. Prangishvili D, Bamford DH, Forterre P, et al. (2017) The enigmatic archaeal virosphere. Nat Rev Microbiol 15: 724–739.
6. Heaton SM (2019) Harnessing host–virus evolution in antiviral therapy and Immunotherapy. Clin Transl Immunol 8: e1067. www.wileyonlinelibrary.com/journal/cti. https://doi.org/10.1002/cti2.1067.
7. Steele EJ, Al-Mufti S , Kenneth A. Augustyn KA et al (2018) Cause of cambrian explosion - Terrestrial or cosmic? Progr Biophys Mol Biol 136: 3e23
8. Yaron JR, Zhang L, Guo Q, et al (2020) Deriving immune modulating drugs from viruses - A new class of biologics. J Clin Med 9:972.
9. Yaron JR, Ambadapadi S, Zhang L, et al (2020). Immune protection is dependent on the gut microbiome in a lethal mouse gamma-herpesviral infection. Sci Rep 10:2371. 10.1038/s41598-020-59269-9; SREP-19-31172

Acknowledgments

I would like to thank all of the scientists and students in my research group who contributed to the work reported in the book, in particular Liqiang Zhang, PhD; Jordan R. Yaron, PhD; Qiuyun Gun, MD, PhD; Michelle Burgin, BSc; Lauren Schutz; Enkidio Awo; Ayman Fath, MD; Michael Juby, MD; and Grant McFadden, PhD. Without their contributions and dedicated work, this book would not have been possible.

Contents

Contributors

SRIRAM AMBADAPADI • *Center for Personalized Diagnostics, Biodesign Institute, Arizona State University, Tempe, AZ, USA; Department of Molecular Genetics and Microbiology, University of Florida, Gainesville, FL, USA*

ENKIDIA A. AWO • *Centers for Personalized Diagnostics and for Immunotherapy, Vaccines and Virotherapy, Biodesign Institute, Arizona State University, Tempe, AZ, USA*

MICHELLE BURGIN • *Centers for Personalized Diagnostics and for Immunotherapy, Vaccines and Virotherapy, Biodesign Institute, Arizona State University, Tempe, AZ, USA*

HAO CHEN • *The Department of Tumor Surgery, Second Hospital of Lanzhou University, Lanzhou, China*

QIANG CHEN • *The Biodesign Institute and School of Life Sciences, Arizona State University, Tempe, AZ, USA*

SAMANTHA M. COTSMIRE • *Center for Immunotherapy, Vaccines and Virotherapy, Biodesign Institute, Arizona State University, Tempe, AZ, USA*

JULIANE DAGGETT • *DACT, Biodesign Institute, Ariizona State University, Tempe, AZ, USA*

KATHLEEN H. DELANEY • *Department of Pathology and Molecular Medicine, McMaster University, Hamilton, ON, Canada*

ANA LEMOS DE MATOS • *Center for Immunotherapy, Vaccines and Virotherapy, The Biodesign Institute, Arizona State University, Tempe, AZ, USA*

ADRIAN ESQUEDA • *The Biodesign Institute and School of Life Sciences, Arizona State University, Tempe, AZ, USA*

PETRA FROMME • *Center for Applied Structural Discovery, The Biodesign Institute, Arizona State University, Tempe, AZ, USA*

CHRISTOPHER GISRIEL • *Department of Chemistry, Yale University, New Haven, CT, USA*

HONOR L. GLENN • *Center for Immunotherapy, Vaccines and Virotherapy, Biodesign Institute, Arizona State University, Tempe, AZ, USA*

HAIWEI GU • *Arizona Metabolomics Laboratory, College of Health Solutions, Arizona State University, Scottsdale, AZ, USA*

QIUYUN GUO • *Centers for Personalized Diagnostics and for Immunotherapy, Vaccines and Virotherapy, Biodesign Institute, Arizona State University, Tempe, AZ, USA; Department of Oncology, Tongji Hospital, Tongji Medical College, Huazhong University of Science and Technology, Wuhan, China*

ROBYN N. HALL • *Black Mountain Laboratories, CSIRO Health & Biosecurity, Acton, ACT, Australia*

CRISTHIAN J. ILDEFONSO • *Department of Ophthalmology, University of Florida College of Medicine, Gainesville, FL, USA*

BERTRAM L. JACOBS • *School of Life Sciences Center for Immunotherapy, Vaccines and Virotherapy, Biodesign Institute, Arizona State University, Tempe, AZ, USA*

PANIZ JASBI • *Arizona Metabolomics Laboratory, College of Health Solutions, Arizona State University, Scottsdale, AZ, USA*

NICOLA C. JONES • *Department of Pharmacology and Toxicology, University of Otago, Dunedin, New Zealand*

SHAHAR KEINAN • *Polaris Quantum Biotech, Durham, NC, USA*

PETER J. KERR • *Black Mountain Laboratories, CSIRO Health & Biosecurity, Acton, ACT, Australia*

JACQUELYN KILBOURNE • *Biodesign Institute, Arizona State University, Tempe, AZ, USA*

WEI KONG • *Center for Immunotherapy, Vaccines and Virotherapy, The Biodesign Institute, Arizona State University, Tempe, AZ, USA*

JACEK M. KWIECIEN • *Department of Pathology and Molecular Medicine, McMaster University, Hamilton, ON, Canada*

KENNETH M. LOWE • *Biodesign Institute, Arizona State University, Tempe, AZ, USA*

ALEXANDRA R. LUCAS • *Center for Personalized Diagnostics, Biodesign Institute, Arizona State University, Tempe, AZ, USA; Center for Immunotherapy, Vaccines and Virotherapy, Biodesign Institute, Arizona State University, Tempe, AZ, USA; Department of Molecular Genetics and Microbiology, University of Florida, Gainesville, FL, USA; St Joseph Hospital, Dignity Health, Creighton University, Phoenix, AZ, USA; Division of Cardiology, Saint Joseph's Hospital, Dignity Health, Phoenix, AZ, USA*

JOSE M. MARTIN-GARCIA • *Center for Applied Structural Discovery, The Biodesign Institute, Arizona State University, Tempe, AZ, USA*

GRANT MCFADDEN • *Center for Immunotherapy, Vaccines and Virotherapy, The Biodesign Institute, Arizona State University, Tempe, AZ, USA; Department of Molecular Genetics and Microbiology, University of Florida, Gainesville, FL, USA*

JEFFREY PATTERSON • *Arizona Metabolomics Laboratory, College of Health Solutions, Arizona State University, Scottsdale, AZ, USA*

MASMUDUR M. RAHMAN • *Center for Immunotherapy, Vaccines and Virotherapy, The Biodesign Institute, Arizona State University, Tempe, AZ, USA*

RAELA B. RIDLEY • *Department of Ophthalmology, University of Florida College of Medicine, Gainesville, FL, USA*

LAUREN N. SCHUTZ • *Centers for Personalized Diagnostics and for Immunotherapy, Vaccines and Virotherapy, Biodesign Institute, Arizona State University, Tempe, AZ, USA*

XIAOJIAN SHI • *Arizona Metabolomics Laboratory, College of Health Solutions, Arizona State University, Scottsdale, AZ, USA*

TANJA STRIVE • *Black Mountain Laboratories, CSIRO Health & Biosecurity, Acton, ACT, Australia*

GABRIELLA S. STUART • *Department of Pharmacology and Toxicology, University of Otago, Dunedin, New Zealand*

MATEUSZ SZCZERBA • *Center for Immunotherapy, Vaccines and Virotherapy, Biodesign Institute, Arizona State University, Tempe, AZ, USA*

LINO E. TORRES-DOMÍNGUEZ • *Center for Immunotherapy, Vaccines and Virotherapy, The Biodesign Institute, Arizona State University, Tempe, AZ, USA*

NANCY VILLA • *Center for Immunotherapy, Vaccines and Virotherapy, Biodesign Institute, Arizona State University, Tempe, AZ, USA*

SARAH E. WALLACE • *Centers for Personalized Diagnostics and for Immunotherapy, Vaccines and Virotherapy, Biodesign Institute, Arizona State University, Tempe, AZ, USA*

ERIN M. WALSH • *Department of Ophthalmology, University of Florida College of Medicine, Gainesville, FL, USA*

LYN M. WISE • *Department of Pharmacology and Toxicology, University of Otago, Dunedin, New Zealand*

JORDAN R. YARON • *Centers for Personalized Diagnostics and for Immunotherapy, Vaccines and Virotherapy, Biodesign Institute, Arizona State University, Tempe, AZ, USA*

LIQIANG ZHANG • *Centers for Personalized Diagnostics and for Immunotherapy, Vaccines and Virotherapy, Biodesign Institute, Arizona State University, Tempe, AZ, USA*

Chapter 1

Viruses for Landscape-Scale Therapy: Biological Control of Rabbits in Australia

Peter J. Kerr, Robyn N. Hall, and Tanja Strive

Abstract

Viral diseases, whether of animals or humans, are normally considered as problems to be managed. However, in Australia, two viruses have been used as *landscape-scale* therapeutics to control European rabbits (*Oryctolagus cuniculus*), the preeminent invasive vertebrate pest species. Rabbits have caused major environmental and agricultural losses and contributed to extinction of native species. It was not until the introduction of Myxoma virus that effective control of this pest was obtained at a continental scale. Subsequent coevolution of rabbit and virus saw a gradual reduction in the effectiveness of biological control that was partially ameliorated by the introduction of the European rabbit flea to act as an additional vector for the virus. In 1995, a completely different virus, Rabbit hemorrhagic disease virus (RHDV), escaped from testing and spread through the Australian rabbit population and again significantly reduced rabbit numbers and environmental impacts. The evolutionary pressures on this virus appear to be producing quite different outcomes to those that occurred with myxoma virus and the emergence and invasion of a novel genotype of RHDV in 2014 have further augmented control. Molecular studies on myxoma virus have demonstrated multiple proteins that manipulate the host innate and adaptive immune response; however the molecular basis of virus attenuation and reversion to virulence are not yet understood.

Key words Myxoma virus, Rabbit hemorrhagic disease virus, Biological control, European rabbit, Myxomatosis, Coevolution

1 Background to Biological Control

The European rabbit (*Oryctolagus cuniculus*) is Australia's major vertebrate pest. In the Australian summer of 1950, a continent-wide experiment in evolution was unwittingly initiated when a poxvirus, Myxoma virus (MYXV), the cause of myxomatosis, spread from experimental field sites where it was being evaluated as a biological control for European rabbits. The coevolution of MYXV and rabbit has become a paradigm of host-pathogen evolution [1]. Some 45 years later, history repeated, or at least rhymed, when a calicivirus, Rabbit hemorrhagic disease virus (RHDV), escaped from an offshore test site onto the Australian mainland. In both cases, the pathogen spread rapidly causing crashes in the

Alexandra R. Lucas (ed.), *Viruses as Therapeutics: Methods and Protocols*, Methods in Molecular Biology, vol. 2225, https://doi.org/10.1007/978-1-0716-1012-1_1,

rabbit population, with case fatality rates in excess of 90%. The two viruses have provided lasting landscape control of the European rabbit. However, initial impacts have not been sustained with recovery of rabbit populations, albeit at lower levels. Evolution of these two very different viruses within the same host species illustrates how different transmission dynamics may drive pathogen virulence, while emergence of resistance in the rabbit population exerts selection pressure on the viruses in an ongoing biological arms race [1–3]. There is only one other example of deliberate use of a virus for control of a vertebrate species: feral cats (*Felis catus*) on Marion Island were partially controlled by the introduction of feline panleukopenia virus [4]. In addition, cyprinid herpesvirus-3 is under evaluation as a biological control for invasive European carp (*Cyprinus carpio*) in Australia [5].

2 The European Rabbit in Australia

The European rabbit was repeatedly introduced into Australia from the time of the first European settlement in 1788. These were domestic breeds and did not spread far beyond settled areas. It was the introduction of a small number of wild rabbits, imported from Britain in 1859, that is credited with initiating the uncontrolled spread of rabbits across the continent, although it is likely that other introductions occurred [6–8]. Rabbits were described as a plague or a grey blanket covering the country with paddocks appearing to move due to the mass of rabbits [6]. Rabbits compete with livestock and native herbivores for grazing, eliminate plant species and prevent recruitment and regeneration, encourage soil erosion due to plant cover loss, and introduce and spread weeds that thrive in the disturbed soil of warrens (reviewed in [9]). Furthermore, rabbits support large populations of introduced predators: European red fox (*Vulpes vulpes*) and feral cats, which also prey on native species and are blamed for causing multiple extinctions [10–12]. Conventional lethal control measures such as shooting, trapping, and poisoning and physical controls such as rabbit-proof fencing and habitat destruction were either ineffective or only useful at a local scale and could not be applied in Australia's vast, sparsely populated rangelands where rabbit populations boom and crash driven by rainfall and food availability.

3 Viruses

3.1 Myxoma Virus and Myxomatosis

Myxomatosis was first described as a novel, lethal disease in laboratory rabbits at the Institute of Hygiene in Montevideo, Uruguay [13]. All the breeds of domestic rabbits used for meat, fur, and pets and in laboratories are derived from European rabbits, which originally evolved in the Iberian Peninsula. These are not native to the Americas although many native American lagomorphs such as cottontails (*Sylvilagus* spp.) and hares (*Lepus* spp.) are called rabbits. It was subsequently demonstrated that the natural host of MYXV is probably the South American tapeti (*Sylvilagus menensis*; also called *S. brasiliensis*, jungle or forest rabbit) [14]. In *S. menensis*, inoculation of MYXV induces a local cutaneous fibroma that persists for some weeks before it is cleared by the host immune response. Closely related viruses circulate in North America in the eastern cottontail (*S. floridanus*) and the brush rabbit (*S. bachmani*). The viruses are passively transmitted by adhering to the mouthparts of biting arthropods such as fleas and mosquitoes when they probe through the virus-rich epidermis overlying the fibroma; there is no replication in the vector [15, 16].

In European rabbits, MYXV causes a lethal, generalized disease characterized by cutaneous skin swellings; swollen, closed eyelids; swollen head, lips, nose, and ears; and severe swelling of the ano-genital region. Mucopurulent discharge from eyelids and nostrils is common in the terminal stages (Fig. 1). Death typically occurs 8-14 days after initial infection, with case fatality rates over 99% in susceptible rabbits infected with virulent virus. Transmission is mainly by fleas and mosquitoes probing through virus-rich lesions in skin, base of ears, or eyelids but can also occur by nasal or conjunctival transfer of virus in discharges or by fighting; oral infection is inefficient (Fig. 2). An amyxomatous form of the disease that lacks the large cutaneous lesions has also emerged [17, 18]. Some strains of virus cause an accelerated fatal course in

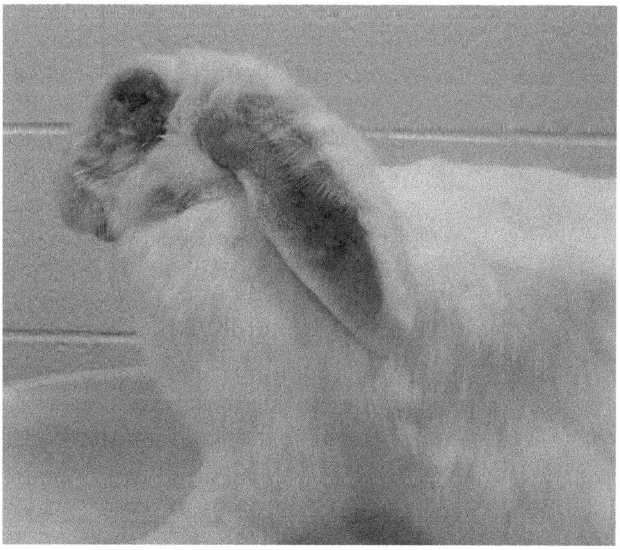

Fig. 1 Laboratory rabbit infected with the Lausanne strain of MYXV. (Reproduced from [20] under creative commons license)

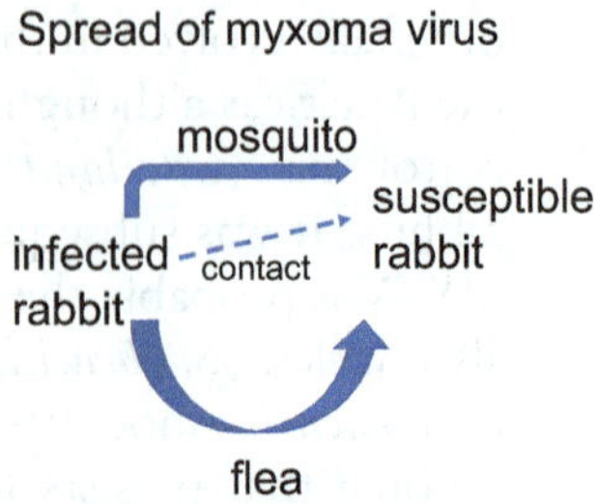

Fig. 2 Transmission of MYXV in European rabbits. Virus adheres to the mouthparts of mosquitoes or fleas probing through virus-rich epidermis and is passively transferred into the epidermis/dermis of susceptible rabbits when the vector attempts to feed. Rabbit fleas move freely between rabbits. Virus can also be directly transmitted during social interactions from conjunctival or nasal discharges

domestic rabbits with few of the clinical signs typically associated with myxomatosis [18–20].

MYXV is a poxvirus, the type species of the *Leporipoxvirus* genus. It is a large, complex virus that like other poxviruses replicates in membrane-bound factories within the cytoplasm of infected cells [21]. The virion is cuboid to oval in shape approximately 300 × 250 × 75 nm and consists of a nucleosome core containing the genome and all enzymes and transcription factors required for early mRNA transcription. The core is enclosed within a lipoprotein membrane together with two lateral bodies of unknown function. This form of the virus, termed the intracellular mature virion (MV), is infectious and released by cell lysis. A second infectious form of the virus, EV or enveloped virus, which has a lipid envelope derived from internal cell membranes with virus-encoded surface proteins, is released from the cell by budding through the plasma membrane [21, 22]. The Lausanne strain reference genome consists of 161.8 kb of double-strand DNA encoding 158 genes, 12 of which are duplicated within terminal inverted repeats at either end of the genome [23, 24].

MYXV has not just been a paradigm for host-pathogen coevolution; it has also been an outstanding model for elucidating how viruses manipulate the host immune response and understanding host antiviral pathways (reviewed in [25]). MYXV encodes multiple proteins involved in suppressing the host innate and adaptive antiviral immune responses. In its natural host, these enable viral persistence in the fibroma but in European rabbits, the virus is able to disseminate throughout the body and overwhelm the immune system [26]. Viral immunomodulators may be secreted from cells or may act within infected cells and include proteins that inhibit types I and II interferon, tumor necrosis factor, NF-κB, and inflammasome formation; downregulate MHC-I expression; and bind chemokines. Serine proteinase inhibitors disrupt proteinase cascades involved in coagulation and inflammation. Many proteins are involved in maintaining cell viability by inhibiting cell death pathways and a secreted epidermal growth factor homolog is important for virulence [25, 27].

While many of these proteins act largely on rabbit pathways some are active across species and may have potential as novel therapeutics. The most advanced example is a serine proteinase inhibitor (serpin), Serp-1, secreted from MYXV-infected cells. Serp-1 inhibits proteases involved in blood clotting, thrombolysis, and inflammation. It has shown efficacy in animal models of vascular disease and other inflammatory conditions, for example [28], and has progressed to phase II human trials [29]. In addition, peptides derived from the functional loop region of Serp-1 have demonstrated potential as therapeutics in animal studies [30, 31]. A second serpin (Serp-2), which acts within infected cells to inhibit the serine protease granzyme B and the cysteine proteases caspase 1 and caspase 8, has also shown potential as an anti-inflammatory mediator in animal models, for example [32, 33]. A secreted type II interferon receptor homolog, MT-7, is able to bind chemokines and inhibit inflammation in addition to its interferon-binding activity. It has shown promising anti-inflammatory activity in animal studies, for example [34–36]. M013, a pyrin domain protein, is an inhibitor of inflammasome formation and NF-κB activation [37] and has shown activity against inflammation in an ocular gene transfer model [38]. In addition, MT-5, that functions as an adaptor in E3 ubiquitin ligase complexes and prevents cell cycle arrest [39, 40], has been used to design peptides with *in vitro* antitumor activity [41, 42].

Perhaps the most intriguing outcome of the molecular studies on MYXV is the development of MYXV as an oncolytic agent for the treatment of human cancer (reviewed in [43]). Although MYXV is unable to replicate in humans, many human cancer cell types are permissive for replication either because these cells have lost the innate immune system pathways that normally prevent MYXV replication or because the serine/threonine protein kinase Akt has been upregulated [43]. This means that in humans, only cancer cells would be permissive for infection. No human trials have commenced, but efficacy has been demonstrated in many animal models. Of particular interest is the ability of the virus to specifically kill residual tumor cells ex vivo in autologous bone marrow grafts and to inhibit graft vs. host disease in allogeneic bone marrow transplants treated ex vivo (reviewed in [44]).

3.2 Rabbit Hemorrhagic Disease Virus and Rabbit Hemorrhagic Disease

The first outbreak of rabbit hemorrhagic disease (RHD) occurred in China in 1984 in domestic rabbits recently imported from Germany. This novel, fatal disease was eventually shown to be caused by a calicivirus, subsequently termed Rabbit hemorrhagic disease virus (RHDV). The virus spread in farmed rabbits in China and Korea and in 1986 emerged in farmed rabbits in Italy and then other parts of Europe where it also spread into wild rabbits [45]. A variant virus, termed RHDVa, emerged in Europe in 1996 and rapidly spread suggesting higher fitness than RHDV [46, 47]. The geographic origin of RHDV is not clear but it most likely evolved from an apathogenic, enteric rabbit calicivirus (RCV) (see below) [48]. A second pathogenic virus, here termed RHDV2 (the original virus will be termed RHDV1), emerged in France in 2010 and has likely evolved independently to RHDV1 from another apathogenic RCV [49].

RHD is characterized by fulminant hepatitis with collapse and death 36–72 h after infection often with minimal clinical signs. Infection is typically oral, but the virus is infectious by most routes of inoculation. Case fatality rates due to RHD may be as high as 90–95%. Virus is shed in the feces and most secretions; carcasses may retain infectivity for many months depending on the ambient temperature [50]. Flies feeding on carcasses are major mechanical vectors with infectious virus regurgitated or passed through the insect gut to be shed on pasture or directly onto rabbits [51, 52] (Fig. 3). Because of its resistance to environmental degradation the virus is readily spread by trade in rabbits and rabbit products. Unusually, RHDV1 does not cause disease in rabbits younger than 4–5 weeks of age although infection occurs, and virus is shed. This age-related resistance is gradually lost with almost all rabbits susceptible to lethal disease by 12 weeks of age [45]. The rapid, massive hepatic necrosis induced by RHDV has led to its use as an animal model for fulminant hepatitis [53–55].

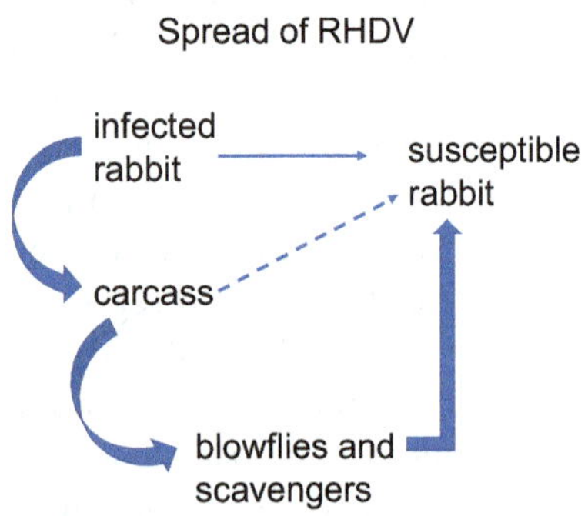

Fig. 3 Transmission of RHDV in rabbits. Infected rabbits shed RHDV into the environment in feces and other secretions. The virus can be ingested by susceptible rabbits in a typical fecal/oral cycle. Carcasses of infected rabbits contain very high concentrations of virus and provide a source of environmental contamination. Carrion feeding flies and scavengers such as foxes can ingest virus from the carcass and infectious virus can pass through the gut of the fly or scavenger to contaminate the environment or to directly contaminate rabbits, for example by flies feeding around the eyes

RHDV is a calicivirus in the *Lagovirus* genus. Small, approximately 40 nm in diameter, and non-enveloped, it has a genome of 7.4 kb of positive-sense single-strand RNA encoding nine genes. During replication within the cytoplasm of infected cells, a subgenomic RNA is transcribed that encodes the two structural genes for the capsid proteins, VP60 and VP10 [56]. This may facilitate recombination between the structural genes and nonstructural genes of lagoviruses. The capsid comprises 180 copies of VP60 organized as dimers into a shell (S) domain and a protruding (P) domain [57]. VP10 probably forms a pore complex in the capsid based on human calicivirus structure [58].

The *Lagovirus* genus currently contains two species: RHDV and European brown hare syndrome virus (EBHSV) [59]. It is proposed that these be collapsed into a single species, *Lagovirus europaeus*, with subdivision into genogroups: GI for RHDV and related rabbit lagoviruses and GII for EBHSV and similar hare lagoviruses. These genogroups are divided into genotypes and variants based on phylogenic relationships [60]. As mentioned above, two genotypes (GI.1 and GI.2) of RHDV are now circulating (here termed RHDV1 and RHDV2) while the apathogenic RCVs are classified into GI.3 and GI.4.

3.3 Testing of MYXV and RHDV as Biological Controls

An effective biological control must significantly reduce the ecological impact of the pest by reducing population size and suppressing population recovery. This requires high mortality rates, although disruption of reproduction would achieve the same outcome. The disease agent must spread efficiently and be maintained within the population in a virulent form. It should also be highly species specific [61]. There are very few pathogens that meet these criteria.

Early studies on MYXV demonstrated that it was highly lethal for laboratory rabbits and that inoculation of other species, including humans, did not lead to disease [62]. In 1933, the Australian Government commissioned research to determine the potential of myxomatosis as a biological control. These studies confirmed the lethality and apparent species specificity of MYXV and demonstrated the potential for fleas and mosquitoes to transmit the virus [61, 63]. However, it remained to be determined whether the virus could establish and spread in wild rabbit populations.

In 1937–1938, trials were undertaken in wild rabbits on Wardang Island off the coast of South Australia. These were followed by trials in enclosed warrens on the adjacent mainland and finally, in 1942–1943, releases in semiarid country (<200 mm annual rainfall) further north. In all these trials, despite high case fatality rates, the virus was poorly transmitted, and it was concluded that MYXV had little potential as a biological control [64].

In the autumn of 1950, further testing was undertaken in higher rainfall areas along the Murray River in south eastern

Australia. This area was chosen because it was expected that mosquitoes would be more prevalent than in the previous tests. Again, the virus appeared to die out. However, with the emergence of large numbers of *Culex annulirostris* mosquitoes in the summer, outbreaks of myxomatosis were reported across wide areas [65, 66].

In contrast to MYXV, it was clear from epizootics in Spain, France, and Portugal that RHDV could establish in and significantly reduce wild rabbit populations. No other species appeared to be affected indicating that RHDV had the potential to be a new biological control agent [45]. In 1991, an isolate of RHDV (Czech v351) was imported into the level-4-biosecurity Australian Animal Health Laboratory for evaluation of lethality, routes of infection, humaneness, and species specificity of the virus. The rapid and apparently "quiet" death, often with no prior disease signs, caused by RHDV was in contrast to myxomatosis, where no assessment of animal welfare seems to have occurred. Although it was demonstrated that both fleas and mosquitoes could passively transmit RHDV, the role of carcass-feeding flies in transmission was not anticipated [67].

In 1995, testing of RHDV commenced in wild rabbits on Wardang Island, the same location used to evaluate MYXV nearly 60 years earlier. In October 1995, virus escaped to wild rabbit populations on the adjacent mainland, probably transferred by blowflies (*Calliphora* spp.) or bush flies (*Musca vetustissima*). Attempts to contain the escape by eliminating rabbits around the outbreak were abandoned when RHD outbreaks were reported hundreds of kilometers away [45].

4 The Early Spread of MYXV and RHDV

MYXV initially spread along the banks of rivers and creeks because this was where the mosquitoes and rabbits were juxtaposed. However, in north-western New South Wales and south-western Queensland, widespread summer flooding had enabled massive breeding of mosquitoes, which allowed the virus to spread widely. This was enhanced by deliberate transport of infected rabbits and inoculation of rabbits with virus. Vast numbers of rabbits provided the basis for an epizootic that must have killed tens of millions over an area that extended for 1600 km by 1800 km [65, 66]. As mosquito activity diminished in the autumn of 1951, this first epizootic came to an end. Rabbit breeding mostly occurs in the winter-spring period and during this time the virus trickled on at a local scale in areas not affected by the initial epizootic. It re-emerged the following spring and spread was augmented with deliberate releases including across the continent in Western Australia [65, 68] and subsequently into Tasmania.

During the initial epizootic there was considerable outcry about the risks to humans from myxomatosis. This concern was heightened by a concurrent outbreak in humans of the mosquito-borne viral disease Murray Valley encephalitis. To allay public concerns, three leading scientists, Frank Macfarlane Burnet (Director of the Hall Institute and Nobel laureate in 1960), Frank Fenner (Professor of Microbiology at the John Curtin School of Medical Research), and Ian Clunies-Ross (Chairman of the Commonwealth Scientific and Industrial Research Organization), injected themselves with MYXV: there were no ill effects [69]. The scientists did not seek publicity and remained anonymous, but the outcome was subsequently announced in the Australian Federal Parliament.

RHDV spread rapidly across large parts of South Australia and emerged elsewhere, probably helped by human transfer of infected carcasses. In some areas, rabbit populations declined by 90–95% [45, 70, 71]. The plans for a systematic evaluation of RHDV prior to any release and the legislative requirements for a release were replaced by the on-ground reality. Following a series of reports, RHDV was officially registered as a pest control agent in September 1996 [45]. This allowed virus to be deliberately released in inoculation campaigns. However, by this time RHDV had already spread across most of Australia. In August 1997, virus from Australia was illegally introduced into New Zealand, where it established in the wild rabbit population [72].

5 The Impact of Biological Control

After the spread of MYXV in the 1950s, rabbit populations dropped to perhaps 1–5% of their previous levels. However, the impact was quite mixed. Regions where summer mosquito activity was low achieved relatively little sustained control despite intensive release campaigns and occasional epizootics [68]. In other parts of the agricultural zone, populations were suppressed and stayed low, sometimes helped by conventional control measures [73]. Economic benefits from increases in agricultural production and cost savings in rabbit control followed, while in the rangelands, tree and shrub regeneration occurred for the first time since the rabbit spread. From the mid 1950s, there was a gradual recovery in rabbit numbers as less virulent virus strains came to predominate in the field and genetic resistance to MYXV emerged in the rabbit population. Before RHDV release, it was estimated that rabbit populations were around 5% of their pre-myxomatosis levels in the agricultural regions and 25% in the rangelands [74].

The impact of RHD was highest in the semiarid rangelands. At some sites, rabbit populations were reduced to 5% of the previous level [70, 75, 76]. At other locations, particularly in higher rainfall areas, relatively minor reductions occurred [77]. In at least one area, RHDV failed to invade rabbit populations despite repeated introductions [78]. Low impact coincided with the unexpected presence of an apathogenic intestinal lagovirus termed rabbit calicivirus-A1 (RCV-A1) [79, 80]. Similar viruses occur in rabbits in Europe, New Zealand, and North America and probably other countries [72, 81, 82]. Infection with RCV-A1 induces antibodies that cross-react with RHDV; recent infection with RCV-A1 may inhibit infection with RHDV or the disease course may be prolonged with consequent lower likelihood of transmission and increased survival rates [83–86]. The presence of the partially cross-protective RCV-A1 meant that, unlike MYXV, RHDV did not invade a completely naïve host population.

Rabbit populations gradually began to recover after the initial impact of RHD [87]. Most older rabbits would have survived RHD and been immune [87–89]. Maternal antibody transferred across the placenta provides protection for young rabbits for many weeks and rabbits infected during this time have a higher likelihood of recovery [90]. Thus, the impact of the disease was gradually reduced due to higher proportions of immune rabbits. Of course, myxomatosis continued to suppress rabbit numbers with the epidemiology of the viruses interacting in complex ways depending on the proportions of susceptible animals in the population [91, 92].

In late 2014, RHDV2 entered Australia by an unknown route. The ability to infect rabbits that are immune to RHDV1 and cause disease in young kittens appears to have enabled successful invasion of the wild rabbit population partially replacing RHDV1 [93–98]. It is estimated that population losses due to RHDV2 averaged around 60% across Australian rabbit populations [99]. RHDV2 is antigenically and genetically distinct from RHDV1 and appears to have evolved from an avirulent lagovirus independently of RHDV1; it has a broader host range, spilling over into multiple species of hare (*Lepus* spp.) [100–102]. It is possible that by killing young animals in which RCV-A1 circulates, RHDV2 may be reducing the prevalence of RCV-A1, which could further enhance control [99]. Having two serotypes in circulation could increase pressure on the rabbit population as one serotype might be able to invade populations immune to the other serotype.

6 Host-Pathogen Coevolution: Biological Arms Race or Detente?

Unlike most disease outbreaks, the epizootics of myxomatosis and RHD followed point releases of a defined novel pathogen into a new host species. Subsequent virus isolates can be compared with the initial virus. In addition, laboratory rabbits and wild rabbits have similar outcomes following infection, so laboratory rabbits provide an unselected population and the short generation interval

Table 1
Virulence classification of myxoma virus isolates[a]

Virulence grade	Case fatality rate	Average survival time
1	>99%	≤13 days
2	95–99%	14–16 days
3	70–95%	17–28 days
4	50–70%	29–50 days
5	<50%	Not applicable

[a]Table is adapted from [19, 106]

(approximately 12 months) of wild rabbits allows natural selection to be studied in real time.

It was predicted, early in 1951, that over time MYXV would become less lethal and the rabbit population more resistant [69]. Professor Frank Fenner seized this opportunity to study host-pathogen coevolution in real time on a continental scale. Over the next 15 years, the studies of Fenner and others in Australia, and later in Europe, developed into a paradigm of host-pathogen coevolution [1, 3, 27, 103].

As predicted, it quickly became apparent that some field isolates of MYXV were slightly attenuated compared to the original virus, standard laboratory strain (SLS). Infected rabbits survived longer and the occasional recovery occurred [104]. These field strains outcompeted introductions of SLS [66, 105]. MYXV isolates were assigned to one of five virulence grades based on case fatality rates and average survival times of infected laboratory rabbits (Table 1) [19, 106]. Over the next 30 years, the great majority of field isolates were of grade 3 virulence (SLS had grade 1 virulence) (Fig. 4) [103]. Rabbits infected with moderately attenuated viruses survived longer in an infectious state for mosquitoes and so the attenuated virus had a higher probability of transmission compared to highly virulent virus, which tended to kill the rabbits days earlier; very attenuated, grade 5, viruses were more readily controlled by the infected rabbit so limiting transmissibility [16]. Interestingly, in transmission experiments, rabbits that survived infection were much less likely to transmit virus at any stage compared to those that had prolonged survival times, but subsequently died. This suggests that moderate virulence could be positively selected compared to evolving towards an attenuated fibroma type of disease [107].

Myxomatosis exerted strong selection pressure on the rabbit population. At many locations, rabbits that had survived myxomatosis constituted most of the breeding population and epizootics occurred annually so favorable alleles would have become

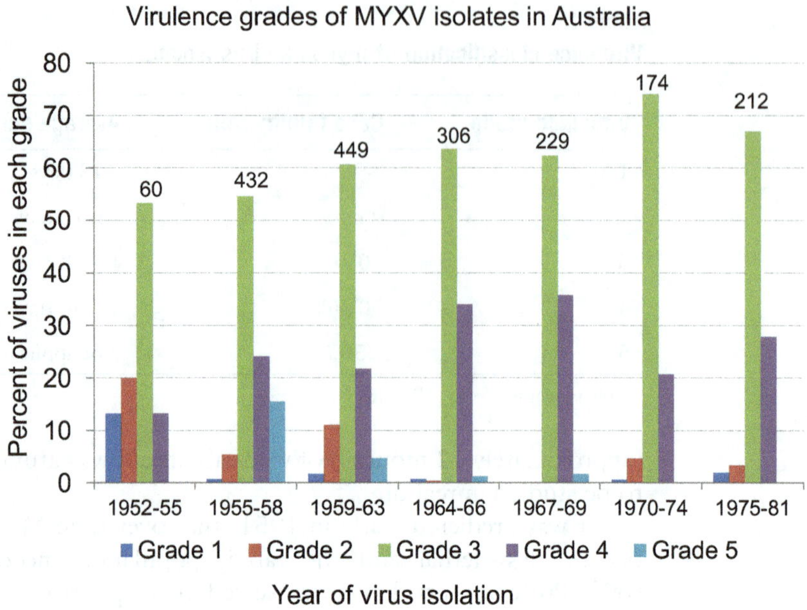

Fig. 4 Virulence grades of field isolates of MYXV. Virus isolates were inoculated into laboratory rabbits (usually 5) and virulence grades assigned based on average survival time and case fatality rates as in Table 1. Numbers above the bars are the number of isolates tested. (Data are from [103]. Figure 4 is reproduced from [27] with permission)

widespread. The best data set is from Lake Urana, where over 7 years or essentially seven epizootics of selection and breeding by the survivors, case fatality rates, when infected with a standard grade 3 virus, fell from 88% to 26% (Fig. 5) [108, 109]. Extreme heat, such as occurs over much of inland Australia in summer, can increase survival rates of rabbits infected with attenuated viruses and may have aided survival of rabbits with some resistance that might otherwise have died [110]. Resistance likely favored selection for somewhat higher virulence in the virus when measured in unselected rabbits. This is possibly why in laboratory studies, grade 4 viruses were the most transmissible [16] but in the field, grade 3 viruses predominated.

Resistance manifests as control of virus replication during the early stages of infection [111, 112]. It can be overcome by skewing of immune responses or challenge with more virulent virus strains [103, 113–115]. The same gene variants have been selected in wild rabbits in Australia and in Europe following the spread of myxomatosis; many of these implicate innate immunity, but the molecular basis of resistance is not known [116]. If resistance is polygenic, it is likely that the increased resistance in the rabbit population is driving further evolution of the virus with an ongoing dynamic between virulence, resistance, and transmissibility. This idea is supported by a novel immunosuppressive disease phenotype seen in laboratory

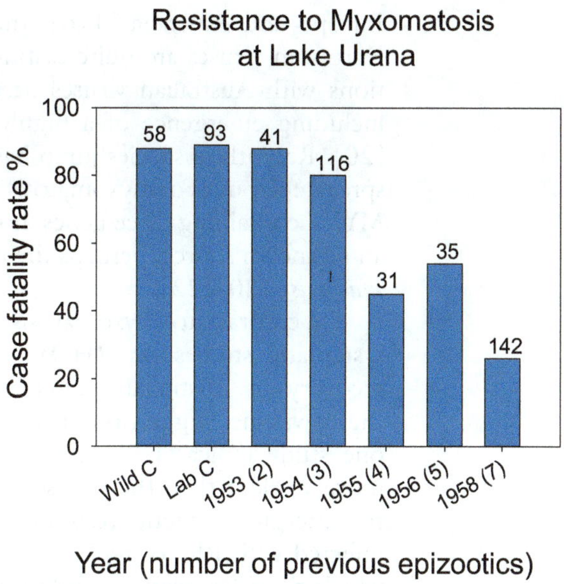

Fig. 5 Emergence of resistance to myxomatosis in wild rabbits. Wild rabbit kittens from each breeding season were trapped at Lake Urana and maintained in an animal house until 4 months of age. Seronegative rabbits were then challenged with a standard grade 3 virulence virus. The number of rabbits tested each year is shown above the bars. The year of birth and the number of epizootics, in brackets, are shown below the bars; cwild: control wild rabbits; clab: control lab rabbits. (Data are from [108, 109]. Figure 3 is reproduced from [27] with permission)

rabbits (with no resistance) infected with more recent viruses, suggesting selection to overcome resistance mediated by the innate immune system in wild rabbits [18, 20]. As already suggested, this implies that evolution towards an innocuous fibroma, as occurs in the natural host, may not readily occur. Attenuated viruses that do not disseminate to cause generalized disease in the infected rabbit will be cleared by the immune response and, in the field, more virulent viruses will tend to outcompete such attenuated viruses. Thus, there is likely to be a continuing arms race between virus virulence and host resistance [117].

The Lausanne strain of MYXV was deliberately released by a landholder in France in 1952. Two wild rabbits were inoculated and from this introduction, MYXV spread through the wild and domestic rabbits of Europe [118]. Although the starting virus strain was different (and more virulent than SLS) and the rabbit flea (*Spilopsyllus cuniculi*) was a major vector in addition to mosquitoes, the evolutionary outcomes were remarkably similar to those in Australia, with the emergence of attenuated viruses and selection for resistant rabbits [27, 103, 119]. Interestingly, while the same gene variants were selected in the rabbit population in

Europe and Australia [116], the evolutionary pathways of the European viruses are quite distinct, with almost no shared mutations with Australian viruses despite convergence of phenotype, including emergence of a highly immunosuppressive phenotype [20]. Recently, a species jump has occurred in Europe with widespread epizootics of myxomatosis in hares caused by a recombinant MYXV containing three genes, homologous to MYXV genes, from an unknown source, perhaps the related leporipoxvirus *Hare fibromavirus* [120–122].

In contrast to myxomatosis, there have been no large-scale systematic studies on RHDV virulence. However, despite the recovery in Australian rabbit populations, and unlike MYXV, there was no suggestion that RHDV was attenuating. Indeed, one study indicates that field isolates have been selected to be more virulent than the released v351 strain, possibly in response to emerging genetic resistance [123]. Rabbits experimentally infected with RHDV typically die within 36–72 h after infection, which provides relatively little time for shedding virus. However, RHDV persists in rabbit carcasses, from where it can be released to contaminate warrens and pasture or is spread by flies and scavengers such as foxes. In experimentally infected rabbits, virus titers in liver peaked at about 30–50 h and then declined coincident with the development of serum IgM [86, 124]. This suggests that in wild rabbits, high virulence with rapid death optimizes transmission by maximizing the amount of infectious virus in the carcass [123].

Early studies with RHDV2 in France showed that more rabbits survived infection or developed chronic disease compared to RHDV1, although this was partially isolate dependent [49, 94]. It was suggested that moderate virulence was advantageous for invading wild rabbit populations [49, 125]. However subsequent experience indicates that some isolates have high virulence [98, 126]. This suggests a possible model whereby RHDV2 emerged from a low-pathogenicity rabbit lagovirus by a switch in tissue tropism from intestinal epithelium to hepatocytes with subsequent selection for higher virulence because of more efficient transmission.

Emerging resistance to RHDV1 has been demonstrated in wild rabbits from some locations in Australia. Resistant rabbits had significantly reduced rates of infection following low oral doses of virus [127]. This could be overcome in some rabbits by intramuscular injection of virus. However, others were still not infected. This implies that resistant rabbits were not expressing one or more factors needed for infection at low doses.

RHDV attaches to polymorphic α1,2-fucosylated glycan molecules called histo-blood group antigens (HBGAs) expressed on epithelial cells in intestine and other tissues [128]. Binding is determined by amino acid residues in the P domain of the VP60 capsid protein [129] with different virus strains preferentially binding to different HGBAs [130]. Rabbits differentially express HBGAs [131]; low expression of particular HGBAs is associated with survival of infection with some strains of RHDV, especially at low virus doses [130]. Following RHDV epizootics, the surviving rabbits were those that expressed low levels of particular HGBAs [130]. Given the high mutation frequency of RHDV (see below), this suggests that there could be rapid selection for virus variants able to bind the most frequent HGBAs within a population followed by selection against rabbits expressing those HGBAs. Since HGBAs are not expressed on hepatocytes, it is likely that these are predominately attachment factors and other receptor molecules must be used in the infection process [128].

Single-nucleotide polymorphisms (SNPs) significantly associated with survival following RHDV infection have been identified in Australian rabbits [132]. Many of these SNPs were associated with genes involved in innate immune responses or virus responses indicating that polymorphisms in multiple genes may contribute to surviving infection. However, the molecular basis of any resistance associated with these SNPS is not understood. Whether RHDV2 overcomes genetic resistance to RHDV1 is not known.

Unlike MYXV, high virulence with rapid death of the infected rabbit appears to be an evolutionarily successful strategy for RHDV. The high mutation rate and frequent recombination (see below) throw up variants for natural selection to act upon and potentially overcome genetic resistance. It is possible that in fragmented, low-density rabbit populations, high virulence could be less successful and lead to local extinction of RHDV. Whether the virus can evolve towards longer survival of the infected rabbit with continued shedding in the face of the adaptive immune response remains to be seen.

7 Genomic Evolution of RHDV and MYXV

Phylogenetic studies on RHDV since its emergence reveal a high rate of evolution and recombination between strains of RHDV and between RHDV and apathogenic rabbit caliciviruses particularly exchanging the nonstructural genes and the structural genes [133–137]. Indeed, the RHDV2 strain that invaded Australia was a recombinant between RHDV1 nonstructural genes and RHDV2 capsid genes similar to a strain originally described in Portugal [95]. Recombinants with nonstructural genes from apathogenic viruses and RHDV capsid genes appear to be virulent and have the host range of the virus that contributed to the capsid genes suggesting that the capsid determines virulence, tissue tropism, and host range. The emergence of RHDVa and then RHDV2 suggests that further variants may arise.

Genomic studies on MYXV have demonstrated a high rate of evolution for a double-stranded DNA virus [138, 139]. The evolutionary rate in Australia was essentially the same as that in Europe [20], but one Australian lineage exhibited a short-lived aberrantly high mutation rate coincident with the invasion of RHDV [140]. It has not been possible to determine key mutations involved in attenuation or reversion to virulence suggesting that in these large, complex genomes there are multiple pathways converging on similar phenotypes [141]. Genes with key roles in virulence, defined by systematic gene knockout experiments [25], were sometimes disrupted in highly virulent viruses suggesting that single amino acid changes in other genes could be critical for epistatic compensation [18, 20].

8 Attempts to Enhance Biocontrol

The most successful deliberate addition to rabbit biocontrol in Australia was the release of an exotic flea species. The European rabbit flea (*Spilopsyllus cuniculi*), introduced in 1969, is an efficient year-round vector for MYXV, unlike more seasonal mosquitoes, and was effective in areas where mosquito activity was low [142, 143]. The spread of rabbit fleas enhanced the impact of myxomatosis at a time when rabbit populations were recovering [144]. Unfortunately, these fleas survive poorly in the hotter arid areas of Australia [145, 146]; to overcome this limitation, Spanish rabbit fleas (*Xenopsylla cunicularis*) were introduced in the early 1990s [147, 148]. The impact of Spanish fleas on myxomatosis was not documented because their release coincided with the spread of RHDV and research attention was directed elsewhere. Anecdotally, the fleas have established and persisted.

Introductions of MYXV, using virulent grade 1 viruses to inoculate rabbits, were made across Australia for 50 years. However, virulent viruses could not compete with attenuated field strains and probably had little impact. The Lausanne strain (released in Europe) was widely released in conjunction with the European rabbit flea as it was thought to be better adapted to flea transmission. Lausanne is genetically distinct from SLS, but no field strains descended from Lausanne have been identified indicating that the virus did not establish [140, 149–151].

As with MYXV, deliberate introductions of RHDV, prepared from the original v351 strain, were made widely. However, establishment and spread of v351 subsequent to the original escape have not occurred [133, 152]. In a bid to enhance the effect of RHDV, a RHDVa strain from Korea, termed K5, was chosen for release based on its ability to infect genetically resistant wild rabbits and rabbits previously infected with RCV-A1. K5 was widely released in Australia in 2017 with the aim of slowing the recovery in rabbit populations. However, K5 had limited impact and failed to establish, probably because of the presence of the more competitive RHDV2 [99, 153]. An earlier incursion by an exotic RHDVa strain with some spread in wild and domestic rabbits was detected in 2014 [136]. Having multiple genetic variants in circulation may provide opportunities for recombination and emergence of novel variants of RHDV [137].

9 Conclusions and Future Prospects

Biological control, initially with MYXV and subsequently with RHDV, has provided continent-wide landscape therapy by dramatically reducing the impact of rabbits. The cumulative value of biological control was estimated at AUS$70 billion (in 2012 dollars) [154]. However, this excludes the ecological impact, which is not readily quantified. Despite its success, biological control is not a panacea for rabbits; over time rabbit populations gradually recover, albeit at lower levels. Indeed, mortality rates of over 90% are needed to suppress rabbit populations under ideal conditions [73]. Searches are underway to identify new biological control agents [153]. However, both MYXV and RHDV were very unusual emerging pathogens with high case fatality rates, rapid spread, and high species specificity. Identification of further agents may be more difficult. Sustained control of rabbits is likely only to be achieved by ongoing biological control combined with broad-based conventional con trols as part of an integrated pest management strategy. Meanwhile the fascinating coevolution of the rabbit and its new pathogens will continue.

References

1. Anderson RM, May RM (1982) Coevolution of hosts and parasites. Parasitology 85:411–426
2. Boots M, Hudson PJ, Sasaki A (2004) Large shifts in pathogen virulence relate to host population structure. Science 303:842–844
3. Di Giallonardo F, Holmes EC (2015) Viral biocontrol: grand experiments in disease emergence and evolution. Trends Microbiol 23:83–90
4. Bester MN, Bloomer JP, van Aarde RJ et al (2002) A review of the successful eradication of feral cats from sub-Antarctic Marion Island, Southern Indian Ocean. S Afr J Wildlife Res 32:65–73
5. McColl KA, Sunarto A, Holmes EC (2016) Cyprinid herpesvirus 3 and its evolutionary future as a biological control for carp in Australia. Virol J 13:206
6. Rolls EC (1969) They all ran wild. Angus and Robertson, Melbourne
7. Stoddart EM, Parer I (1988) Colonisation of Australia by the rabbit *Oryctolagus cuniculus* (L): CSIRO Australia, Division of Wildlife and Ecology: Project Report no. 6. CSIRO Australia
8. Iannella A, Peacock D, Cassey P et al (2019) Genetic perspectives on the historical introduction of the European rabbit (*Oryctolagus cuniculus*) to Australia. Biol Invasions 21:603–614
9. Cooke BD (2012) Rabbits: manageable environmental pests or participants in new Australian ecosystems? Wildlife Res 39:279–289

10. Dickman CR (1996) Impact of exotic generalist predators on the native fauna of Australia. Wildlife Biol 2:185–195
11. Woinarski JCZ, Burbidge AA, Harrison PL (2015) Ongoing unraveling of a continental fauna: decline and extinction of Australian mammals since European settlement. Proc Natl Acad Sci U S A 112:4531–4540
12. Murphy BP, Woolley L-A, Geyle HM et al (2019) Introduced cats (*Felis catus*) eating a continental fauna: the number of mammals killed in Australia. Biol Conserv 237:28–40
13. Sanarelli G (1898) Das myxomatogene Virus. Beitrag zum Stadium der Krankheitserreger ausserhalb des Sichtbaren. Zbl Bakt 23:865–873
14. Aragão HDB (1943) O virus do mixoma no coelho do mato (*Sylvilagus menensis*) sua transmissao pelos *Aedes scapularis e aegypti*. Mem Inst Oswaldo Cruz 38:93–99
15. Day MF, Fenner F, Woodroofe GM et al (1956) Further studies on the mechanism of mosquito transmission of myxomatosis in the European rabbit. J Hyg (Camb) 54:258–283
16. Fenner F, Day MF, Woodroofe GM (1956) Epidemiological consequences of the mechanical transmission of myxomatosis by mosquitoes. J Hyg (Camb) 54:284–303
17. Joubert L, Duclos P, Toaillen P (1982) La myxomatose des garennes dans le sud-est. La myxomatose amyxomateuse. Rev Méd Vét 133:739–753
18. Kerr PJ, Cattadori IM, Liu J et al (2017) Next step in the ongoing arms race between myxoma virus and wild rabbits in Australia is a novel disease phenotype. Proc Natl Acad Sci U S A 114:9397–9402
19. Fenner F, Marshall ID (1957) A comparison of the virulence for European rabbits (*Oryctolagus cuniculus*) of strains of myxoma virus recovered in the field in Australia, Europe and America. J Hyg (Camb) 55:149–191
20. Kerr PJ, Cattadori IM, Rogers MB et al (2017) Genomic and phenotypic characterization of myxoma virus from Great Britain reveals multiple evolutionary pathways distinct from those in Australia. PLoS Pathog 13:e1006252
21. Duteyrat J-L, Gelfi J, Bertagnoli S (2007) Ultrastructural study of myxoma virus morphogenesis. Arch Virol 151:2160–2180
22. Condit RC, Moussatche N, Traktman P (2006) In a nutshell: structure and assembly of the vaccinia virion. Adv Virus Res 66:31–124
23. Cameron C, Hota-Mitchell S, Chen L et al (1999) The complete DNA sequence of myxoma virus. Virology 264:298–318
24. Morales M, Ramirez MA, Cano M et al (2009) Genome comparison of a nonpathogenic myxoma virus field strain with its ancestor, the virulent Lausanne strain. J Virol 83:2397–2403
25. Spiesschaert B, McFadden G, Hermans K et al (2011) The current status and future directions of myxoma virus, a master in immune evasion. Vet Res 42:76
26. Jeklova E, Leva L, Matiosovic J et al (2008) Characterisation of immunosuppression in rabbits after infection with myxoma virus. Vet Microbiol 129:117–130
27. Kerr PJ (2012) Myxomatosis in Australia and Europe: a model for emerging infectious diseases. Antiviral Res 93:387–415
28. Zhang L, Yaron JR, Tafoya A et al (2019) A virus-derived immune modulating serpin accelerates would closure with improved collagen remodeling. J Clin Med 8:1626
29. Tardif J-C, L'Allier PL, Grégoire J et al (2010) A randomized controlled, phase 2 trial of the viral serpin Serp-1 in patients with acute coronary syndromes undergoing percutaneous coronary intervention. Circ Cardiovasc Interv 3:543–548
30. Ambadapadi S, Munuswamy-Ramanujam G, Zheng D et al (2016) Reactive center loop (RCL) peptides derived from serpins display independent coagulation and immune modulating activities. J Biol Chem 29:2874–2887
31. Mahon BP, Ambadapadi S, Yaron JR et al (2018) Crystal structure of cleaved Serp-1, a myxoma virus-derived immune modulating serpin: structural design of serpin reactive centre loop peptides with improved therapeutic function. Biochemistry 57:1096–1107
32. Viswanathan K, Bot I, Liu L et al (2012) Viral cross-class serpin inhibits vascular inflammation and T lymphocyte fratricide; a study in rodent models in vivo and human cell lines in vitro. PLoS One 7:e44694
33. Yaron JR, Chen H, Ambadapadi S et al (2019) Serp-2 a virus derived apoptosis and inflammasome inhibitor, attenuates liver ischemia-reperfusion injury in mice. J Inflamm (Lond) 16:12
34. Dai E, Liu L-Y, Wang H et al (2010) Inhibition of chemokine-glycosaminoglycan interactions in donor tissue reduces mouse allograft vasculopathy and transplant rejection. PLoS One 5(5):e10510
35. Lucas AR, Verma RK, Dai E et al (2014) Myxoma virus anti-inflammatory chemokine

binding protein reduces the increased plaque growth induced by chronic *Porphyromonas gingivalis* oral infection after balloon angioplasty aortic injury in mice. PLoS One 9: e111353

36. Kwiecien JM, Dabrowski W, Marzec-Kotarska B et al (2019) Myxoma virus derived immune modulating proteins, M-T7 and Serp-1, reduce early inflammation after spinal cord injury in the rat model. Folia Neuropathol 57:41–50
37. Garg RR, Jackson CB, Khan AR et al (2019) Myxoma virus M013 protein antagonizes NF-κB and inflammasome pathways via distinct structural motifs. J Biol Chem 294:8480–8489
38. Ildefonso CJ, Jaime H, Rahman MM et al (2015) Gene delivery of a viral anti-inflammatory protein to combat ocular inflammation. Hum Gene Ther 26:59–68
39. Johnston JB, Wang G, Barrett JW et al (2005) Myxoma virus M-T5 protects infected cells from the stress of cell cycle arrest through its interaction with host cell cullin-1. J Virol 79:10750–10763
40. Zhang L, Villa NY, McFadden G (2009) Interplay between poxviruses and the cellular ubiquitin/ubiquitin-like pathways. FEBS Lett 583:607–614
41. Istivan TS, Pirogova E, Gan E et al (2011) Biological effects of a de novo designed myxoma virus peptide analogue: evaluation of cytotoxicity on tumor cells. PLoS One 6: e24809
42. Almansour NM, Pirogova E, Coloe PJ et al (2012) A bioactive peptide analogue for myxoma virus protein with a targeted cytotoxicity for human skin cancer in vitro. J Biomed Sci 19:65
43. Torres-Domínguez LE, McFadden G (2019) Poxvirus oncolytic therapy. Expert Opin Biol Ther 19:561–573
44. Villa NY, McFadden G (2018) Virotherapy as potential adjunct therapy for graft-vs-host disease. Curr Pathobiol Rep 6:247–263
45. Cooke BD, Fenner F (2002) Rabbit haemorrhagic disease and the biological control of wild rabbits, *Oryctolagus cuniculus*, in Australia and New Zealand. Wildlife Res 29:689–706
46. Capucci L, Fallacara F, Grazioli S et al (1998) A further step in the evolution of rabbit hemorrhagic disease virus: the appearance of the first consistent antigenic variant. Virus Res 58:115–126
47. Schirrmeier H, Reimann I, Köllner B et al (1999) Pathogenic, antigenic and molecular properties of rabbit haemorrhagic disease virus (RHDV) isolated from vaccinated rabbits: detection and characterization of antigenic variants. Arch Virol 144:719–735
48. Mahar JE, Nicholson L, Eden J-S et al (2016) Benign rabbit caliciviruses exhibit evolutionary dynamics similar to those of their virulent relatives. J Virol 90:9317–9329
49. Le Gall RG, Zwinglestein F, Boucher S et al (2011) Detection of a new variant of Rabbit Haemorrhagic Disease Virus in France. Vet Record 168:137–138
50. McColl KA, Morrissy CJ, Collins BJ et al (2002) Persistence of rabbit haemorrhagic disease virus in decomposing rabbit carcasses. Aust Vet J 80:298–299
51. Asgari S, Hardy JRE, Sinclair RG et al (1998) Field evidence for mechanical transmission of rabbit haemorrhagic disease virus (RHDV) by flies (Diptera: Calliphoridae) among wild rabbits in Australia. Virus Res 54:123–132
52. McColl KA, Merchant JC, Hardy J et al (2002) Evidence for insect transmission of rabbit haemorrhagic disease virus. Epidemiol Infect 129:655–633
53. Tuñón MJ, Sanchez-Campos S, Garcia-Ferreras J et al (2003) Rabbit hemorrhagic viral disease: characterization of a new animal model of fulminant liver failure. J Lab Clin Med 141:272–278
54. Tuñón MJ, San Miguel B, Crespo I et al (2011) Melatonin attenuates apoptotic liver damage in fulminant hepatic failure induced by the rabbit haemorrhagic disease virus. J Pineal Res 50:38–45
55. Tuñón MJ, San Miguel B, Crespo I et al (2011) Cardiotrophin-1 promotes a high survival rate in rabbits with lethal fulminant hepatitis of viral origin. J Virol 85:13124–13132
56. Abrantes J, van der Loo W, Le Pendu J et al (2012) Rabbit haemorrhagic disease (RHD) and rabbit haemorrhagic disease virus (RHDV): a review. Vet Res 43:12
57. Wang X, Xu F, Liu J et al (2013) Atomic model of rabbit hemorrhagic disease virus by cryo-electron microscopy and crystallography. PLoS Pathog 9:e1003132
58. Conley MJ, McElwee M, Azmi L et al (2019) Calicivirus VP2 forms a portal-like assembly following receptor engagement. Nature 565:377–381
59. VInjé J, Estes MK, Esteves P, Green KY et al (2019) ICTV virus taxonomy profile: *Caliciviridae*. J Gen Virol. https://doi.org/10.1099/jgv.0.001332
60. Le Pendu J, Abrantes J, Bertagnoli S et al (2017) Proposal for a unified classification

system and nomenclature of lagoviruses. J Gen Virol 98:1658–1666

61. Martin CJ (1936) Observations on myxomatosis cuniculi (Sanarelli) made with a view to the use of the virus in the control of rabbit plagues. CSIR Australia Research Bulletin, No. 96:1-28
62. Aragão HB (1927) Myxoma of rabbits. Mem Inst Oswaldo Cruz 20:237–247
63. Bull LB, Dickinson CG (1937) The specificity of the virus of rabbit myxomatosis. J Counc Sci Ind Res 10:291–294
64. Bull LB, Mules MW (1944) An investigation of *myxomatosis cuniculi* with special reference to the possible use of the disease to control rabbit populations in Australia. J Counc Sci Ind Res 17:79–93
65. Ratcliffe FN, Myers K, Fennessy BV et al (1952) Myxomatosis in Australia. A step towards the biological control of the rabbit. Nature 170:7–19
66. Myers K (1954) Studies in the epidemiology of infectious myxomatosis of rabbits. II. Field experiments, August–November 1950, and the first epizootic of myxomatosis in the Riverine plain of south-eastern Australia. J Hyg (Camb) 52:47–59
67. Lenghaus C, Westbury H, Collins B et al (1994) Overview of the RHD project in the Australian Animal Health Laboratory. In: Munro RK, Williams RT (eds) Rabbit haemorrhagic disease: issues in assessment for biological control. Bureau of Resource Sciences, Canberra, pp 104–129
68. Calaby JH, Gooding CD, Tomlinson AR (1960) Myxomatosis in Western Australia. CSIRO Wildlife Res 5:89–101
69. Burnet FM (1968) Changing patterns. Heinemann, Melbourne
70. Mutze G, Cooke B, Alexander P (1998) The initial impact of rabbit hemorrhagic disease on European rabbit populations in South Australia. J Wildlife Dis 34:221–227
71. Kovaliski J (1998) Monitoring the spread of rabbit haemorrhagic disease virus as a new biological control agent for control of wild European rabbits in Australia. J Wildlife Dis 34:421–428
72. O'Keefe JS, Tempero JE, Motha MXJ et al (1999) Serology of rabbit haemorrhagic disease virus in wild rabbits before and after release of the virus in New Zealand. Vet Microbiol 66:29–40
73. Myers K (1962) A survey of myxomatosis and rabbit infestation trends in the eastern Riverina, N.S.W. 1951–1960. CSIRO Wildlife Res 7:1–12
74. Williams K, Parer I, Coman B et al (1995) Managing vertebrate pests: rabbits. Bureau of Resource Sciences, Australian Government Publishing Services, Canberra
75. Edwards GP, Dobbie W, Berman D (2002) Population trends in European rabbits and other wildlife of central Australia in the wake of rabbit haemorrhagic disease. Wildlife Res 29:557–565
76. Bruce JS, Twigg LE (2004) Rabbit haemorrhagic disease virus: serological evidence of a non-virulent RHDV-like virus in south-western Australia. Wildlife Res 31:602–612
77. Saunders G, Choquenot D, McIlroy J et al (1999) Initial effects of rabbit haemorrhagic disease on free-living rabbit (*Oryctolagus cuniculus*) populations in central-western New South Wales. Wildlife Res 26:69–74
78. Richardson BJ, Philips S, Hayes RA et al (2007) Aspects of the biology of the European rabbit (*Oryctolagus cuniculus*) and rabbit haemorrhagic disease virus (RHDV) in coastal eastern Australia. Wildlife Res 34:398–407
79. Strive T, Wright JD, Robinson AJ (2009) Identification and partial characterisation of a new lagovirus in Australian wild rabbits. Virology 384:97–105
80. Liu J, Fordham DA, Cooke BD et al (2014) Distribution and prevalence of the Australian non-pathogenic rabbit calicivirus is correlated with rainfall and temperature. PLoS One 9: e113976
81. Capucci L, Fusi P, Lavazza A et al (1996) Detection and preliminary characterization of a new calicivirus related to rabbit haemorrhagic disease virus but non pathogenic. J Virol 70:8614–8623
82. Bergin IL, Wise AG, Bolin SR et al (2009) Novel calicivirus identified in rabbits, Michigan, USA. Emerg Infect Dis 15:1955–1962
83. Strive T, Wright J, Kovaliski J et al (2010) The non-pathogenic Australian lagovirus RCV-A1 causes a prolonged infection and elicits partial cross-protection to rabbit haemorrhagic disease virus. Virology 398:125–134
84. Strive T, Elsworth P, Liu J et al (2013) The non-pathogenic Australian rabbit calicivirus RCV-A1 provides temporal and partial cross-protection to lethal rabbit haemorrhagic disease virus infection which is not dependent on antibody titres. Vet Res 44:51
85. Cox T, Liu J, van de Ven R et al (2017) Different serological profiles to co-occurring pathogenic and nonpathogenic caliciviruses in wild European rabbits (*Oryctolagus cuniculus*) across Australia. J Wildlife Dis 53:472–481

86. Cooke BD, Duncan RP, McDonald I et al (2018) Prior exposure to non-pathogenic calicivirus RCV-A1 reduces both infection rate and mortality from rabbit haemorrhagic disease in a population of wild rabbits in Australia. Transbound Emerg Dis 65: e470–e477
87. Mutze G, Bird P, Peacock D et al (2014) Recovery of South Australian rabbit populations from the impact of rabbit haemorrhagic disease. Wildlife Res 41:552–559
88. Bruce JS, Twigg LE (2005) The reintroduction, and subsequent impact of rabbit haemorrhagic disease virus in a population of wild rabbits in south-western Australia. Wildlife Res 32:139–150
89. McPhee SR, Butler KL, Kovaliski J et al (2009) Antibody status and survival of Australian wild rabbits challenged with rabbit haemorrhagic disease virus. Wildlife Res 36:447–456
90. Robinson AJ, So PTM, Muller WJ et al (2002) Statistical models for the effect of age and maternal antibodies on the development of rabbit haemorrhagic disease in Australian wild rabbits. Wildlife Res 29:663–671
91. Wells K, Brook BW, Lacy RC et al (2015) Timing and severity of immunizing diseases in rabbits is controlled by seasonal matching of host and pathogen dynamics. J R Soc Interface 12:21041184
92. Wells K, Fordham DA, Brook BW et al (2018) Disentangling synergistic disease dynamics: implications for the viral control of rabbits. J Anim Ecol 87:1418–1428
93. Dalton KP, Nicieza I, Balseiro A et al (2012) Variant rabbit haemorrhagic disease virus in young rabbits in Spain. Emerg Infect Dis 18:2009–2012
94. Dalton KP, Nicieza I, Abrantes J et al (2014) Spread of new variant RHDV in domestic rabbits on the Iberian Peninsula. Vet Microbiol 169:67–73
95. Hall RN, Mahar JE, Haboury S et al (2015) Emerging rabbit hemorrhagic disease virus 2 (RHDVb) in Australia. Emerg Infect Dis 21:2276–2278
96. Peacock D, Kovaliski J, Sinclair R et al (2017) RHDV2 overcoming RHDV immunity in wild rabbits (*Oryctolagus cuniculus*) in Australia. Vet Record. https://doi.org/10.1136/vr.104135
97. Mahar JE, Hall RN, Peacock D et al (2018) Rabbit hemorrhagic disease virus 2 (RHDV2; G1.1) is replacing endemic strains of RHDV in the Australian landscape within 18 months of its arrival. J Virol 92:e01374–e01317
98. Neimanis A, Larsson Pettersson U, Huang N et al (2018) Elucidation of the pathology and tissue distribution of *Lagovirus europaeus* G1.2/RHDV2 (rabbit haemorrhagic disease virus 2) in young and adult rabbits (*Oryctolagus cuniculus*). Vet Res 49:46
99. Ramsey DSL, Cox T, Strive T et al (2019) Diverse impacts of two novel strains of rabbit haemorrhagic disease virus on wild rabbit populations in Australia. J App Ecol. https://doi.org/10.1111/1365-2664.13548
100. Lopes AM, Marques S, Silva E et al (2014) Detection of RHDV strains in the Iberian hare (*Lepus granatensis*): earliest evidence of rabbit lagovirus cross-species infection. Vet Res 45:94
101. Hall RN, Peacock DE, Kovaliski J et al (2017) Detection of RHDV2 in European brown hares (*Lepus europaeus*) in Australia. Vet Record 180:121
102. Neimanis A, Ahola H, Larsson Pettersson U et al (2018) Overcoming species barriers: an outbreak of *Lagovirus europaeus* G1.2/RHDV2 in an isolated population of mountain hares (*Lepus timidus*). BMC Vet Res 14:367
103. Fenner F (1983) Biological control as exemplified by smallpox eradication and myxomatosis. Proc R Soc Lond B 218:259–285
104. Fenner F, Marshall ID, Woodroofe GM (1953) Studies in the epidemiology of infectious myxomatosis of rabbits. I. Recovery of Australian wild rabbits (*Oryctolagus cuniculus*) from myxomatosis under field conditions. J Hyg (Camb) 51:225–244
105. Marshall ID, Dyce AL, Poole WE et al (1955) Studies in the epidemiology of infectious myxomatosis of rabbits. IV. Observations of disease behaviour in two localities near the northern limit of rabbit infestation in Australia, May 1952 to April 1953. J Hyg (Camb) 53:12–25
106. Fenner F, Chapple PJ (1965) Evolutionary changes in myxoma virus in Britain. An examination of 222 naturally occurring strains obtained from 80 counties during the period October–November 1962. J Hyg (Camb) 63:175–185
107. Mead-Briggs AR, Vaughan JA (1975) The differential transmissibility of myxoma virus strains of differing virulence grades by the rabbit flea *Spilopsyllus cuniculi* (Dale). J Hyg (Camb) 75:237–247
108. Marshall ID, Fenner F (1958) Studies in the epidemiology of infectious myxomatosis of rabbits. V. Changes in the innate resistance

of wild rabbits exposed to myxomatosis. J Hyg (Camb) 56:288–302

109. Marshall ID, Douglas GW (1961) Studies in the epidemiology of infectious myxomatosis of rabbits. VIII. Further observations on changes in the innate resistance of Australian wild rabbits exposed to myxomatosis. J Hyg (Camb) 59:117–122
110. Marshall ID (1959) The influence of ambient temperature on the course of myxomatosis in rabbits. J Hyg (Camb) 57:484–497
111. Best SM, Kerr PJ (2000) Coevolution of host and virus: the pathogenesis of virulent and attenuated strains of myxoma virus in resistant and susceptible European rabbits. Virology 267:36–48
112. Best SM, Collins SV, Kerr PJ (2000) Coevolution of host and virus: cellular localization of myxoma virus infection of resistant and susceptible European rabbits. Virology 277:76–91
113. Douglas GW (1962) The Glenfield strain of myxoma virus. Its use in Victoria. J Agric (Victoria) 60:511–516
114. Kerr PJ, Perkins HD, Inglis B et al (2004) Expression of rabbit IL-4 by recombinant myxoma viruses enhances virulence and overcomes genetic resistance to myxomatosis. Virology 324:117–128
115. Silvers L, Inglis B, Labudovic A et al (2006) Virulence and pathogenesis of the MSW and MSD strains of Californian myxoma virus in European rabbits with genetic resistance to myxoma virus compared to rabbits with no genetic resistance. Virology 348:72–83
116. Alves JM, Carneiro M, Cheng JY et al (2019) Parallel adaptation of rabbit populations to myxoma virus. Science 363:1319–1326
117. Kerr PJ, Liu J, Cattadori IM et al (2015) Myxoma virus and the leporipoxviruses: an evolutionary paradigm. Viruses 7:1020–1061
118. Fenner F, Ratcliffe FN (1965) Myxomatosis. Cambridge University Press, Cambridge, England
119. Arthur CP, Louzis C (1988) La myxomatose du lapin en France: une revue. Rev Sci Tech Off Int Epiz 7:937–957
120. Dalton KP, Martin JM, Nicieza I et al (2019) Myxoma virus jumps species to the Iberian hare. Transbound Emerg Dis 66:2218–2226
121. Águedo-Pinto A, Lemos de Matos A, Abrantes M et al (2019) Genetic characterization of a recombinant myxoma virus in the Iberian hare (*Lepus granatensis*). Viruses 11:530
122. García-Bocanegra I, Camacho-Sillero L, Risalde MA et al (2019) First outbreak of myxomatosis in Iberian hares (*Lepus granatensis*). Transbound Emerg Dis. https://doi.org/10.1111/tbed.13289
123. Elsworth P, Cooke BD, Kovaliski J et al (2014) Increased virulence of rabbit haemorrhagic disease virus associated with genetic resistance in wild Australian rabbits (*Oryctolagus cuniculus*). Virology 464–465:415–423
124. Cooke BD, Berman D (2000) Effect of inoculation route and ambient temperature on the survival time of rabbits, *Oryctolagus cuniculus* (L.) infected with Rabbit Haemorrhagic Disease Virus. Wildlife Res 27:137–142
125. Fouchet D, Le Pendu J, Guitton J-S et al (2009) Evolution of microparasites in spatially and genetically structured host populations: the example of RHDV infecting rabbits. J Theor Biol 257:212–227
126. Capucci L, Cavadini P, Schiavitto M et al (2017) Increased pathogenicity in rabbit haemorrhagic disease virus type 2 (RHDV-2). Vet Record 180:426
127. Elsworth PG, Kovaliski J, Cooke BD (2012) Rabbit haemorrhagic disease: are Australian rabbits (*Oryctolagus cuniculus*) evolving resistance to infection with Czech CAPM 351 RHDV? Epidemiol Infect 140:1972–1981
128. Ruvoën-Clouet N, Ganière JP, André-Fontaine G et al (2000) Binding of rabbit haemorrhagic disease virus to antigens of the ABH histo-blood group family. J Virol 74:11950–11954
129. Leuthold MM, Dalton KP, Hansman GS (2015) Structural analysis of a rabbit haemorrhagic disease virus binding histo-blood group antigens. J Virol 89:2378–2387
130. Nystrom K, Le Gall Reculé G, Grassi P et al (2011) Histo-blood group antigens act as attachment factors of rabbit hemorrhagic disease virus infection in a virus strain-dependent manner. PLoS Pathog 7:e1002188
131. Nyström K, Abrantes J, Lopes AM et al (2015) Neofunctionalization of the Sec1 α1,2-fucosyltransferase paralogue in leporids contributes to glycan polymorphism and resistance to rabbit haemorrhagic disease. PLoS Pathog 11:e1004759
132. Schwensow NI, Detering H, Pederson S et al (2017) Resistance to RHD virus in wild Australian rabbits: comparison of susceptible and resistant individuals using a genome-wide approach. Mol Ecol 26:4551–4561
133. Eden J-S, Kovaliski J, Duckworth JA et al (2015) Comparative phylodynamics of rabbit hemorrhagic disease virus in Australia and New Zealand. J Virol 89:9548–9558

134. Hu B, Wang F, Fan Z et al (2017) Recombination between G2 and G6 strains of rabbit hemorrhagic disease virus (RHDV) in China. Arch Virol 162:269–272
135. Lopes AM, Silvério D, Magalhães MJ et al (2017) Characterization of old RHDV strains by complete genome sequencing identifies a novel genetic group. Sci Rep 7:13599
136. Mahar JE, Read AJ, Gu X et al (2018) Detection and circulation of a novel rabbit hemorrhagic disease virus in Australia. Emerg Infect Dis 24:22–31
137. Hall RN, Mahar JE, Read AJ et al (2018) A strain-specific multiplex RT-PCR for Australian rabbit haemorrhagic disease viruses uncovers a new recombinant virus variant in rabbits and hares. Transbound Emerg Dis 65: e444–e456
138. Kerr PJ, Ghedin E, DePasse JV et al (2012) Evolutionary history and attenuation of myxoma virus on two continents. PLoS Pathog 8: e1002950
139. Kerr PJ, Rogers MB, Fitch A et al (2013) Genome scale evolution of myxoma virus reveals host pathogen adaptation and rapid geographic spread. J Virol 87 (23):12900–12915
140. Kerr PJ, Eden J-S, Di Giallonardo F et al (2019) Punctuated evolution of myxoma virus: rapid and disjunct evolution of a recent viral lineage in Australia. J Virol 93: e01994–e01918
141. Liu J, Cattadori IM, Sim DG et al (2017) Reverse engineering field isolates of myxoma virus demonstrates that some gene disruptions or losses of function do not explain virulence changes observed in the field. J Virol 91:e01289–e01217
142. Sobey WR, Menzies W (1969) Myxomatosis: the introduction of the European rabbit flea *Spilopsyllus cuniculi* (Dale) into Australia. Aust J Sci 31:404–405
143. Sobey WR, Conolly D (1971) Myxomatosis: the introduction of the European rabbit flea *Spilopsyllus cuniculi* (Dale) into wild rabbit populations in Australia. J Hyg (Camb) 69:331–346
144. Cooke BD (1983) Changes in the age-structure and size of populations of wild rabbits in South Australia, following the introduction of European rabbit fleas, *Spilopsyllus cuniculi* (Dale), as vectors of myxomatosis. Aust Wildlife Res 10:105–120
145. Cooke BD (1984) Factors limiting the distribution of the European rabbit flea *Spilopsyllus cuniculi* (Dale) (Siphonoptera) in inland South Australia. Aust J Zool 32(4):493–506
146. Cooke BD, Skewes MA (1988) The effects of temperature and humidity on the survival and development of the European rabbit flea, *Spilopsyllus cuniculi* (Dale). Aust J Zool 36:649–659
147. Fenner F, Fantini B (1999) Biological control of vertebrate pests. The history of myxomatosis – an experiment in evolution. CAB International, New York
148. Saunders G, Cooke B, McColl K et al (2010) Modern approaches for the biological control of vertebrate pests: an Australian perspective. Biol Control 52:288–295
149. Saint KM, French N, Kerr P (2001) Genetic variation in Australian isolates of myxoma virus: an evolutionary and epidemiological study. Arch Virol 146:1105–1123
150. Berman D, Kerr PJ, Stagg R et al (2006) Should the 40-year-old practice of releasing virulent myxoma virus to control rabbits (*Oryctolagus cuniculus*) be continued? Wildlife Res 33:549–556
151. Kerr PJ, Hone J, Perrin L et al (2010) Molecular and serological analysis of the epidemiology of myxoma virus in rabbits. Vet Microbiol 143:167–178
152. Kovaliski J, Sinclair R, Mutze G et al (2014) Molecular epidemiology of rabbit haemorrhagic disease virus in Australia: when one became many. Mol Ecol 23:408–420
153. Strive T, Cox T (2019) Lethal biological tools for landscape-scale mitigation of rabbit impacts in Australia. Aust Zool. https://doi.org/10.7882/AZ.2019.7016
154. Cooke BD, Chudleigh P, Simpson S et al (2013) The economic benefits of the biological control of rabbits in Australia, 1950–2011. Aust Econ Hist Rev 53:1–17

Chapter 2

Development and Expression of Subunit Vaccines Against Viruses in Plants

Adrian Esqueda and Qiang Chen

Abstract

Various systems exist for the robust production of recombinant proteins. However, only a few systems are optimal for human vaccine protein production. Plant-based transient protein expression systems offer an advantageous alternative to costly mammalian cell culture-based systems and can perform posttranslational modifications due to the presence of an endomembrane system that is largely similar to that of the animal cell. Technological advances in expression vectors for transient expression in the last two decades have produced new plant expression systems with the flexibility and speed that cannot be matched by those based on mammalian or insect cell culture. The rapid and high-level protein production capability of transient expression systems makes them the optimal system to quickly and versatilely develop and produce vaccines against viruses such as 2019-nCoV that have sudden and unpredictable outbreaks. Here, expression of antiviral subunit vaccines in *Nicotiana benthamiana* plants via transient expression is demonstrated.

Key words Plant-produced vaccines, Agroinfiltration, Transient expression, Vaccines, *Nicotiana benthamiana*

1 Introduction

Flaviviruses such as the yellow fever and dengue viruses are responsible for causing illness in a large portion of the global population [1–3]. The 2015 outbreak of the flavivirus Zika (ZIKV) in the Americas and Caribbean was linked to devastating congenital conditions in newborns such as microcephaly and Guillain-Barre syndrome in adults [1–3]. Unfortunately, neither a vaccine nor a specific treatment is available, calling for the urgent development of ZIKV vaccines. Like all flaviviruses, the virally encoded envelope (E) protein covers the entire virion surface, consequently making it the main antigenic target of the immune system and neutralizing antibodies [4]. Thus, the ability to quickly express and evaluate the ZIKV E protein and its various domains becomes an important issue, highlighting the need for flexible and fast systems for vaccine protein production [5].

Alexandra R. Lucas (ed.), *Viruses as Therapeutics: Methods and Protocols*, Methods in Molecular Biology, vol. 2225,
https://doi.org/10.1007/978-1-0716-1012-1_2,

A wide range of recombinant protein production systems exist and include systems based on mammalian, insect, bacterial, and yeast cell culture [6]. However, these cell-based systems command a large investment in capital required to establish the necessary infrastructure for protein production. Furthermore, the output of proteins can reach a cap very quickly, hindering large-scale production of biologic proteins. Plant-based protein expression systems offer numerous advantages over traditional cell culture-based systems, such as greater scalability, faster protein production times, lower startup costs, and potential of great biomass production by employing commercial agriculture practice [6]. Furthermore, plants possess the eukaryotic endomembrane system which enables plants to perform posttranslational modifications that are important in the protein folding, structure, stability, and function [7]. In addition, plants also offer the advantage of a lower risk of harboring animal pathogens, thereby reducing the risk of inadvertent contamination of the final drug products [8, 9].

Recombinant protein expression in plants can be achieved from two sources: the nuclear genome and the plastid genome. Because posttranslational modifications are important in the function of therapeutic proteins, expression of proteins from the nucleus is preferred [10]. Nuclear or plastid stable expression of transgenes can be heritable through succeeding generations in the form of seeds, which allows transgenic plants to be grown on agricultural scales [10]. Despite the advantages of stable transformation, production of a plant line that stably expresses a transgene is laborious and extremely time consuming, and biosafety regulations are in place to restrict growing transgenic plants in open fields [11].

Transient expression is another option for the expression of recombinant proteins in plants. Here, the gene of interest (GOI) is expressed during its transient presence in the host cell [12]. Decades of research have led to the development of systems that use the ability of *Agrobacterium tumefaciens* (*A. tumefaciens*), a Gram-negative soil bacterium, in transferring its plasmid DNA to plant cells. In agroinfiltration, *A. tumefaciens* bacteria carrying the GOI are infiltrated into the interstitial space of the mesophyll of leaves to deliver the GOI into plant cells to drive target protein production [13, 14]. When compared to other methods, transient protein production via agroinfiltration reaches greater protein accumulation levels in much less production lead times. Furthermore, the risk of inadvertent environmental contamination is virtually eliminated since agroinfiltration is done indoors and infiltrated plants are grown in enclosed areas [13]. Due to the underlying genetic mechanisms of transient expression, infiltration must be repeated for each batch of recombinant protein desired. However, transient expression offers a significantly improved yield, biosafety, and production speed of recombinant proteins, which makes it the best selection for multiple applications including basic research and industrial-scale production of plant-made vaccine recombinant proteins.

The last two decades have seen advances in the development of numerous vectors for transient expression of recombinant proteins in plants. The more recently developed "deconstructed viral" vectors have gained favor due to their superiority in the production of high levels of recombinant protein production with unprecedented speed [15, 16]. Deconstructed viral vectors also offer advantages such as improved biosafety and reduced environmental risks due to the elimination of infectious viral functions [17, 18]. In particular, the MagnICON and geminiviral deconstructed viral vectors are described here. The MagnICON vectors are RNA replicons based on the genomes of the tobacco mosaic virus and the noncompeting potato virus X [19]. On the other hand, the geminiviral vectors are a DNA replicon system based on the bean yellow dwarf virus and it is comprised of three parts that can be combined to form a single vector: the pBY-GOI segment containing the replicon of the target gene, the pREP110 construct coding for the Rep/RepA protein necessary for replication, and the p19 that codes for a post-transcription silencing suppressor [20, 21]. Noteworthy, one or more replicons can be encoded in a single geminiviral vector to produce several different proteins or a protein with multiple hetero-subunits [21]. Additionally and in contrast to MagnICON vectors, geminiviral vectors have a broader range of plant hosts. Both systems have been successfully used to produce plant-made pharmaceuticals [22, 23]. Results from our laboratory and others have shown that transient expression with "deconstructed" plant viral vectors allows for the production of up to 5 mg of antivirus vaccines per gram of leaf fresh weight within 4–10 days of vector inoculation [16, 24–33]. Therefore, the rapid and high-level protein production capability of these new transient expression systems makes them the optimal system to produce and evaluate vaccine candidates for viral diseases with unpredictable and sudden outbreaks. To transiently express vaccine proteins in plants, the GOI has to be cloned into an appropriate transient expression vector and this GOI-carrying "deconstructed" plant viral vector is then introduced into *A. tumefaciens*.

The introduction of GOI into plant cells can be done by agroinfiltration during which the transient expression vector is transferred into plant cells by *A. tumefaciens*. Syringe infiltration is the most common method of agroinfiltration in the laboratory for small-scale GOI delivery as it allows multi-GOI infiltration on different parts of the plant or combinations thereof.

Nicotiana benthamiana (*N. benthamiana*) has been the plant of choice for transient protein production since it offers advantages such as a thoroughly characterized biology, high biomass yield, abundant availability of compatible expression vectors, and amenability to agroinfiltration [8]. Agroinfiltrated plants will return to greenhouses to allow target protein production. With MagnICON or geminiviral vectors, transiently expressed target proteins generally reach their peak accumulation between 5 and 10 days post-agroinfiltration. Leaves will be harvested and total soluble proteins will be extracted from the plant material [34–38]. The target protein will then be purified by a target-protein-specific process [39, 40]. Purified target proteins will be further characterized by biochemical and immunological assays and ultimately tested in animal models. Therefore, before expressing a target protein in plants, a thorough purification strategy should be in place to remove unwanted compounds so that the final recombinant protein preparation is conducive to downstream applications and assays. Here, a protocol designed to produce a subunit vaccine based on ZIKV E protein by transient expression via syringe agroinfiltration of *N. benthamiana* plants is presented.

2 Materials

Prepare all media using double-distilled water and sterilize via autoclave or a filtration method that is capable of excluding bacterial contamination. Please observe biological and chemical waste disposal regulations when disposing waste material.

2.1 Plant Growth Materials

1. T.O. Plastics (Clearwater, MN) SFT-1020-OPEN-NH standard flat carry tray.
2. T.O. Plastics STI-804 standard insert.
3. Humidity dome.
4. Peat pellets (Jiffy-7727W/Hole, Canada).
5. Handheld seeder (e.g., Vibro Hand Seeder VHS-1).
6. Water-soluble fertilizer: 15–16–17.
7. *N. benthamiana* seeds.
8. Linear fluorescent bulbs Sylvania FO32/25 W/835/XP/SS/ECO3 (Wilmington, MA).

2.2 Bacterial Culture

1. *A. tumefaciens* strain GV3101 carrying the GOI.
2. 100 mg/mL (1000×) Carbenicillin disodium (*see* **Note 1**).
3. 16.65 mg/mL (333×) Rifampicin.
4. Yeast extract nutrient broth (YenB) medium: 1% Nutrient broth, 0.5% yeast extract. When preparing agar plates, add 1.5% agar to the YenB medium.
5. Shaker incubator.
6. Spectrophotometer capable of reading absorbance at 600 nm.

2.3 Infiltration

1. 15 mL Syringe.
2. Sterile thumbtack.
3. Rubber bands.
4. Tray.
5. Absorbing pad.
6. 2× 100 mL Plastic beakers.
7. Goggles.
8. Gloves.
9. 2-(*N-morpholino*) ethanesulfonic acid (MES) buffer (1×: 10 mM MES hydrate (FW 195.24), 10 mM $MgSO_4$ hydrate (FW 246.47)): Mix in deionized water and adjust pH to 5.5 with sodium hydroxide (NaOH). Sterilize by passing the MES buffer through a 0.2 μM filter.
10. Phenylmethylsulfonyl fluoride (PMSF) 1 M stock.

2.4 Harvesting

1. Razor blade(s).
2. Backing board.
3. Blender.
4. Centrifuge bottles/caps.
5. Funnel.
6. PBS buffer, pH 5.2.
7. PMSF.
8. Ice container.
9. Ice.
10. Cheese cloth.
11. 0.5–0.2 μM Vacuum filtration device.

2.5 Immobilized Metal Affinity Chromatography

1. Ring stand.
2. Chromatography column.
3. Nickel-nitrilotriacetic acid (Ni-NTA) agarose beads.
4. Imidazole buffers (50, 100, 300, 500 mM).
5. 50 and 15 mL conical tubes.

3 Methods

Plants are grown inside a climate-controlled room at 25 °C, 60–80% humidity, and with a 16/8-h day/night cycle under artificial fluorescent lighting (85–100 Lux).

Fig. 1 Peat pellets and growth trays with a humidity-promoting dome. After sowing with *N. benthamiana* seeds the tray should be covered to promote germination

3.1 Plant Growth

1. Place 8 standard inserts and 32 peat pellets inside the flat carry tray. Add 1–2 L water and allow the pellets to expand to their full size.
2. Using a handheld seeder, add 2–4 *N. benthamiana* seeds to the center of each pellet. Add enough deionized water to cover the bottom quarter of the peat pellet (*see* **Note 2**).
3. Cover the tray with a humidity dome (Fig. 1) and allow the seeds to germinate for 1 week in the growth chamber. Beginning with week 2, shift the dome so that a small opening is created (*see* **Note 3**).
4. After 2 weeks, remove the dome and add 2 L of 1.48 g/L fertilizer to the tray. Continue to grow plants under the same environmental conditions, adding fertilizer every 2 days or as necessary (*see* **Notes 4** and **5**).
5. Starting with week 5, remove all but five of the largest leaves on the plant. They should ideally be evenly spaced.
6. Continue to fertilize the plants and remove any new sprouting leaves, leaving a total of just five leaves to fully grow, up until the end of week 6, at which point the plants are ready for infiltration (Fig. 2).

3.2 **A. tumefaciens** *Culturing and Preparation for Syringe Infiltration*

1. Streak 50–100 μL (5.0×10^7–1.0×10^8 CFU) *A. tumefaciens* GV3101 strains containing ICON vector modules carrying the gene of interest (e.g., ZIKV E protein) on LB-agar plates with 100 μg/mL carbenicillin and 333 μg/mL of rifampicin. Grow for 48 h at 30 °C [5].

Fig. 2 *N. benthamiana* plants at the beginning of week 6. At week 6, the plants are ready for agroinfiltration. The plants should have five large and healthy leaves uniform in color

2. Inoculate a 10 mL liquid YenB starter culture from a single colony for each strain. Grow at 30 °C in a shaker (200 RPM) for 16–20 h, until the OD600 reaches at least 1.0 (measure OD600 with a spectrophotometer).
3. Use the culture to inoculate a new 100 mL YenB subculture containing the appropriate antibiotics for each strain and incubate at 30 °C and 200 rpm for 16–20 h to reach an approximate OD600 of 1.5–1.8.
4. Centrifuge the cultures at room temperature and 6000 × *g* for 15 min. Remove the supernatant to obtain a bacterial pellet.
5. Resuspend the pellet in 10 mL MES buffer and measure the OD600 (*see* **Note 6**).
6. If multiple vector modules are used, mix equal amounts of each vector-carrying *Agrobacterium* strain together in 200 mL of MES buffer. Use the following formula to calculate the volume of each of the strains to transfer. The final OD600 of each strain should be equally divided to obtain a total OD600 of 0.21 (*see* **Note 7**):

$$V\text{subculture} = \frac{(\text{OD600 } 0.21)(200\text{ mL})}{\text{OD600}}$$

3.3 Syringe Agroinfiltration

1. Gather a tray, paper towels, adsorbent pad, syringe, thumbtack, and rubber band.
2. Assemble the syringe/thumbtack infiltration setup using a rubber band (*see* Fig. 3 and **Note 8**).
3. Pour approximately 200 mL of infiltration media into a plastic 300 mL beaker (*see* **Note 9**).
4. With the syringe, siphon approximately 10 mL of the infiltration media.

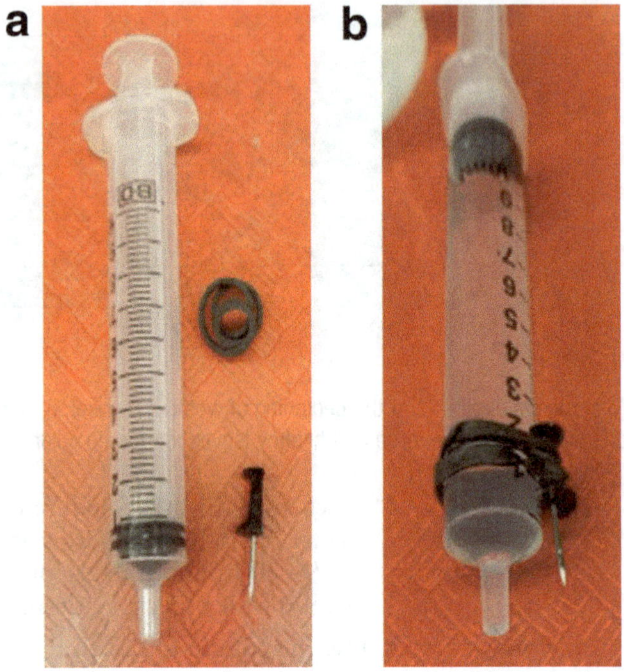

Fig. 3 The assembly of the syringe/thumbtack apparatus allows for the quicker agroinfiltration of plant leaves. (**a**) The syringe/thumbtack apparatus consists of a 15 mL syringe, rubber band, and a thumbtack. To increase the ease of use and stability of the thumbtack on the syringe, it is best to cut off one side of the thumbtack's plastic handle using cutting pliers. (**b**) Secure the thumbtack to the syringe with a couple of loops using a rubber band

5. Pick a leaf and target infiltration area. With the hand not holding the syringe, place a tip of a finger against the back side of the leaf to act as a support for infiltration.
6. With the hand holding the syringe, lightly pierce a small hole with the thumbtack and place the nozzle of the syringe over the hole.
7. Slowly push on the syringe plunger and observe infiltration media spread throughout the leaf (Fig. 4) (*see* **Note 10**).
8. Pick a different spot on the leaf and infiltrate again. Once the entire leaf takes on a darker color due to the infiltration media, the leaf is done.
9. Infiltrate all leaves in plants and place back on tray.
10. Once all plants on tray have been infiltrated, fill the tray with tap water up to 1/3 of the peat pellet, return tray to growth chamber, and monitor expression (*see* note: leaves should be harvested in approximately 1 week).

Fig. 4 Syringe agroinfiltration of *N. benthamiana* leaves. Slowly depress the plunger and allow the leaf to take up the infiltration media while noting pressure and relative rate of flow/spreading of the infiltrate. Once the rate of flow slows down, pick a different part of the leaf and begin the infiltration process again. Infiltration is complete once the entire leaf has changed to a slightly darker shade of green

3.4 Harvesting, Isolation, and Characterization of Recombinant Zika Envelope Protein

1. Harvest the plant leaves on the sixth day post-infiltration by cutting the leaf stem from the plant.
2. Remove the central vein from the leaf using a razor blade leaving behind just the two halves of the leaf. Discard the central vein.
3. Prepare the homogenization buffer using the following calculation:

$$\text{Homogization buffer} = (\times \text{g of leaves}) * \left(\frac{1.5\ \text{mL}}{\text{g of leaves}}\right)$$

4. The homogenization buffer should contain phenylmethylsulfonyl fluoride at a final concentration of 2 mM.
5. Place bottles on ice and funnel on top of them with the cheesecloth over the funnel.

6. Add the homogenization buffer to the blender and add the leaves.
7. Blend at low speed until the leaves have largely been broken down. Thereafter, increase the blender speed to high for about 1 min.
8. Pour the homogenate on the cloth on top of the funnel. Allow the homogenate to flow through the cloth for about a minute.
9. Carefully wrap the homogenate with the cloth and gently squeeze to allow filtration of the homogenate to flow through the funnel and into the bottle(s) (*see* **Note 11**).
10. Spin the sample at 14.8K × *g* for 30 min at 4 °C, twice.
11. Pour the supernatant onto a large plastic container (on ice) and add a stir bar.
12. Adjust the pH to 5.2 and let sit overnight at 4 °C or a minimum of 4 h. The isoelectric point of the ribulose-1,5-bisphosphate, an enzyme involved in the first major step of carbon fixation, is reached at a pH of 5.2, thus promoting the removal of this most abundant plant leaf protein via isoelectric point precipitation.
13. Transfer the sample to centrifuge bottles and spin again at 14.8K × *g* for 30 min at 4 °C, twice to remove the precipitated protein.
14. Transfer the sample to a plastic container and adjust the pH to 7.0.
15. Transfer the sample to centrifuge bottles and spin at 14.8K × *g* at 4 °C.
16. Transfer the sample to a vacuum-filtering device to remove fine particles that can potentially and quickly foul the column.
17. Using immobilized metal affinity chromatography, flow the sample through Ni^{2+} His-Bind column in accordance with the manufacturer's instructions (MilliporeSigma, Billerica, MA).
18. Analyze purified ZIKA envelop protein by SDS-PAGE and Western blot under reducing conditions [5]. The expected molecular weight of plant-produced Zika E protein is approximately 52.7 kDa (*see* Fig. 5).

4 Notes

1. All antibiotics must be filter sterilized.
2. Use 2–4 seeds to maximize the possibility of germination.
3. The use of a dome significantly aids in maintaining a high level of humidity and minimal airflow. Beginning with week 2, the

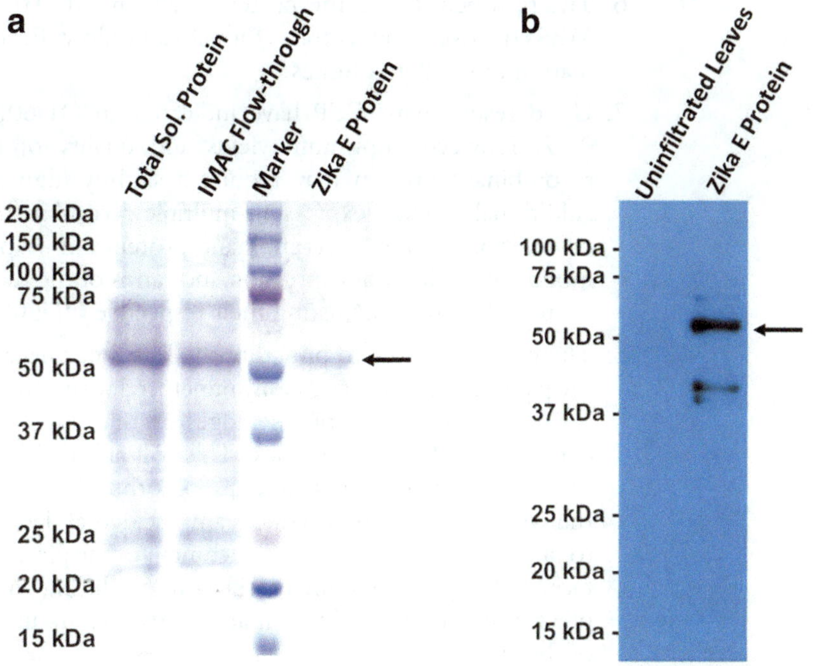

Fig. 5 SDS-PAGE and Western blot analysis of plant-produced Zika E expressed in *N. benthamiana* plants and purified via Ni^{2+}-immobilized metal affinity chromatography. (**a**) Plant-produced Zika envelope recombinant protein was analyzed on a 12% SDS-PAGE Coomassie-stained gel. The far-right lane shows the plant-produced Zika envelope recombinant protein slightly above the 50 kDa marker (arrow), as expected for the calculated MW of 52.7 kDa. The far-left lane contains total leaf-soluble protein and the second lane from the right contains Ni^{2+}-immobilized metal affinity chromatography flow-through. (**b**) Western blot analysis of plant-produced ZIKV E protein. The left lane contains extract from uninfiltrated leaves, whereas the right lane contains extract from the leaves agroinfiltrated with the plant-produced Zika envelope recombinant protein construct. The plant-produced Zika envelope recombinant protein is noted slightly above the 50 kDa mark, as expected

dome is shifted to unseat it, thereby allowing airflow and a subsequent decrease in humidity.

4. The condition and overall quality of the host plant are the utmost important factors in the final yield and quality of the target recombinant protein. Thus, it is imperative to monitor the color of the plants and avoid the use of any leaves that are yellowing, necrotic, beginning to wilt, or thin to the touch (may be dehydrated). These changes or abnormalities can translate to significant changes in recombinant protein yield and quality.
5. Avoid overwatering the plants since it will lead to the overgrowth of roots and cover the trays. Ideally, it is best to re-fertilize when the tray has become close to dry and the peat pellets are still visibly moist.

6. The resuspension of the bacterial cells in a 1/10 volume of MES effectively concentrates the cells and allows for the manipulation of smaller volumes.
7. Good results with GFP have indicated an OD600 value of 0.07. However, optimum yields and quality of the target recombinant protein may require a slightly higher OD600. Additionally, when expressing multiple recombinant proteins or multiple subunit recombinant proteins, it is possible to modulate the level of expression and ratios of the recombinant proteins by simply adjusting their respective OD600s.
8. The use of a needle to puncture the leaf is not necessary since the plant tissue is soft and easily punctured with a thumbtack. It is also helpful to trim off the edges on one side of the plastic component of the thumbtack. This will aid in the stability of the assembled apparatus. The use of a thumbtack attached to the syringe significantly reduces any accidental deep punctures to fingers. Furthermore, the attachment of the piercing thumbtack to the syringe eliminates the constant lifting and releasing of a thumbtack needed to initiate infiltration on different parts of the leaves.
9. Prior to starting the infiltration process, it is helpful to place a few paper towels under the front part of the tray so as to elevate it and allow infiltrate overflow to run to the back of the tray. This significantly helps keep the process tidy.
10. Avoid initiating infiltration on any leaf vein, especially the central vein. It is extremely difficult to infiltrate from the central vein and counterproductive. It is important to start the infiltration process slowly and gauge the pressure and rate of spread of the infiltration media. If too much pressure is applied a hole may be blown on the leaf tissue. Furthermore, blistering of the leaf may occur at which point one should back off and reinitiate the infiltration process at a different part of the leaf.
11. When squeezing the homogenate in the cheesecloth, it is best if all sides and angles are squeezed in alternating fashion. This will ensure that all areas of the homogenate are pressed and will also minimize the possibility of bursting the cheesecloth and spreading the homogenate all over the work area.

References

1. Attar N (2016) ZIKA virus circulates in new regions. Nat Rev Microbiol 14(2):62
2. Dowd KA, Ko S-Y, Morabito KM et al (2016) Rapid development of a DNA vaccine for Zika virus. Science 354:237
3. Hickman HD, Pierson TC (2016) Zika in the brain: new models shed light on viral infection. Trends Mol Med 22(8):639–641
4. Zhao H, Fernandez E, Dowd KA et al (2016) Structural basis of Zika virus-specific antibody protection. Cell 166(4):1016–1027
5. Yang M, Sun H, Lai H et al (2018) Plant-produced Zika virus envelope protein elicits neutralizing immune responses that correlate with protective immunity against Zika virus in mice. Plant Biotechnol J 16(2):572–580

6. Chen Q (2008) Expression and purification of pharmaceutical proteins in plants. Biol Eng 1 (4):291–321
7. Chen Q, Mason HS, Mor T et al (2009) Subunit vaccines produced using plant biotechnology. In: Levine MM (ed) New Generation Vaccines, 4th edn. Informa Healthcare, Boca Raton, pp. 306–315. https://doi.org/10.3109/9781420060744.030
8. Chen Q (2011) Expression and manufacture of pharmaceutical proteins in genetically engineered horticultural plants. In: Scorza R, Baiquan M (eds) Transgenic Horticultural Crops Challenges and Opportunities. Taylor & Francis, pp 83–124. https://doi.org/10.1201/b10952-5
9. Faye L, Gomord V (2010) Success stories in molecular farming—a brief overview. Plant Biotechnol J 8(5):525–528
10. Davies HM (2010) Review article: Commercialization of whole-plant systems for biomanufacturing of protein products: evolution and prospects. Plant Biotechnol J 8(8):845–861
11. Chen Q (2011) Turning a new leaf. Eur Biopharm Rev 2:64–68
12. Komarova TV, Baschieri S, Donini M et al (2010) Transient expression systems for plant-derived biopharmaceuticals. Expert Rev Vaccines 9(8):859–876
13. Chen Q, Lai H, Hurtado J et al (2013) Agroinfiltration as an effective and scalable strategy of gene delivery for production of pharmaceutical proteins. Adv Tech Biol Med 1(1):103
14. Leuzinger K, Dent M, Hurtado J et al (2013) Efficient agroinfiltration of plants for high-level transient expression of recombinant proteins. J Vis Exp 77:50521
15. Gleba Y, Marillonnet S, Klimyuk V (2004) Engineering viral expression vectors for plants: the 'full virus' and the 'deconstructed virus' strategies. Curr Opin Plant Biol 7(2):182–188
16. Lico C, Chen Q, Santi L (2008) Viral vectors for production of recombinant proteins in plants. J Cell Physiol 216(2):366–377
17. Buswell S, Medina-Bolivar F, Chen Q et al (2005) Expression of porcine prorelaxin in transgenic tobacco. Ann N Y Acad Sci 1041 (1):77–81
18. Santi L, Batchelor L, Huang Z et al (2008) An efficient plant viral expression system generating orally immunogenic Norwalk virus-like particles. Vaccine 26(15):1846–1854
19. Marillonnet S, Thoeringer C, Kandzia R et al (2005) Systemic Agrobacterium tumefaciens--mediated transfection of viral replicons for efficient transient expression in plants. Nat Biotechnol 23(6):718–723
20. Chen Q, He J, Phoolcharoen W et al (2011) Geminiviral vectors based on bean yellow dwarf virus for production of vaccine antigens and monoclonal antibodies in plants. Hum Vaccin 7(3):331–338
21. Huang Z, Phoolcharoen W, Lai H et al (2010) High-level rapid production of full-size monoclonal antibodies in plants by a single-vector DNA replicon system. Biotechnol Bioeng 106 (1):9–17
22. Diamos AG, Hunter JGL, Pardhe MD et al (2020) High level production of monoclonal antibodies using an optimized plant expression system. Front Bioeng Biotechnol 7:472–472
23. Rattanapisit K, Phakham T, Buranapraditkun S et al (2019) Structural and in vitro functional analyses of novel plant-produced anti-human PD1 antibody. Sci Rep 9(1):15205
24. Chen Q (2016) Glycoengineering of plants yields glycoproteins with polysialylation and other defined N-glycoforms. Proc Natl Acad Sci U S A 113(34):9404–9406
25. He J, Peng L, Lai H et al (2014) A plant-produced antigen elicits potent immune responses against West Nile virus in mice. Biomed Res Int 2014:10. https://doi.org/10.1155/2014/952865
26. Huang Z, Chen Q, Hjelm B et al (2009) A DNA replicon system for rapid high-level production of virus-like particles in plants. Biotechnol Bioeng 103(4):706–714
27. Lai H, He J, Engle M et al (2012) Robust production of virus-like particles and monoclonal antibodies with geminiviral replicon vectors in lettuce. Plant Biotechnol J 10(1):95–104
28. Lai H, Paul AM, Sun H et al (2018) A plant-produced vaccine protects mice against lethal West Nile virus infection without enhancing Zika or dengue virus infectivity. Vaccine 36 (14):1846–1852
29. Loos A, Steinkellner H (2014) Plant glyco-biotechnology on the way to synthetic biology. Front Plant Sci 5:523
30. Phoolcharoen W, Bhoo SH, Lai H et al (2011) Expression of an immunogenic Ebola immune complex in Nicotiana benthamiana. Plant Biotechnol J 9(7):807–816
31. Phoolcharoen W, Dye JM, Kilbourne J et al (2011) A nonreplicating subunit vaccine protects mice against lethal Ebola virus challenge. Proc Natl Acad Sci U S A 108 (51):20695–20700
32. Shaaltiel Y, Tekoah Y (2016) Plant specific N-glycans do not have proven adverse effects in humans. Nat Biotechnol 34(7):706–708
33. Yang M, Lai H, Sun H et al (2017) Virus-like particles that display Zika virus envelope

protein domain III induce potent neutralizing immune responses in mice. Sci Rep 7(1):7679

34. Dent M, Hurtado J, Paul AM et al (2016) Plant-produced anti-dengue virus monoclonal antibodies exhibit reduced antibody-dependent enhancement of infection activity. J Gen Virol 97(12):3280–3290
35. He J, Lai H, Engle M et al (2014) Generation and analysis of novel plant-derived antibody-based therapeutic molecules against West Nile virus. PLoS One 9(3):e93541. https://doi.org/10.1371/journal.pone.0093541
36. Hurtado J, Acharya D, Lai H et al (2019) In vitro and in vivo efficacy of anti-chikungunya virus monoclonal antibodies produced in wild-type and glycoengineered Nicotiana benthamiana plants. Plant Biotechnol J 18(1):266–273
37. Lai H, Engle M, Fuchs A et al (2010) Monoclonal antibody produced in plants efficiently treats West Nile virus infection in mice. Proc Natl Acad Sci U S A 107(6):2419–2424
38. Lai H, He J, Hurtado J et al (2014) Structural and functional characterization of an anti-West Nile virus monoclonal antibody and its single-chain variant produced in glycoengineered plants. Plant Biotechnol J 12(8):1098–1107
39. He J, Lai H, Brock C et al (2012) A novel system for rapid and cost-effective production of detection and diagnostic reagents of West Nile virus in plants. J Biomed Biotechnol 2012:1–10. https://doi.org/10.1155/2012/106783
40. Yang M, Dent M, Lai H et al (2017) Immunization of Zika virus envelope protein domain III induces specific and neutralizing immune responses against Zika virus. Vaccine 35(33):4287–4294

Chapter 3

Development of Antiviral Vaccine Utilizing Self-Destructing *Salmonella* for Antigen and DNA Vaccine Delivery

Wei Kong

Abstract

Vaccines are the most effective means to prevent infectious diseases, especially for viral infection. The key to an excellent antiviral vaccine is the ability to induce long-term protective immunity against a specific virus. Bacterial vaccine vectors have been used to impart protection against self, as well as heterologous antigens. One significant benefit of using live bacterial vaccine vectors is their ability to invade and colonize deep effector lymphoid tissues after mucosal delivery. The bacterium *Salmonella* is considered the best at this deep colonization. This is critically essential for inducing protective immunity. This chapter describes the methodology for developing genetically modified self-destructing *Salmonella* (GMS) vaccine delivery systems targeting influenza infection. Specifically, the methods covered include the procedures for the development of GMSs for protective antigen delivery to induce cellular immune responses and DNA vaccine delivery to induce systemic immunity against the influenza virus. These self-destructing GMS could be modified to provide effective biological containment for genetically engineered bacteria used for a diversity of purposes in addition to vaccines.

Key words Self-destructing, Genetically modified *Salmonella*, Codon optimization, DNA Vaccine, Influenza

1 Introduction

Geneticallymodified *Salmonella* strains that harbor a plasmid to synthesize protective antigens or DNA vaccine vectors have been, and are currently, being developed to provide novel needle-free and low-cost protection against various diseases. Oral administration of GMSs enables *Salmonella* to attach to and invade mucosal associated lymphoid tissues (*see* **Notes 1–3**). The use of *Salmonella* GMS leads to the induction of mucosal, systemic, and cellular immune responses, which result in long-term protective immunity [1–6]. The methodology of antiviral vaccine delivery described in this chapter is to use a biological containment system designed to cause programmed bacterial cell lysis with no survivors. The system is composed of two parts. The first component is *Salmonella* strain

Alexandra R. Lucas (ed.), *Viruses as Therapeutics: Methods and Protocols*, Methods in Molecular Biology, vol. 2225, https://doi.org/10.1007/978-1-0716-1012-1_3, © Springer Science+Business Media, LLC, part of Springer Nature 2021

with deletions of two genes required for peptidoglycan synthesis and additional mutations to enhance complete lysis and antigen delivery. The second component is a plasmid, which encodes arabinose-regulated peptidoglycan synthesis. These GMS strains are attenuated yet capable of synthesizing protective antigens. The regulated delayed attenuation and programmed self-destructing features are designed into these GMS strains to enable them to efficiently colonize host tissues and allow the release of the bacterial cell contents after lysis [2]. To turn such an antigen-delivering GMS strain into a universal DNA vaccine-delivery vehicle, the approach is to enable GMS strain to display a hyperinvasive phenotype to maximize *Salmonella* host entry and host cell internalization, to enable *Salmonella* endosomal escape to release a DNA vaccine into the cytosol and to decrease *Salmonella*-induced pyroptosis/apoptosis that allows the DNA vaccine time to traffic to the nucleus for efficient synthesis of encoded protective antigens [3]. The development and application of the GMS strains are described in detail in the following sections.

1.1 Self-Destructing GMS for Antigen Delivery

To eliminate the use of plasmid vectors with non-permitted drug resistance genes and to stabilize plasmid vectors in *Salmonella* strains in vivo, a balanced lethal *Salmonella* host-vector system with deletion of the aspartate-semialdehyde dehydrogenase (*asdA*) gene is developed to impose an obligate requirement for diaminopimelic acid (DAP) [7] and a plasmid vector with the wild-type *asdA* gene [8, 9]. Muramic acid is another essential constituent of peptidoglycan. The UDP-N-acetylglucosamine enolpyruvyl transferase (*murA*) gene encodes the first enzyme in muramic acid synthesis [10]. The *asdA* and *murA* genetic modifications are combined, which provides redundant mechanisms to ensure *Salmonella* cell lysis. A regulated delayed lysis system is devised for antigen delivery after colonization of host lymphoid tissues that rely on using a more tightly regulated *araC* P_{BAD} activator-promoter (encodes a positive regulatory protein and a promoter required for L-arabinose utilization in *Escherichia coli*) [2] than the original sequence from *E. coli* B/r [11] for the arabinose-dependent synthesis of the AsdA and MurA enzymes. This system is composed of two parts. The first component is the GMS strain χ8937 (genotype: Δ*asdA*::TT *araC* P_{BAD}*c2* TT ΔP_{murA}::TT *araC* P_{BAD}*murA* Δ(*gmd-fcl*) Δ*relA* Δ*endA*, symbols "Δ", "::", and "P" stand for gene deletion, gene insertion, and promoter, respectively) with deletion of the *asdA* gene, arabinose-regulated expression of *murA*, and additional mutations to enhance complete lysis and antigen or DNA vaccine delivery. Unlike strains with *asdA* deletions, which can be grown by the addition of DAP to the growth medium, strains with *murA* deletions are lethal due to an inability to supply the phosphorylated substrate for the MurA enzyme. Therefore, a conditional lethal arabinose-dependent *murA* mutation is created by replacing the chromosomal *murA* promoter with the *araC* P_{BAD} activator-promoter. Although arabinose is present in plant foods, most are in a complex form unavailable to support the growth of strains with ΔP_{murA}::TT *araC* P_{BAD}*murA* deletion-insertion mutations. Thus a strain with the ΔP_{murA}::TT *araC* P_{BAD}*murA* mutation undergoes about two cell divisions and then commences lysing in media without arabinose [2].

The GDP-mannose 4,6-dehydratase and GDP-L-fucose synthase mutation Δ(*gmd-fcl*) deletes genes encoding enzymes for GDP-fucose synthesis, thereby precluding the formation of colanic acid, a polysaccharide made in response to stress associated with cell wall damage [12]. This mutation is included because we have observed that under some conditions, *asdA* mutants can survive if they produce copious amounts of colanic acid [13]. Therefore, by deleting the genes required for colanic acid synthesis, this possibility is circumvented. The ribosome-dependent (p)ppGpp synthetase mutation Δ*relA* uncouples cell wall-less death from dependence on protein synthesis to further ensure that the bacteria do not survive in vivo or after excretion and to allow for maximum antigen production in the face of amino acid starvation resulting from a lack of aspartate semi-aldehyde dehydrogenase synthesis due to the *asdA* mutation [14, 15]. The second component is plasmid pYA3681 (Fig. 1) [2], which encodes arabinose-regulated *murA* and *asdA* expression and bacteriophage P22 repressor protein (C2)-regulated synthesis of antisense *asdA* and *murA* mRNA transcribed from the P22 P_R promoter with opposite polarity at the 3′ end of the *asdA* gene. The mRNA translation efficiency is reduced for both the *murA* and *asdA* genes by changing their start codons

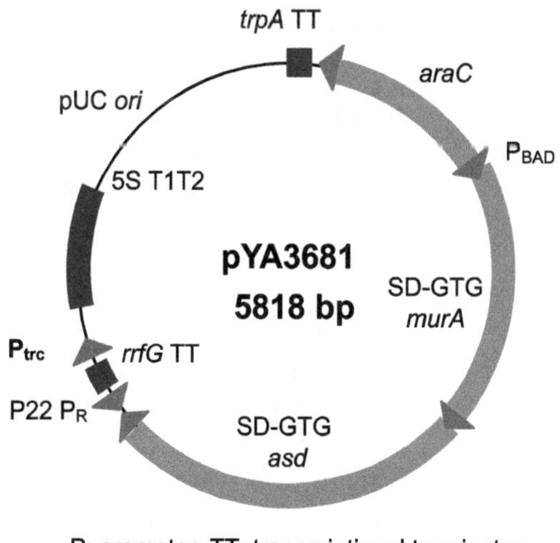

Fig. 1 Map of plasmid pYA3681. Plasmid sequences include the *trpA, rrfG,* and 5S ribosomal RNA transcriptional terminators; the P_{BAD}, P_{trc}, and P22 P_R promoters; the *araC* gene; and the start codon-modified *murA* and *asdA* genes

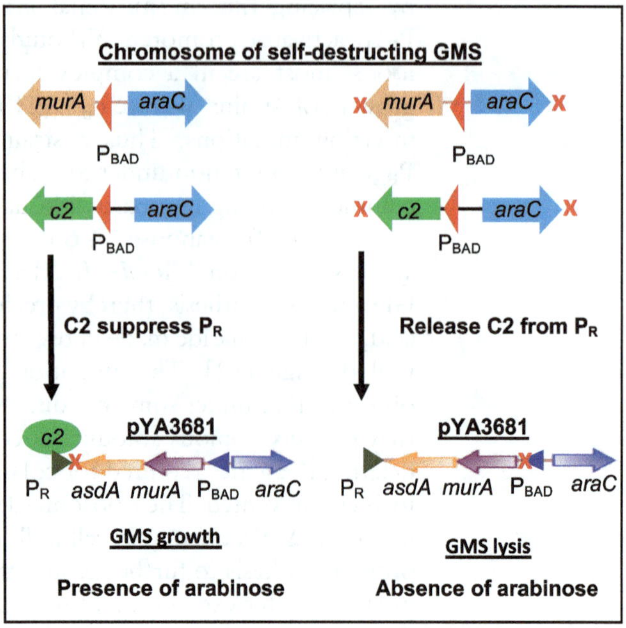

Fig. 2 Schematic illustration of regulatory interactions in self-destructing GMS for peptide delivery. In the presence of arabinose, *asdA, murA,* and *c2* are transcribed, allowing for bacterial growth and repression of the P22 P_R promoter by C2. In the absence of arabinose, such as animal host, no further synthesis of Asd and MurA or C2. The concentrations of Asd, MurA, and C2 decrease due to cell division, leading to reduced synthesis of DAP and muramic acid, and imbalanced synthesis of the rigid layer of the cell wall. As the C2 concentration drops, P22 P_R is de-repressed, resulting in P_R-directed synthesis of antisense mRNA, which blocks the translation of residual *asdA* and *murA* mRNA. These concerted activities lead to cell lysis to release synthesized peptides

from ATG to GTG. An arabinose-regulated *c2* gene is present in the chromosome due to the Δ*asdA*::TT *araC* P_{BAD}*c2* TT deletion-insertion. The cloning of a sequence encoding a protective antigen is under P_{trc} control. Transcription terminators (TT) flank all of the domains for controlling lysis, replication, and expression so that the expression of a function in one domain does not affect the activities of another domain. Upon invasion of host tissues, an arabinose-free environment, transcription of *asdA*, *murA*, and *c2* ceases, and concentrations of their gene products decrease due to cell division. The drop in C2 concentration results in the activation of P_R, driving the synthesis of antisense mRNA to block translation of any residual *asdA* and *murA* mRNA (Fig. 2) [2, 3]. This host-vector grows in LB broth with 0.2% L-arabinose as well as the wild-type strain χ3761, but is unable to grow in or on media devoid of arabinose since it undergoes cell wall-less death by lysis [2]. Vaccine strains with this regulated lysis system are avirulent at oral doses in excess of 10^9 CFU to BALB/c mice and, by the release of a bolus of protective antigen upon lysis, induce excellent immune responses [2]. These self-destructing GMS host-vector systems are ideal for the delivery of antigens that are difficult to secrete due to structural attributes and/or DNA vaccinevectors. In addition, they provide complete biological containment with no persistence in vivo and no survival if excreted [2].

1.2 Self-Destructing GMS Enabling Immediate Escape from the Endosome Compartment into the Cytosol

Salmonella invasion into host cells results in the bacterium residing in an endosomal compartment, often termed the *Salmonella*-containing vacuole. This problem is solved by deleting the *sifA* gene, which encodes a type III secreted effector protein SifA that governs the conversion of the *Salmonella*-containing vacuoles into filaments. Thus inactivation of *sifA* leads to the escape of *Salmonella* into the cytosol [16]. Including a Δ*sifA* mutation enables self-destructing GMS immediately upon invasion into a host cell, exits the endosome, and rapidly multiplies in the cytoplasm. Mice immunized with an improved GMS with the Δ*sifA* mutation, harboring lysis vector pYA3681 expressing the codon-optimized nucleoprotein (NP) gene of influenza WSN strain, are fully protected from challenge with 100 times the LD_{50} of rWSN [6].

1.3 Self-Destructing GMS for DNA Vaccine Delivery

A second regulated delayed lysis host-vector system is devised that harbors DNA vaccine vector with the same regulatory domain that contributes to the lysis phenotype encoded on pYA3681 but with a eukaryotic expression promoter. Although the use of nonviral DNA vaccine vectors offers advantages, such as decreasing inflammatory responses, gene expression in vivo remains much lower than observed with their viral counterparts. One reason for such low-level expression is that bacterial plasmids have not evolved mechanisms to target the nucleus in nondividing cells [17–19]. Plasmid nuclear import is dependent on DNA nuclear-targeting sequences (DTS) [20–22]. The DTS frequently contain transcription factor-binding sites, which allow transcription factors to bind to the plasmid in the cytoplasm and escort it to the nucleus by the nuclear localization signal-mediated machinery. The simian virus 40 (SV40) enhancer, which is known to bind to over ten distinct transcription factors, is an excellent DTS [23]. The minimum requirement for this function is a single copy of a 72 bp element of the SV40 enhancer, in combination with the cytomegalovirus (CMV) immediate-early gene enhancer/promoter (CMV E/P) [24]. In addition, the synthesis of eukaryotic transcription factors, e.g., nuclear factor kappa-light-chain enhancer of activated B cells (NF-κB) and activating enhancer-binding protein 2 (AP-2), are stimulated by *Salmonella* infection [25–27]. These transcription factors can bind to nonviral DNA vaccinevectors, allowing the nuclear-locating signal to mediate the import of plasmid DNA into the nucleus. However, nuclease degradation of DNA

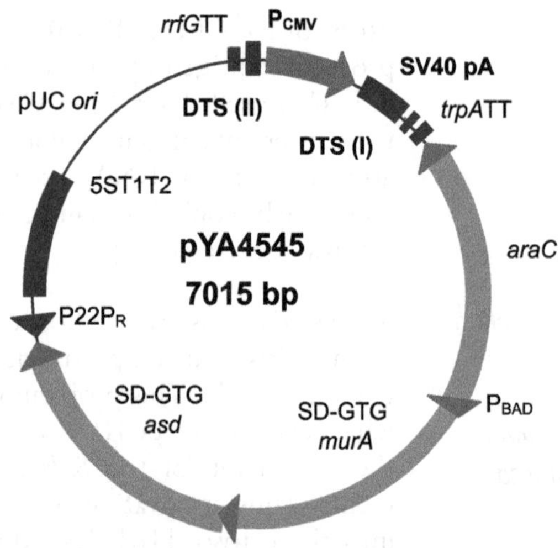

Fig. 3 Map of improved DNA vaccine vector pYA4545. Plasmid sequences include the *rrfG, trpA,* and 5S ribosomal RNA transcriptional terminators; the P_{BAD}, P22 P_R, and P_{CMV} promoters; the *araC* gene; and start codon-modified *murA* and *asdA* genes, DTS (I), DTS (II), and SV40 late poly(A)

vaccinevectors after delivery and during trafficking to the nucleus is another barrier that leads to inefficient DNA vaccination. Homopurine-rich tracts in the bovine growth hormone polyadenylation signal (BGH poly A) are identified as labile sequences, and replacement of BGH poly A with SV40 late poly A has improved resistance to attack from mammalian nucleases [28, 29]. Therefore, to increase the efficiency of the DNA vaccine vector delivery system, the 72 bp DTS (I) of the SV40 enhancer is inserted into plasmid pYA3650, and the BGH poly A is replaced with the SV40 late poly A. Also, the artificial DNA-binding sites DTS (II) for NF-κB and AP-2 are designed and inserted upstream of CMV E/P as an additional DNA nuclear targeting/enhancer sequence to yield the improved DNA vaccine vector pYA4545 (Fig. 3) [3]. The plasmid pYA4545 allows the rapid nuclear import and high-level synthesis of encoded genes. One additional beneficial feature of pYA4545 is that it possesses 24 immune-enhancing CpG oligodeoxynucleotide motifs that may contribute to the enhancement of innate inflammatory response (*see* **Note 4**).

1.4 Hyper-Invasive Self-Destructing GMS Displaying Reduced Pyroptosis/Apoptosis-Inducing Phenotype

One of the major mechanisms of *S.* Typhimurium invasion of animal hosts is by entering and traversing the epithelial monolayer through microfold (M) cells [30–33]. The *hilA* (hyper-invasion locus) regulator encodes an OmpR/ToxR family transcriptional regulator that activates the expression of invasion genes in response to both environmental and genetic regulatory factors [34, 35]. To improve M cell-mediated *Salmonella* invasion, the *hilA* promoter is replaced with the P_{trc} promoter to enable the constitutive synthesis of HilA. The GMS strain χ9971 (ΔP_{hilA}::P_{trc} $_{\Delta lacO}$*hilA*) is able to invade human intestinal Int-407 cells and colonize mouse tissues in significantly higher numbers than the wild-type strain. It has been reported that variation in the initial dose of infection with *Salmonella* has a profound effect on the response of *Salmonella* flagellin-specific CD4 T cells in vivo and that low-dose infection could evade *Salmonella* flagellin-specific T-cell activation completely [36]. Using hyper-invasive *Salmonella* vaccine strain may reduce the vaccine dose required for efficient DNA delivery and achieve an immune evasion of *Salmonella* carrier.

Invasive *Salmonella* induces pyroptosis/apoptosis in a fraction of infected macrophages [37, 38], and this is likely to diminish the transcription of a DNA vaccine after trafficking to the nucleus. *Salmonella* expressing the *Salmonella* pathogenicity island (SPI)-1-encoded type 3 secretion system (T3SS) (SPI-1 T3SS) and flagellin trigger caspase-1-dependent pyroptosis of infected macrophages and strains harboring mutations in the genes encoding the SPI-1 T3SS are not cytotoxic [39–42]. On the other hand, caspase-1-mediated delayed macrophage death requires the SPI-2 T3SS [43, 44]. SseL, a *Salmonella* deubiquitinase, is required for *Salmonella*-induced cytotoxicity of macrophages, and strains with the Δ*sseL* mutation are defective for the delayed cytotoxic effect [45–47]. TlpA, a *Salmonella* temperature-sensing protein, is encoded on the virulence plasmid of *S.* Typhimurium [48], and characterized by a remarkably long a helical coiled-coil motif [49]. Strains with the Δ*tlpA* mutation exhibited reduced *Salmonella*-induced host cell apoptosis [3]. The GMS strains χ9923 (Δ*tlpA*) and χ9924 (Δ*sseL*) significantly increased the ability of replication in Int-407 and colonized in mouse tissues in significantly greater numbers than the wild-type strain at day 6 postinfection [3]. In addition, the release of a DNA vaccine vector due to programmed lysis of a *Salmonella* cell within *Salmonella*-containing vacuole would be unlikely to stimulate an immune response, due to difficulties that the DNA vaccine molecule would encounter in getting to the nucleus for transcription of the protective antigen-encoding gene. Therefore, the Δ*sifA* mutation is included into self-destructing GMS for efficient DNA vaccine delivery. The GMS strain χ11218 with Δ*tlpA*, Δ*sseL*, ΔP_{hilA}::P_{trc} $_{\Delta lacO}$*hilA*, and Δ*sifA* mutations harboring improved DNA vaccine vector expressing hemagglutinin (HA) of influenza A/WSN/33 virus induced much higher serum IgG responses to WSN HA in mice after oral immunization. Mice immunized with such vaccine strain are fully protected from challenge with 100 times the LD_{50} of rWSN [3].

2 Materials

1. *E. coil.*
   χ7213: *thi-1 thr-1 leuB6 fhuA21 lacY1 glnV44 asdA4 recA1* RP4 2-Tc::Mu[*pir*]

2. *Salmonella enterica* serovar Typhimurium UK-1 strains.
   χ3761: Wild type,

   χ8806: Δ*asdA*::TT *araC* P_{BAD}*c2* TT ΔP_{murA}::TT *araC* P_{BAD}*murA*,

   χ8807: Δ*asdA*::TT *araC* P_{BAD}*c2* TT ΔP_{murA}::TT *araC* P_{BAD}-*murA* Δ*araBAD* Δ*araE*,

   χ8888: Δ*asdA*::TT *araC* P_{BAD}*c2* TT ΔP_{murA}::TT *araC* P_{BAD}-*murA* Δ*araBAD* Δ*araE*,

   Δ*endA* Δ(*gmd-fcl*) Δ*relA*.

   χ8926: Δ*sifA*,

   χ9354: Δ*asdA*::TT *araC* P_{BAD}*c2* TT ΔP_{murA}::TT *araC* P_{BAD}-*murA* Δ*araBAD* Δ*araE*,

   Δ*endA* Δ(*gmd-fcl*) Δ*relA* Δ*sifA*.

   χ9923: Δ*tlpA*,

   χ9924: Δ*sseL*,

   χ9971: ΔP_{hilA}::$P_{trc\Delta lacO}$*hilA*,

   χ11017: Δ*asdA*::TT *araC* P_{BAD}*c2* Δ*araBAD* Δ(*gmd-fcl*) Δ*pmi* Δ*relA*::*araC* P_{BAD}*lacI*,

   TT ΔP_{murA}::TT *araC* P_{BAD}*murA*.

   χ11214: Δ*asdA*::TT *araC* P_{BAD}*c2* TT ΔP_{murA}::TT *araC* P_{BAD}-*murA* Δ*araBAD* Δ*araE*,

   Δ*endA* Δ(*gmd-fcl*) Δ*relA* Δ*sifA* ΔP_{hilA}::$P_{trc\Delta lacO}$*hilA*.

   χ11215: Δ*asdA*::TT *araC* P_{BAD}*c2* TT ΔP_{murA}::TT *araC* P_{BAD}-*murA* Δ*araBAD* Δ*araE*,

   Δ*endA* Δ(*gmd-fcl*) Δ*relA* Δ*sifA* Δ*tlpA* Δ*sseL*.

   χ11218: Δ*asdA*::TT *araC* P_{BAD}*c2* TT ΔP_{murA}::TT *araC* P_{BAD}-*murA* Δ*araBAD* Δ*araE*,

   Δ*endA* Δ(*gmd-fcl*) Δ*relA* Δ*sifA* Δ*tlpA* Δ*sseL* ΔP_{hilA}::$P_{trc\Delta lacO}$*hilA*.

   χ11246: Δ*asdA*::TT *araC* P_{BAD}*c2* Δ*araBAD* Δ(*gmd-fcl*) Δ*pmi* Δ*relA*::*araC* P_{BAD}*lacI*,

   TT ΔP_{murA}::TT *araC* P_{BAD}*murA* Δ*sifA*.

3. Plasmids.

   pCAGGS-NP: Vector containing nucleoprotein (NP) gene from A/WSN/33),

   pUC57-NP-WSN: Commercial vector pUC-57 containing codon-optimized A/WSN/33 NP, pYA3681: Lysis vector containing P_{trc} promoter,

pYA4702: NP gene expressed from P_{trc} promoter cloned into lysis vector,

pYA3681 with pBR *ori*,

pYA4858: Codon-optimized NP gene expressed from P_{trc} promoter cloned into pYA3681 with pBR *ori*,

pYA4651: *ply* gene from *S. pneumonia* expressed from P_{trc} promoter cloned into pYA3681 with pBR *ori*,

pYA3716: Δ*sifA* flanking sequences in pRE112,

pVAX-1: Nonoptimized first-generation DNA vaccine vector (Invitrogen),

pYA3587: pVAX-1 with *rrfG*TT at HincII site,

pYA3650: pUC *ori araC* P_{BAD} SD-GTG *murA* SD-GTG *asdA* P22 P_R antisense mRNA eukaryotic expression cassette,

pYA3681: pBR *ori araC* P_{BAD} SD-GTG *murA* SD-GTG *asdA* P22 P_R antisense mRNA prokaryotic expression cassette,

pYA3836: pUC *ori araC* P_{BAD} SD-GTG *murA* SD-GTG *asdA* P22 P_R antisense mRNA eukaryotic expression cassette with DTS (I),

pYA4050: pUC *ori araC* P_{BAD} SD-GTG *murA* SD-GTG *asdA* P22 P_R antisense mRNA eukaryotic expression cassette with DTS (I) and SV 40 polyA,

pYA4545: pUC *ori araC* P_{BAD} SD-GTG *murA* SD-GTG *asdA* P22 P_R antisense mRNA eukaryotic expression cassette of with DTS (I), DTS (II), and SV 40 polyA,

pYA4611: pUC *ori araC* P_{BAD} SD-GTG *murA* SD-GTG *asdA* P22 P_R antisense mRNA eukaryotic expression cassette of WSN-HA with DTS (I) and SV 40 polyA,

pYA4859: pUC *ori araC* P_{BAD} SD-GTG *murA* SD-GTG *asdA* P22 P_R antisense mRNA eukaryotic expression cassette of WSN-HA with DTS (I), DTS (II), and SV 40 polyA,

pYA3484: Suicide vector for deletion of the Δ*araBAD*,

pYA3485: Suicide vector for deletion of the Δ*araE*,

pYA3629: Suicide vector for deletion of the Δ(*gmd-fcl*),

pYA3652: Suicide vector for deletion of the Δ*endA*,

pYA3679: Suicide vector for deletion of the Δ*relA*,

pYA3716: Suicide vector for deletion of the Δ*sifA*,

pYA4620: Suicide vector for deletion of the Δ*tlpA*,

pYA4621: Suicide vector for deletion of the Δ*sseL*,

pYA4641: Suicide vector for deletion of the ΔP_{hilA}::$P_{trc\Delta lacO}$*hilA*.

4. **Bacteriophage:** P22HT*int* (*see* **Note 5**).
5. **Petri dishes:** Polystyrene-disposable sterilized Petri dishes, size 100 x 15 mm (VWR, Radnor, PA).
6. **Luria-Bertani (LB) broth:** 10 g Tryptone (BD, Franklin Lake, NJ), 10 g NaCl (Sigma Aldrich, St. Louis, MO), 5 g yeast extract (BD, Franklin Lake, NJ), and 950 mL H_2O. Combine the reagents and shake until the solutes have dissolved. Adjust the final volume of the solution to 1 L with H_2O. Sterilize by autoclaving for 20 min at 15 psi (1.05 kg/cm^2) on the liquid cycle.
7. **LB agar:** 10 g Tryptone (BD, Franklin Lake, NJ), 10 g NaCl (Sigma-Aldrich, St. Louis, MO), 5 g yeast extract (BD, Franklin Lake, NJ), 15 g agar (BD, Franklin Lake, NJ), and 1000 mL H_2O; combine the reagents. Sterilize by autoclaving for 20 min at 15 psi (1.05 kg/cm^2) on the liquid cycle.
8. **2,6-Diaminopimelic acid (DAP):** To prepare a standard stock solution of 0.5 mg/mL, dissolve 50 mg DAP (Sigma-Aldrich, St. Louis, MO) in the 100 mL of 0.1 M pH 2.0 citrate buffer. Filter the arabinose solution with a 0.22 μm filter. The working concentration of DAP is 10 mg/L.
9. **Arabinose:** Dissolve 20 g arabinose (Sigma-Aldrich, St. Louis, MO) in the 90 mL of H_2O to make a stock solution (20%). Bring up the volume to 100 mL with H_2O. Filter the arabinose solution with a 0.22 μm filter. The working concentration of arabinose is 0.2%.
10. **Sucrose:** Dissolve 50 g sucrose (Sigma-Aldrich, St. Louis, MO) in the 80 mL of H_2O to make a stock solution (50%). Bring up the volume to 100 mL with H_2O. Filter the sucrose solution with a 0.22 μm filter. The working concentration of sucrose is 5%.
11. **Synthetic peptideNP147–155 (TYQRTRALV)** (GenScript, Piscataway, NJ).
12. **Electroporation cuvettes:** 0.2 cm gap Gene Pulser®/Micro-Pulser™ Electroporation Cuvettes (BIO-RAD, Hercules, CA).
13. **Gene Pulser:** Gene Pulser Xcell modular electroporation system (BIO-RAD Laboratories, Hercules, CA).
14. **Polymerase chain reaction**: Phusion high-fidelity heat-stable DNA polymerase and supplied buffers (New England Biolabs, Ipswich, MA) or Taq DNA polymerase (Sigma-Aldrich, St. Louis, MO).
15. **T4 DNA ligase** (New England Biolabs, Ipswich, MA).
16. **Restrictive enzymes** (New England Biolabs, Ipswich, MA).
17. **Bioscreen-C Automated Growth Curve Analysis System** (Growth Curves USA, Piscataway, NJ).

18. **37 °C Warm room.**
19. **PAGE gel apparatus:** Mini-PROTEAN 3 system (BIO-RAD Laboratories, Hercules, CA).
20. **10% Sodium dodecyl sulfate (SDS):** Weigh out 10 g SDS and add to a 100 mL Duran bottle. SDS is a fine powder and should be weighed using a fume mask or under a fume hood to avoid inhalation. Add 80 mL of distilled water to slowly dissolve SDS. Once the powder has fully dissolved, top up the solution to 100 mL using distilled water. Do not autoclave. Filter the solution to remove any particles that have failed to dissolve.
21. **4 x SDS-PAGE loading buffer:** 250 mM Tris–HCl (pH 6.8), 8% (w/v) sodium dodecyl sulfate (SDS), 0.2% (w/v) bromophenol blue, 40% (v/v) glycerol, and 20% (v/v) β-mercaptoethanol.
22. **Nitrocellulose membrane:** 0.45 μm Roll (Bio-Rad Laboratories, Hercules, CA).
23. **Transfer buffer (Tris-glycine-SDS) for western blotting**: 25 mM Tris–HCl, pH 7.6, 192 mM glycine, 20% methanol.
24. **Tris-buffered saline (TBS):** 10 mM Tris–Cl, 15 mM NaCl, pH 7.5.
25. **Tween-20** (Sigma-Aldrich, St. Louis, MO).
26. **TBST:** TBS containing 0.05% Tween 20.
27. **Blocking solution:** 3% BSA in TBST.
28. **Rabbitpolyclonal anti-influenza A NP antibody** (Abcam, Cambridge, MA).
29. **Nitroblue tetrazolium-5-bromo-4-chloro-3-indolyl phosphate (BCIP)** (Sigma-Aldrich, St. Louis, MO).
30. **Phosphate-buffered saline (PBS):** Combine 8 g NaCl, 1.15 g Na_2HPO_4, 0.2 g KH_2PO_4, and 0.2 g KCl. Dissolve in 0.9 L ddH_2O. Adjust pH to 7.4 and make up to 1.0 L. Autoclave.
31. **PBS-T20:** PBS containing 0.05% Tween 20.
32. **Goat anti-rabbit IgG alkaline phosphatase conjugate** (Abcam, Cambridge, MA).
33. **BSG**: Buffered saline containing 0.01% gelatin (Sigma-Aldrich, St. Louis, MO).
34. **BALB/c mice:** 6–8-Week-old female (Charles River, Wilmington, MA).
35. **96-Well flat-bottom polystyrene microtiter plates** (Dynatech Laboratories, El Paso, TX).
36. **Pipette:** Fixed and adjustable volume single channel and multichannel (METTLERTOLEDO, Columbus, OH).

37. Lipopolysaccharide (LPS): 100 ng LPS (Sigma-Aldrich, St. Louis, MO) per well suspended in carbonate-coating buffer (pH 9.5).
38. **Biotinylated goat anti-mouse IgG** (Southern Biotechnology, Birmingham, AL).
39. **Streptavidin-alkaline phosphatase conjugate** (Southern Biotechnology, Birmingham, AL).
40. **p-Nitrophenyl phosphate** (Sigma-Aldrich, St. Louis, MO).
41. **ELISA plate reader** (SpectraMax; Molecular Devices, San Jose, CA).
42. **ELISPOT kit** (Sigma-Aldrich, St. Louis, MO).
43. **Fluorescencecell viability assay kit** (Abcam, Cambridge, MA).
44. **RPMI-1640 containing 2 μg/mL acetyl-trypsin** (Sigma-Aldrich, St. Louis, MO).
45. **Ketamine cocktail:** 21.0 mg Ketamine, 2.4 mg xylazine, and 0.3 mg acepromazine.

3 Methods

3.1 Development of GMS for Protective Antigen Delivery to Induce Cellular Immune Responses Against Influenza Virus

Influenza infections are primarily and effectively controlled by vaccines that elicit neutralizing antibodies against the surface proteins hemagglutinin (HA) and neuraminidase (NA). However, influenza vaccines are frequently reformulated to match the circulating strains. Therefore, the development of a vaccine that elicits broad-based protection against different strains is desirable. Influenza nucleoprotein (NP) is over 90% conserved between influenza A strains and generates cellular immunity [50]. $CD8^+$ and $CD4^+$ T-cell NP epitopes are well defined, and induced T cells afford NP-specific protection [51] (*see* **Note 1**). GMS (χ11246) ($SifA^-$) exits the endosome soon after the invasion of a host cell, multiplies in the cytoplasm, and delivers by lysis the synthesized NP protein from the plasmid (pYA4858) to the cell cytosol to induce cellular immunity (*see* **Note 2**).

3.1.1 Bacterial Cultures and General DNA Procedures

1. Grow *E. coli* and *S.* Typhimurium cultures in Luria-Bertani (LB) broth [52], or on LB agar plates at 37 °C.
2. Use LB agar without NaCl and with 5% sucrose for *sacB* gene-based counterselection in allelic exchange experiments.
3. Use bacteriophage P22HT*int* for generalized transduction [53] (*see* **Note 5**).
4. Add DAP at a concentration of 50 μg/mL for the growth of Asd^- strains.

5. Supplement LB with 0.2% arabinose for host-regulated delayed lysis vector combinations [2].
6. Transform *E. coli* and *S.* Typhimurium strains by electroporation.
7. Select transformant containing Asd$^+$ plasmids on LB agar plates without DAP
8. Determine plasmid stability as described previously [26].

3.1.2 Strain Construction and Characterization

1. The Δ*sifA* mutation is a defined in-frame deletion of the *sifA* gene. Introduce Δ*sifA* into strain χ11017 using phage P22 transduction from (χ8926::pYA3716) to generate strain χ11246. Verify the presence of the mutation by PCR.
2. Perform plasmid stability assays according to the standard test protocol. Briefly, inoculate a 25 μL volume of overnight static cultures (LB broth with mannose and arabinose) of GMSs into 25 mL of LB broth with mannose, DAP, and arabinose (non-selective medium) for 12 h of incubation by rotation at 37 °C (T_0). Then transfer 25 μL of T_0 culture to 25 mL fresh nonselective medium described above and incubate for further 12 h with rotation at 37 °C (T_1). Repeat the processes every 12 h for a total of five passages (~50 generations). Prior to each transfer, dilute the cultures serially and spread onto LB agar plates containing mannose, DAP, and arabinose. After incubating overnight at 37 °C, isolate 100 colonies picked randomly and patch onto LB agar plates with mannose and arabinose (selective agar plate). Growth on these plates would occur only if the plasmid is still present. Use the numbers of colonies that grew on selective agar plate out of the 100 colonies picked from the nonselective agar plate to calculate the plasmid loss for each passage. Then select ten representative colonies from the T_5 plates and evaluate for the ability to synthesize antigens.
3. Determine the growth curves of GMSs using the Bioscreen-C Automated Growth Curve Analysis System. Dilute overnight cultures of GMS strains (1:200) into fresh LB broth containing mannose and arabinose. Incubate cultures at 37 °C for 24 h.
4. To confirm arabinose-dependent growth of the GMS lysis strains, dilute overnight standing cultures (LB + mannose + arabinose) of strains in a tenfold series and plate on LB agar plates supplemented with mannose and with or without arabinose, respectively. Examine plates after 24-h incubation at 37 °C.

3.1.3 Codon Optimization of the NP Gene and Synthesis of Peptide

1. Codon-optimize the sequence of the nucleoprotein (NP) gene of influenza virus strain A/WSN/33 for maximal expression in *Salmonella*. Clone commercially codon-optimized NP cDNA sequence into pUC-57 to yield pUC57-NP-WSN. Change the

average G + C content from 46.66 for the non-optimized gene to 54.59 after codon optimization. Disrupt the stem-loop structures, which impact ribosomal binding and stability of mRNA. Increase the codon usage bias for *E. coli* from a codon adaptation index of 0.57–0.98.

2. Dissolve synthetic peptide NP147–155 (TYQRTRALV) in water according to the manufacturer's instruction, aliquot, and store at −20 °C until use.

3.1.4 Vectors for Antigen Delivery via Regulated Delayed Lysis System

1. Amplify the NP gene from plasmid pCAGGS-NP by PCR. Digest the PCR product using *Nco*I and *Xma*I restrictive enzyme sites and clone into plasmid pYA3681 yielding pYA4702.
2. Amplify the codon-optimized NP gene from pUC-57-WSN-NP by PCR. Digest the PCR product with *Nco*I and clone into pYA3681 yielding pYA4858. Confirm the orientation of the NP gene by restriction digestion with *Pst*I and by sequencing. Sequence all derivatives of pYA3681.
3. Construct negative control vector pYA4651 encoding the *ply* gene from *S. pneumonia* cloned in pYA3681.
4. Transfer all vectors into appropriate *S.* Typhimurium strains by electroporation. Confirm all DNA constructs by sequencing.

3.1.5 SDS-PAGE and Immunoblots

1. To evaluate NP protein synthesis from plasmids in *E. coli* and *S.* Typhimurium strains, grow bacterial cells overnight at 37 °C in LB containing 0.2% arabinose. Take aliquots (1 mL), centrifuge at low speed, and resuspend them in 2 × SDS-PAGE loading buffer and boil for 10 min. Centrifuge the samples for 10 min, dilute (1:10) in 2 × sample-loading buffer, and load 10 μL onto 12.5% SDS-PAGE gels for separation by electrophoresis as previously described [54].
2. For analyzing NP synthesis, incubate blots with rabbit polyclonal anti-influenza A NP antibody for 1 h with constant shaking. After washing with TBST-T20, incubate the membranes with goat anti-rabbit IgG alkaline phosphatase conjugate for 1 h and develop with nitroblue tetrazolium-5-bromo-4-chloro-3-indolyl phosphate (BCIP). Wash membranes with water and air-dry.

3.1.6 Immunization of Mice

1. Deprive food and water from each group of 6–8-week-old female BALB/c mice for 4 h prior to oral immunization. Grow *S.* Typhimurium strains χ11017(pYA4858) (SifA$^+$), χ11246(pYA4858) (SifA$^-$), χ11017(pYA3681), χ11246 (pYA3681), and χ11246(pYA4651) individually in LB broth with 0.2% arabinose and 0.2% mannose to an OD_{600} of 0.85. Centrifuge the cultures at 4000 × *g* for 15 min at room

temperature and suspend in buffered saline containing 0.01% gelatin (BSG) to a final concentration of 5×10^9 CFU/mL. Titrate bacteria on LB agar supplemented with arabinose.

2. Orally administer 1×10^9 CFU *S.* Typhimurium strains in 20 μL of BSG to the mice at the back of the mouth with a pipette tip. Return food and water to orally immunized mice 30 min after vaccine administration. Use vectors without any expressed antigen gene (pYA3681) or that expressed the *ply* gene from *S. pneumonia* (pYA4651) and BSG-immunized mice as negative controls.

3.1.7 Experimental Design

1. To evaluate the effect of the *sifA* deletion (χ11246) on the immunogenicity and protective immunity conferred by GMS strains encoding the codon-optimized NP (pYA4858) of influenza virus, orally immunize BALB/c mice ($n = 8$) with parent χ11017(pYA4858) (SifA$^+$), mutant χ11246(pYA4858) (SifA$^-$), and vector controls χ11017(pYA3681) (SifA$^+$) and χ11246(pYA3681) (SifA$^-$) or with BSG at week zero and boost three times at weeks 1, 4, and 7 post-primary immunization (PPI).
2. Collect blood at weeks 3 and 6 PPI by cheek-pouch bleeding and monitor for the presence of antibodies against NP or *S.* Typhimurium LPS by ELISA.
3. For assaying antigen-specific T cells, collect spleens aseptically at week 8 PPI from three mice, pool, and process for ELISPOT [55] (*see* **Note 3**).
4. Challenge the remaining mice ($n = 5$) in each group with rWSN (100 LD_{50}) at week 8 PPI (14–16 weeks of age) and observe for morbidity and mortality for an additional 3 weeks.

3.1.8 ELISA

1. Determine IgG responses against NP or LPS in sera by ELISA. Briefly, coat 96-well flat-bottom polystyrene microtiter plates with 2 μg/mL of purified NP protein or LPS suspend in carbonate-coating buffer (pH = 9.5) and incubate at 4 °C overnight. Block free binding sites with PBS–0.05% Tween 20 containing 3% bovine serum albumin (BSA) for 2 h at room temperature. Dilute sera twofold serially in PBS/3% BSA and incubate 100 μL in duplicate wells for 1 h at room temperature. Wash plates thrice with PBS-Tween 20 and incubate for 1 h with a 1:1000 dilution of either biotinylated goat anti-mouse IgG or IgG1 or IgG2a. After washing as above, incubate the plates for 1 h with streptavidin-alkaline phosphatase conjugate, develop by incubating with *p*-nitrophenyl phosphate for 30 min, and read by an automated ELISA plate reader at 405 nm.

2. Express endpoint titers as the reciprocal log2 value of the last positive sample dilution. Consider absorbance two times higher than pre-immune serum, used as baseline values, as positive.

3.1.9 Cell Proliferation

1. Perform lymphocyte proliferation assays to assess influenza peptide-specific (NP147–155) cell-mediated responses. Plate single-cell suspensions prepared from spleens at a concentration of 5×10^5 cells/well and stimulate with the NP 147–155 peptideTYQRTRALV (20 μg/mL) for 7 days.
2. Add vision blue dye™ from the fluorescence cell viability assay kit according to the manufacturer's instructions, and read plates at excitation 530 nm and emission 590 nm.

3.1.10 Virus Strain Propagation, Purification, and Titration

1. The rWSN virus is a mouse-adapted strain created by reverse genetics and is lethal to mice in doses above 10^3 $TCID_{50}$ [56]. Propagate the virus and titrate in Madin-Darby canine kidney cells cultured in RPMI-1640 containing 2 μg/mL acetyl-trypsin.
2. Pass the virus through a 30% (w/v) sucrose cushion at 11,620 × g for 3 h using an ultracentrifuge. Resuspend the resulting pellet in phosphate-buffered saline (PBS) pH 7.2 and centrifuge at 11,620 × *g* for 1 h. Finally, dissolve the viral pellet in 500 μL of PBS and keep frozen at −80 °C until use.

3.1.11 Virus Challenge

1. Anesthetize mice with 0.05 mL/20 g body weight of a ketamine cocktail (21.0 mg ketamine, 2.4 mg xylazine, and 0.3 mg acepromazine) administered intraperitoneally.
2. Infect groups of mice with 30 μL of the serially diluted purified rWSN virus from 1×10^7–1×10^2 $TCID_{50}$ at 8 weeks of age and LD_{50}. To rule out any age-dependent variation in LD_{50} doses, perform similar experiments with mice at 10 and 14 weeks of age. No difference is observed in terms of virus-associated morbidity and mortality in mice at 8, 10, or 14 weeks of age.
3. Infect intranasally (IN) sedated mice with a 100 LD_{50} (1×10^5 $TCID_{50}$) of rWSN in a total volume of 30 μL, 15 μL per nostril for all experiments with pipette tips. Back-titrate an aliquot of the virus used for challenge on MDCK to ascertain the exact dose given to mice.
4. Inspect the challenged mice daily for signs of infection such as ruffled fur and hunched posture, and weigh on alternate days until 21 days to monitor the progression of the infection. Calculate percent weight loss for individual mice in each group by comparing their daily weight to their pre-challenge weight. Promptly remove mice that succumbed to infection or have to be euthanized.

3.1.12 Statistical Analyses

1. Determine differences in antibody titers between groups, cell proliferation, and quantitative difference in numbers of IFN-γ-secreting cells between the groups using analysis of variance (ANOVA).
2. Further analyze statistically different means using Bonferroni's test or by Tukey's method.
3. Analyze survival analysis using the log-rank test (GraphPad Prism).

3.2 Development of GMS for DNA Vaccine Delivery to Induce Systemic Immunity Against Influenza Virus

Yearly epidemics are caused by antigenic drift, which is the accumulation of mutations in the HA or NA genes in the currently circulating virus, such as it is able to avoid or overcome immunity generated by the original virus. The more infrequent pandemics can be caused by antigenic shift, the replacement of either HA or NA subtypes with novel ones, resulting in influenza subtypes to which the general population is totally naive. This situation presents humans with a novel antigenic experience that often results in the highest morbidity and mortality. Unfortunately, infection and recovery from influenza do not result in lifelong immunity to all strains of influenza because influenza viruses frequently undergo antigenic changes. Increasing the speed of producing a matching vaccine is key in the context of response to an influenza epidemic. This method describes the development of a GMS-improved DNA vaccine delivery platform, which is vectoring the WSN HA protein to exploit the immune response to *Salmonella* infection yet to release a DNA vaccine vector that will solely specify synthesis of WSN HA antigen in the immunized host to induce a protective immune response against WSN virus challenge.

3.2.1 Bacterial Cultures, General DNA Procedures, and Strain Characterization

See Subheadings 3.1.1 and 3.1.1 above.

3.2.2 Construction of the Self-Destructing S. Typhimurium Vaccine Host-Vector System

1. Choose the *S.* Typhimurium host strain, χ8806 (Δ*asdA*::TT *araC* P_{BAD}*c2* TT ΔP_{murA}::TT *araC* P_{BAD}*murA*), which features arabinose-dependent programmed lysis phenotype, as a starting strain.
2. Add the Δ*araBAD* and Δ*araE* mutations sequentially using suicide vectors resulting in vaccine strain χ8807.
3. Add three additional mutations, Δ(*gmd-fcl*), Δ*relA*, and Δ*endA* mutations, sequentially using suicide vectors resulting in vaccine strain χ8888 (Δ*asdA*::TT *araC* P_{BAD}*c2* TT ΔP_{murA}:: TT *araC* P_{BAD}*murA* Δ*araBAD* Δ*araE* Δ(*gmd-fcl*), Δ*relA*, Δ*endA*).
4. Then introduce the Δ*sifA* mutation, a defined in-frame deletion of the *sifA* gene, into strain χ8888 by conjugation with suicide vector pYA3716 to generate strain χ9354.

5. Verify the presence of each mutation by PCR.
6. Confirm the presence of the Δ*asdA* mutation in *Salmonella* by its dependence on DAP for growth.
7. Verify the presence of the ΔP_{murA}::TT *araC* P_{BAD}*murA* mutation by arabinose dependency for growth.
8. To construct a DNA vaccine vector, create a proper eukaryotic expression cassette first in plasmid pVAX-1 by inserting a transcriptional terminator (*rrfG* TT: 5′-AAC TGC AGT CTA GAT TAT GCG AAA GGC CAT CCT GAC GGA TGG CCT TTT TGT TTA AAC GGA TCC GC) at the HincII site so that expression in pUC origin (*ori*) domain does not affect the transcriptional activities from P_{CMV} promoter. Designate the resulting plasmid as pYA3587.
9. Insert a DNA fragment including pUC *ori*, *rrfG*TT, and a eukaryotic expression cassette from plasmid pYA3587 into the lysis vector pYA3681 [2] to replace the original pBR *ori* and a prokaryotic expression cassette using primers pVAX-SmaI-SalI and pVAX-SmaI-CalI. This process resulted in a DNA vaccine vector pYA3650.
10. Introduce pYA3650 into the regulated delayed lysis host strain χ8888.

3.2.3 Construction of Hyper-invasive GMS Strains

1. To construct the hyper-invasive *Salmonella* strains, design a promoter $P_{trc\Delta lacO}$ and build into a suicide vector. The resulting suicide vector is pYA4641.
2. Conjugate wild-type strain χ3761 with *E. coli* strain χ7213 harboring suicide vector pYA4641 to generate mutant strain χ9971.

3.2.4 Construction of RASV Strains to Reduce Host Cell Pyroptosis/Apoptosis

1. Construct two *S.* Typhimurium mutant strains χ9923 and χ9924 using suicide vectors pYA4620 and pYA4621.
2. Introduce mutations Δ*tlpA* and/or Δ*sseL* into *S.* Typhimurium lysis strain χ9354 by conjugation, to achieve strains χ11215 and χ11218.

3.2.5 Construction of Improved DNA Vaccine Vector

1. Insert the simian virus 40 (SV40) 72 bp repeat enhancer into pYA3650 to create plasmid pYA3836.
2. Replace the BGH poly(A) in pYA3650 with the SV40 late poly (A); the resulting plasmid is pYA4050.
3. Design a synthetic DTS (II) based on conserved mammalian NF-κB- and AP-2-binding sites.
4. Insert a DTS (II) fragment into pYA4050 to yield the improved DNA vaccine vector pYA4545.

3.2.6 Construction of Improved DNA Vaccine Vector Encoding WSN HA

1. Use plasmid pPolI-Wsn as the template for the construction of pYA4545 encoding WSN HA.
2. Insert a DNA fragment encoding the WSN HA antigen with Kozak sequence downstream of the CMV promoter of the improved DNA vaccine vector pYA4545 at the KpnI/XhoI site to obtain plasmid pYA4859.
3. Introduce plasmid pYA4859 into χ9354 to yield χ9354 (pYA4859).

3.2.7 Examination of Cell Lysis In Vitro

1. Dilute (1:400) overnight cultures of strains grown in LB broth supplemented with 0.002% arabinose (supply 0.002% arabinose to avoid the accumulation of arabinose in bacteria to be able to detect cell lysis during the short incubation period) into fresh pre-warmed LB broth supplemented with or without 0.02% arabinose.
2. Assay β-galactosidase activity in the supernatant and cell-pellet fractions at indicated time points.

3.2.8 Colonization of Mice with the Salmonella DNA Vaccine Strains

1. Deprive food and water from each group of three inbred 6–8-week-old female BALB/c mice for 4 h before oral administration of *Salmonella* vaccine strains.
2. Grow GMS strains with aeration in LB broth supplemented with 0.2% arabinose to an optical density at 600 nm (OD_{600}) of 0.85 from a non-aerated static overnight culture.
3. Orally administer 1.0×10^9 CFU of GMS strains in 20 μL of PBS containing 0.01% gelatin (BSG) to mice at the back of the mouth with a pipette tip.
4. Return food and water to mice 30 min later.
5. Euthanize mice at indicated times and collect their Peyer's patches, spleens, and livers aseptically. Homogenize tissues and plate on LB agar with 0.2% arabinose to evaluate colonization and persistence, and onto LB agar plates without arabinose to confirm arabinose-dependent growth.

3.2.9 Immunization of Mice and Sample Collection

1. Grow *S.* Typhimurium DNA vaccine strains individually as described above.
2. Orally vaccinate groups of eight 6–8-week-old female BALB/c mice at week zero with the following dose: 1.0×10^9 CFU χ8888(pYA4050) (vector control) and 1.0×10^9 CFU χ8888 (pYA4611) (pYA4050 encoding WSN HA) or 1.0×10^9 CFU χ9354(pYA4545) (vector control), 1.0×10^9 CFU χ9354 (pYA4859) (pYA4545 encoding WSN HA), 1.0×10^9 CFU χ11214(pYA4859), 1.0×10^9 CFU χ11215(pYA4859), and 1.0×10^9 CFU χ11218(pYA4859).

3. Give booster immunizations to all immunization groups three times later at weeks 1, 4, and 7 PPI.
4. Monitor the immunized mice for 8 weeks for any evidence of illness by observing them daily for evidence of diarrhea, ruffled (ungroomed) fur, or irritability. None of these symptoms of infection should be observed in any of the mice. Monitor blood collected at week 5 PPI by cheek-pouch bleeding for the presence of antibodies against HA or *S.* Typhimurium LPS by ELISA.

3.2.10 ELISA, Virus Strain Propagation, Purification, and Titration, and Virus Challenge

See Subheadings 3.1.8, 3.1.11, and 3.1.12 above.

4 Notes

1. Conventional approaches to influenza vaccination, using a live attenuated or an inactivated virus vaccine, are effective at inducing neutralizing antibodies but must be reformulated each year as new antigenic variant strains arise. The live attenuated virus vaccine also elicits cellular immunity, although it does not appear to provide additional protection in humans. Perhaps of greater importance is the generation of more broadly reactive vaccines. Influenza NP could potentially provide broad-based protection: T-cell epitopes in NP are well defined and both $CD8^+$ and $CD4^+$ T cells play an important role in the protection afforded by NP.
2. In Subheading 3.1, a novel bacterial delivery system for an influenza virus NP utilizing an orally administered GMS displaying a self-destructing in vivo phenotype is described. An irreversible chromosomal deletion of the *sifA* gene is introduced into the regulated delayed lysis strain χ11017 yielding χ11246 carrying mutations to enhance complete lysis and antigen delivery. The resultant strain invaded cells, reached the SCVs, and escaped into the cytosol by virtue of the *sifA* deletion, and eventually lysed due to unavailability of DAP and arabinose in vivo, releasing its payload of antigenic peptides/proteins. The antigen delivered should be able to be synthesized in GMS strains in a functional form.
3. Unexpectedly, we could not detect any influenza NP (antigen or peptide NP147–155 specific) IFN-γ production in the splenocytes of the immunized mice 1 week after the last booster dose by ELISPOT assay. We might need to evaluate more mice at additional time points post-immunization to examine this point. However, IFN-γ is not required for recovery from

primary infection with influenza virus and for the induction of secondary influenza virus-specific CTL to lethal influenza A infection in mediastinal lymph nodes, a site in which the production of antigen-specific CTL correlates with protection. Therefore, MLN, in addition to spleens, should be tested for the presence of CTLs and the role of cytokines like TNF-α, IL-2, and IL-12 in future experiments.

4. Current means for delivery of DNA vaccines by intramuscular injection or by use of gene guns do not result in delivery of the vaccine to mucosal as well as to a diversity of internal lymphoid tissues important in yielding a high-level sustained (i.e., memory) protective immunity to pathogens. In Subheading 1, a much-improved DNA vaccine vector was constructed with enhanced transmission to the nucleus for gene expression. Using this system to deliver the WSN HA antigen successfully induce protective immunity to influenza virus challenge in mice immunized with the self-destructing GMS strains. To develop universal flu vaccine, conserved influenza antigen or domains should be selected.
5. In addition, bacteriophage P22HT*int* is used for generalized transduction [57]. *E. coli* χ6212 [9] is used as an intermediate host strain for cloning. Transformations of *E. coli* and *S.* Typhimurium strains are done by electroporation. Transformants containing Asd$^+$ plasmids are selected on LB agar plates without DAP. When necessary, arabinose is provided in LB agar for lysis vector construction. Suicide vector technology is used to generate precise deletion/deletion-insertion mutations. Conjugational transfer of suicide vectors was performed using the suicide vector donor *E. coli* strain χ7213 [58] that serves as a host strain of suicide vectors to create *Salmonella* chromosomal mutations.

References

1. Curtiss R 3rd, Xin W, Li Y et al (2010) New technologies in using recombinantattenuated *Salmonella* vaccine vectors. Crit Rev Immunol 30(3):255–270
2. Kong W, Wanda SY, Zhang X et al (2008) Regulated programmed lysis of recombinant *Salmonella* in host tissues to release protective antigens and confer biological containment. Proc Natl Acad Sci U S A 105(27):9361–9366
3. Kong W, Brovold M, Koeneman BA et al (2012) Turning self-destructing *Salmonella* into a universal DNA vaccine delivery platform. Proc Natl Acad Sci U S A 109 (47):19414–19419
4. Ameiss K, Ashraf S, Kong W et al (2010) Delivery of woodchuck hepatitis virus-like particle presented influenza M2e by recombinant attenuated *Salmonella* displaying a delayed lysis phenotype. Vaccine 28(41):6704–6713
5. Juarez-Rodriguez MD, Yang J, Kader R et al (2012) Live attenuated *Salmonella* vaccines displaying regulated delayed lysis and delayed antigen synthesis to confer protection against Mycobacterium tuberculosis. Infect Immun 80 (2):815–831
6. Ashraf S, Kong W, Wang SF et al (2011) Protective cellular responses elicited by vaccination with influenza nucleoprotein delivered by a live recombinant attenuated *Salmonella* vaccine. Vaccine 29(23):3990–4002
7. Black S, Wright NG (1955) Aspartic beta-semialdehyde dehydrogenase and aspartic beta-semialdehyde. J Biol Chem 213(1):39–50

8. Galan JE, Nakayama K, Curtiss IIIR 3rd (1990) Cloning and characterization of the asd gene of *Salmonella typhimurium*: use in stable maintenance of recombinant plasmids in *Salmonella* vaccine strains. Gene 94 (1):29–35
9. Curtiss R 3rd, Galan JE, Nakayama K et al (1990) Stabilization of recombinant avirulent vaccine strains in vivo. Res Microbiol 141 (7–8):797–805
10. Brown ED, Vivas EI, Kolter R et al (1995) MurA (MurZ), the enzyme that catalyzes the first committed step in peptidoglycan biosynthesis, is essential in *Escherichia coli.* J Bacteriol 177(14):4194–4197
11. Guzman LM, Belin D, Carson MJ et al (1995) Tight regulation, modulation, and high-level expression by vectors containing the arabinose P_{BAD} promoter. J Bacteriol 177 (14):4121–4130
12. Whitfield C (2006) Biosynthesis and assembly of capsular polysaccharides in *Escherichia coli.* Annu Rev Biochem 75:39–68
13. Curtiss R 3rd, Beer RF Jr, Bassett EG (1976) Biological containment: The subordination of *Escherichia coil* K-12. In: Beers RF Jr, Bassett EG (eds) Recombinant Molecules: Impact on Science and Society. Raven Press, New York, NY, pp 45–56
14. Török I, Kari C (1980) Accumulation of ppGpp in a relA mutant of *Escherichia coli* during amino acid starvation. J Biol Chem 255:3838–3840
15. De Groote MA, Testerman T, Xu Y et al (1996) Homocysteine antagonism of nitric oxide-related cytostasis in *Salmonella typhimurium.* Science 272:414–417
16. Brumell JH, Tang P, Zaharik ML et al (2002) Disruption of the *Salmonella*-containing vacuole leads to increased replication of *Salmonella enterica* serovar *typhimurium* in the cytosol of epithelial cells. Infect Immun 70:3264–3270
17. Brunner S, Sauer T, Carotta S et al (2000) Cell cycle dependence of gene transfer by lipoplex, polyplex and recombinant adenovirus. Gene Ther 7:401–407
18. Capecchi MR (1980) High efficiency transformation by direct microinjection of DNA into cultured mammalian cells. Cell 22(2 Pt 2):479–488
19. Mirzayans R, Aubin RA, Paterson MC (1992) Differential expression and stability of foreign genes introduced into human fibroblasts by nuclear versus cytoplasmic microinjection. Mutat Res 281(2):115–122
20. Dean DA (1997) Import of plasmid DNA into the nucleus is sequence specific. Exp Cell Res 230(2):293–302
21. Dean DA, Dean BS, Muller S et al (1999) Sequence requirements for plasmid nuclear import. Exp Cell Res 253(2):713–722
22. Vacik J, Dean BS, Zimmer WE et al (1999) Cell-specific nuclear import of plasmid DNA. Gene Ther 6:1006–1014
23. Black RE, Morris SS, Bryce J (2003) Where and why are 10 million children dying every year? Lancet 361(9376):2226–2234
24. Alba BM, Zhong HJ, Pelayo JC et al (2001) *degS* (*hhoB*) is an essential *Escherichia coli* gene whose indispensable function is to provide sigma (E) activity. Mol Microbiol 40 (6):1323–1333
25. Vitiello M, Isanto MD, Galdiero M et al (2004) Interleukin-8 production by THP-1 cells stimulated by Salmonella enterica serovar Typhimurium porins is mediated by AP-1, NF-kappaB and MAPK pathways. Cytokine 27(1):15–24
26. Chatfield SN, Charles IG, Makoff AJ et al (1992) Use of the nirB promoter to direct the stable expression of heterologous antigens in *Salmonella* oral vaccine strains: development of a single-dose oral tetanus vaccine. Biotechnology 10:888–892
27. Tallant T, Deb A, Kar N et al (2004) Flagellin acting via TLR5 is the major activator of key signaling pathways leading to NF-kappa B and proinflammatory gene program activation in intestinal epithelial cells. BMC Microbiol 4 (33):1471–2180
28. Carswell S, Alwine JC (1989) Efficiency of utilization of the simian virus 40 late polyadenylation site: effects of upstream sequences. Mol Cell Biol 9(10):4248–4258
29. Ribeiro SC, Monteiro GA, Prazeres DM (2004) The role of polyadenylation signal secondary structures on the resistance of plasmid vectors to nucleases. J Gene Med 6 (5):565–573
30. Takeuchi A (1967) Electron microscope studies of experimental *Salmonella* infection. I Penetration into the intestinal epithelium by *Salmonella typhimurium.* Am J Pathol 50 (1):109–136
31. Jones BD, Ghori N, Falkow S (1994) *Salmonella typhimurium* initiates murine infection by penetrating and destroying the specialized epithelial M cells of the Peyer's patches. J Exp Med 180(1):15–23
32. Francis CL, Ryan TA, Jones BD et al (1993) Ruffles induced by *Salmonella* and other stimuli direct macropinocytosis of bacteria. Nature 364(6438):639–642
33. Jepson MA, Clark MA (2001) The role of M cells in *Salmonella* infection. Microbes Infect 3:1183–1190

34. Bajaj V, Hwang C, Lee CA (1995) *hilA* is a novel *ompR/toxR* family member that activates the expression of *Salmonella typhimurium* invasion genes. Mol Microbiol 18(4):715–727
35. Bajaj V, Lucas RL, Hwang C et al (1996) Co-ordinate regulation of *Salmonella typhimurium* invasion genes by environmental and regulatory factors is mediated by control of *hilA* expression. Mol Microbiol 22 (4):703–714
36. Srinivasan A, Foley J, Ravindran R et al (2004) Low-dose *Salmonella* infection evades activation of flagellin-specific CD4 T cells. J Immunol 173(6):4091–4099
37. Fink SL, Cookson BT (2007) Pyroptosis and host cell death responses during *Salmonella* infection. Cell Microbiol 9(11):2562–2570
38. Lundberg U, Vinatzer U, Berdnik D et al (1999) Growth phase-regulated induction of *Salmonella*-induced macrophage apoptosis correlates with transient expression of SPI-1 genes. J Bacteriol 181(11):3433–3437
39. Chen LM, Kaniga K, Galan JE (1996) *Salmonella spp.* are cytotoxic for cultured macrophages. Mol Microbiol 21:1101–1115
40. Jesenberger V, Procyk KJ, Yuan J et al (2000) *Salmonella*-induced caspase-2 activation in macrophages: a novel mechanism in pathogen-mediated apoptosis. J Exp Med 192 (7):1035–1046
41. Mariathasan S, Newton K, Monack DM et al (2004) Differential activation of the inflammasome by caspase-1 adaptors ASC and Ipaf. Nature 430(6996):213–218
42. Monack DM, Raupach B, Hromockyj AE et al (1996) *Salmonella typhimurium* invasion induces apoptosis in infected macrophages. Proc Natl Acad Sci U S A 93(18):9833–9838
43. Monack DM, Detweiler CS, Falkow S (2001) *Salmonella* pathogenicity island 2-dependent macrophage death is mediated in part by the host cysteine protease caspase-1. Cell Microbiol 3(12):825–837
44. van der Velden AW, Lindgren SW, Worley MJ et al (2000) *Salmonella* pathogenicity island 1-independent induction of apoptosis in infected macrophages by *Salmonella enterica* serotype *typhimurium*. Infect Immun 68:5702–5709
45. Coombes BK, Lowden MJ, Bishop JL et al (2007) SseL is a *Salmonella*-specific translocated effector integrated into the SsrB-controlled *Salmonella* pathogenicity island 2 type III secretion system. Infect Immun 75 (2):574–580
46. Le Negrate GFB, Welsh K, Loeffler M et al (2008) *Salmonella* secreted factor L deubiquitinase of *Salmonella typhimurium* inhibits NF-kappaB, suppresses IkappaBalpha ubiquitination and modulates innate immune responses. J Immunol 180(7):5045–5056
47. Rytkönen A, Poh J, Garmendia J et al (2007) SseL, a *Salmonella* deubiquitinase required for macrophage killing and virulence. Proc Natl Acad Sci U S A 104(9):3502–3507
48. Gulig PA, Curtiss R 3rd (1987) Plasmid-associated virulence of *Salmonella typhimurium*. Infect Immun 55(12):2891–2901
49. Koski P, Saarilahti H, Sukupolvi S et al (1992) A new alpha-helical coiled coil protein encoded by the *Salmonella typhimurium* virulence plasmid. J Biol Chem 267(17):12258–12265
50. Epstein SL, Tumpey TM, Misplon JA et al (2002) DNA vaccine expressing conserved influenza virus proteins protective against H5N1 challenge infection in mice. Emerg Infect Dis 8(8):796–801
51. Ulmer JB, Fu TM, Deck RR et al (1998) Protective CD4+ and CD8+ T cells against influenza virus induced by vaccination with nucleoprotein DNA. J Virol 72(7):5648–5653
52. Bertani G (1951) Studies on lysogenesis. I The mode of phage liberation by lysogenic *Escherichia coli*. J Bacteriol 62(3):293–300
53. Kang HY, Dozois CM, Tinge SA et al (2002) Transduction-mediated transfer of unmarked deletion and point mutations through use of counterselectable suicide vectors. J Bacteriol 184(1):307–312
54. Laemmli UK (1970) Cleavage of structural proteins during the assembly of the head of bacteriophage T4. Nature 227 (5259):680–685
55. Sedgwick JD, Holt PG (1983) A solid-phase immunoenzymatic technique for the enumeration of specific antibody-secreting cells. J Immunol Methods 57(1–3):301–309
56. McCown MF, Pekosz A (2005) The influenza A virus M2 cytoplasmic tail is required for infectious virus production and efficient genome packaging. J Virol 79(6):3595–3605
57. Schmieger H, Backhaus H (1976) Altered cotransduction frequencies exhibited by HT-mutants of *Salmonella*-phage P22. Mol Gen Genet 143(3):307–309
58. Roland K, Curtiss R 3rd, Sizemore D (1999) Construction and evaluation of a delta *cya* delta *crp Salmonella typhimurium* strain expressing avian pathogenic *Escherichia coli* O78 LPS as a vaccine to prevent airsacculitis in chickens. Avian Dis 43(3):429–441

Chapter 4

Methods for the Construction of Recombinant Oncolytic Myxoma Viruses

Lino E. Torres-Domínguez, Ana Lemos de Matos, Masmudur M. Rahman, and Grant McFadden

Abstract

Myxoma virus (MYXV) has proven to be an effective candidate for oncolytic virotherapy in many preclinical cancer models. As a nonhuman pathogen, MYXV does not need to overcome any preexisting antiviral immunity, and its DNA cannot integrate into the host genome, making it an extremely safe vector. Moreover, the large dsDNA genome of MYXV allows the insertion of multiple transgenes and the design of engineered recombinant oncolytic viruses (OVs) with enhanced immunostimulatory or other desired properties. In this chapter, we describe detailed protocols for the generation and characterization of transgene-armed recombinant MYXV vectors.

Key words Myxoma virus, Poxvirus, Oncolytic virus, Recombinant virus, Viral vector, Oncolytic virotherapy, Engineered virus, Armed oncolytic virus

1 Introduction

Oncolytic virotherapy is among the new biological strategies added to the portfolio of cancer treatments in the last decade. Many oncolytic virus (OV) candidates are currently in different stages of preclinical and clinical studies; among them is myxoma virus (MYXV) [1–4]. MYXV is a member of the Poxviridae family, the prototype for the genus *Leporipoxvirus*, which has a natural and pathogenically narrow host range within the Lagomorpha order [5]. This restrictive natural tropism guarantees its safety for oncolytic virotherapy in humans and in non-rabbit cancer models. Although MYXV is nonpathogenic to humans and mice, the virus is able to infect and kill most human or murine cancer cells and is highly immunogenic [1–5]. Therefore, oncolytic MYXV has been tested in multiple murine tumor models, including melanoma, gliomas, gallbladder cancer, medulloblastomas, pancreatic cancer, and hematological malignancies [1–4]. These preclinical studies

Alexandra R. Lucas (ed.), *Viruses as Therapeutics: Methods and Protocols*, Methods in Molecular Biology, vol. 2225, https://doi.org/10.1007/978-1-0716-1012-1_4,

have demonstrated the efficacy of oncolytic MYXV against diverse solid and hematologic malignant diseases.

Some OVs have been genetically engineered by inserting a plethora of heterologous transgenes to further improve their cancer cell killing and/or immunostimulatory properties. There are multiple OVs armed with proteins that enhance virus-mediated direct cancer cell killing, like TNF-related apoptosis-inducing ligand (TRAIL), apoptin, and Fas ligand fusion molecules. Many other engineered OVs carry a variety of immunostimulatory genes which promote virus-directed immune responses against cancer cells, including cytokines: interleukin-2, interleukin-12, interleukin-15, and tumor necrosis factor (TNF); chemokines: chemokine C-C motif ligand 2 (CCL2), chemokine CXC motif ligand 9 (CXCL9), and CXCL10; co-stimulatory proteins: cluster of differentiation 40 (CD40) and OX40L; T-cell engagers: bi-specific T-cell engagers (BiTEs); and checkpoint inhibitors: programmed cell death protein 1 (PD1), PDL1, and cytotoxic T-lymphocyte-associated protein 4 (CTLA4) [6, 7]. A notable example is the insertion of granulocyte-macrophage colony-stimulating factor (GM-CSF) into the genome of the engineered herpes simplex virus type 1 T-Vec (the only OV so far approved by FDA) [8]. The large dsDNA MYXV genome (~160 kb) can be genetically modified with many exogenous transgenes, being suitable to construct single or multiple armed OV vectors with enhanced immunotherapeutic potential [5]. In the literature there are several reports showing significant increase in survival in different animal models treated by recombinant armed MYXVs, when compared to the parental unarmed virus [9, 10]. When using MYXV as a gene delivery platform, the insertion of therapeutic genes may improve the efficacy of those vectors, constituting the future generation of transgene-armed oncolytic viral immunotherapy. With this chapter, we provide detailed protocols for designing, generating, and characterizing recombinant MYXV vectors that encode exogenous transgenes, to be used as an oncolytic virotherapy platform.

2 Materials

2.1 Tissue Culture

1. Rabbit RK13 cell line (ATCC # CCL-37).
2. African green monkey kidney VERO cell line (ATCC® CCL-81™).
3. *Homo sapiens* cervix adenocarcinoma HeLa cell line (CRM-CCL-2™).
4. Dulbecco's minimum essential medium (DMEM) supplemented with 10% fetal bovine serum (FBS), 2 mM glutamine, and 100 μg of penicillin-streptomycin (pen/strep); this supplemented medium is referred to as DMEM complete medium.

5. 0.25% Trypsin-EDTA phenol red.
6. Phosphate-buffered saline (PBS): 137 mM NaCl, 2.7 mM KCl, 4.3 mM Na_2HPO_4, 1.47 mM KH_2PO_4 pH 7.4.
7. 37 °C Incubators with supply of 5% CO_2.
8. Tissue culture multiwell plates, dishes, and other plasticwares.

2.2 Cloning and Virus Construction

(The kits routinely used in our work are listed below and only reflect our commonly used kits.)

1. MultiSite Gateway™ Pro cloning kit (we use Thermo Fisher Scientific).
2. Gateway™ BP Clonase™ II and LR Clonase™ II enzyme mix (we use Thermo Fisher Scientific).
3. Mach1 and DH5a bacterial competent cells (we use Thermo Fisher Scientific).
4. AccuPrime Pfx DNA Polymerase (we use Thermo Fisher Scientific).
5. 100 mg/mL Ampicillin (in water).
6. 50 mg/mL Kanamycin (in water).
7. Luria-Bertani (LB) liquid medium and LB agar plates.
8. Primers for amplification of MYXV genes and transgene (*see* **Note 1**).
9. Agarose and low-melting-point agarose.
10. Fluorescence microscope with adapters for multiwell plates.
11. Cup sonicator with ice-water bath.
12. GoTaq green (Promega).
13. Water bath.
14. Effectene Transfection Reagent (Qiagen).
15. NanoDrop™ spectrophotometer.
16. Cell scrapers.
17. Master cycler.
18. 10 mM Tris–HCL, pH 8.0 (sterile).
19. Plasmid Miniprep Kit (Qiagen).
20. Mini Blood DNA Purification Kit (Qiagen).
21. CellTiter 96® AQ_{ueous} One Solution Cell Proliferation Assay (MTS).
22. 96-Well plate reader.

2.3 Vectors

1. pDONR221 p1-p4.
2. pDONR221 p4r-p3r.
3. pDONR221 p3-p2.
4. pDest40 or pcDNA6.2.

2.4 Western Blot and Immunocytochemistry

1. Nitrocellulose membranes.
2. Tris-buffered saline (TBS): 10 mM, 10×.
3. Tween-20.
4. Commercial transfer buffer 5×.
5. TBS containing 0.05% Tween-20 (TBST).
6. Blocking solution: 5% Nonfat dry milk in TBS.
7. Diluent solution: 5% Nonfat dry milk in TBST.
8. Mini PROTEAN® 3 System glass plates.
9. Primary antibody (specific for the transgene).
10. Secondary antibody conjugated to horseradish peroxidase (HRP).
11. Chemiluminescent HRP substrate.
12. Gel imager.
13. 2% Paraformaldehyde (PFA) in PBS.
14. Phosphate-buffered saline (PBS).
15. 0.2% Triton X-100 in PBS.
16. Blocking buffer: 0.5% Bovine serum albumin (BSA) in PBS.

3 Methods

3.1 Construction of Recombinant Oncolytic MYXV-Expressing Transgene and/or Reporter Fluorescent Protein(s)

The first step for making a transgene expressing oncolytic MYXVs is the construction of a recombination plasmid cassette having the desired transgene, often in tandem combination with a reporter fluorescent protein-encoding gene and flanked by MYXV recombination sites. Both the transgene and the reporter gene can be expressed under a poxvirus early-late promoter (for constitutive expression of the transgene) or a poxvirus late promoter (for transgene expression only after viral DNA replication has commenced).

The fluorescent proteins (such as enhanced green fluorescent protein, eGFP, dsRed, or TdTomato) allow for easy visualization of recombinant virus foci under fluorescent microscope, thus facilitating selection and purification of transgene expressing recombinant MYXV. To construct the recombination plasmid cassette, we routinely use the MultiSite Gateway®Pro 3-Fragment recombination system. In this system, three entry vectors need to be constructed using pDONR vectors: (a) pDONR221 p1-p4 (containing 500–1000 nt (nucleotide) upstream of chosen MYXV genome insertion site, for example, M135 locus); (b) pDONR221 p4r-p3r (containing the heterologous gene of interest (GOI) under the control of an early-late promoter or a late viral promoter); and (c) pDONR221 p3-p2 (containing 500–1000 nt downstream of the genome insertion site locus, such as the M136 locus, fused to a

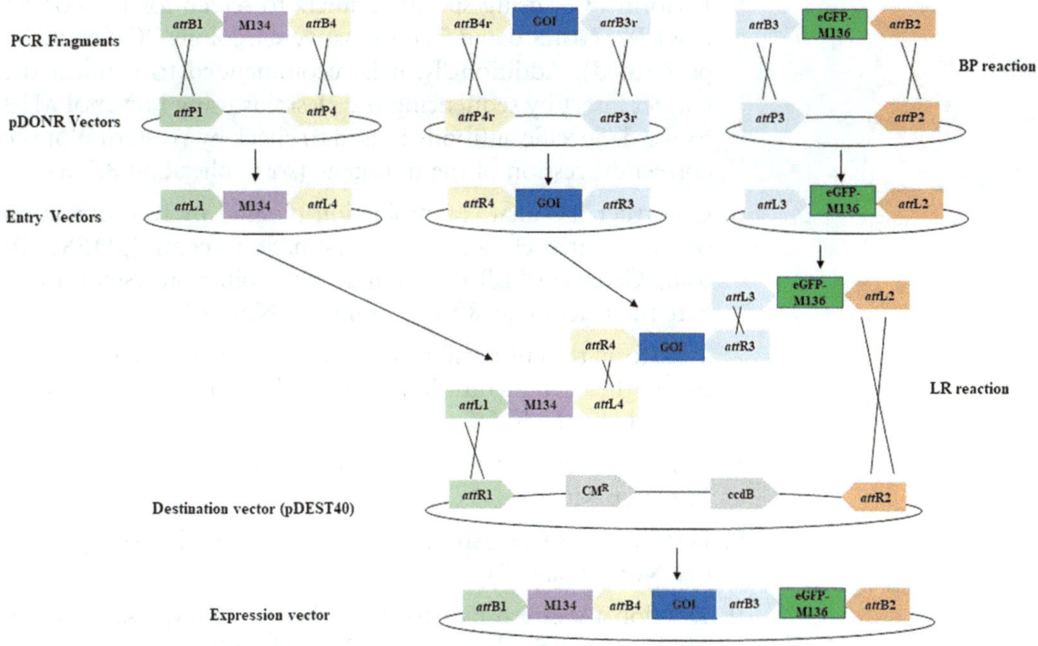

Fig. 1 Schematic representation of the Gateway cloning system for the construction of expression vectors. *BP reactions* recombination between *att*B and *att*P sites, *LR reactions* recombination between *att*L and *att*R sites, *Cm* chloramphenicol gene, *ccdB* survival gene to competent cells

reporter fluorescent gene, like the green fluorescent protein (eGFP), also under the control of an early-late promoter or a late viral promoter) (*see* **Note 2**). Next, these three entry vectors are recombined with a fourth vector (destination vector) to generate the final recombination plasmid cassette, which will allow for the insertion of the transgene and the reporter fluorescent marker into the MYXV genome at the chosen integration site by double-homologous recombination (*see* Fig. 1) (*see* **Note 3**).

3.1.1 Molecular Cloning of Recombinant Transgenes

1. Amplify individual genetic fragments by polymerase chain reaction (PCR) from MYXV genome or from a vector including the transgene of interest, using appropriate primers and the AccuPrime Pfx DNA Polymerase (*see* **Note 4**). Check PCR products on agarose gels with DNA ladder to confirm the size and purity of the amplified PCR products.
2. Perform a BP clonase reaction with the amplified PCR products and the corresponding pDONR vectors at RT for 2–4 h (*see* **Note 5**).
3. Transform the entire BP reaction in Mach1-competent cells and spread the bacteria on LB-kanamycin plates. Incubate the plates at 37 °C for 16 h.
4. Pick 4–6 colonies and amplify the bacteria to purify plasmids using a Plasmid Miniprep Kit.

5. Perform PCR using specific primers to screen for the correct clone (*see* **Notes 6** and 7) (alternatively, a colony PCR could be performed). Additionally, it is recommended to confirm the correct insert by sequencing the plasmids using universal M13 forward. Specific antibody is used to check by Western blot the correct expression of the transgene (*see* Subheading 3.2.3).
6. Construct the final recombination plasmid by recombination of three entry clones with a destination vector (pDEST40) using Gateway™ LR Plus clonase recombination system. Incubate the reaction at RT overnight (*see* **Note 8**).
7. Transform the entire LR reaction in DH5a-competent cells and spread the bacteria on LB-ampicillin plates. Incubate the plates at 37 °C for 16 h.
8. Amplify the bacteria to purify plasmids using Plasmid Miniprep Kit.
9. Perform PCR using specific primers to screen the correct clone (*see* **Notes 6** and 7).
10. Develop a Western blot to check the correct expression of the transgene (*see* Subheading 3.2.3, *see* **Note 9**).

3.1.2 Construction of Recombinant MYXV by Infection and Transfection of the Recombination Plasmid

1. Seed RK13 or Vero cells (using DMEM complete medium) with 70–80% confluency in 6-well plates and allow cells to adhere overnight at 37 °C in a CO_2 incubator.
2. Infect RK13 or Vero cells with WT-MYXV (Lausanne strain) for 1 h with a multiplicity of infection (MOI) of 1.
3. 15–30 min after infection prepare the transfection mixture with the final recombination expression vector using Effectene or any other transfection reagent (it is recommended to use several clones).
4. Remove virus-containing medium after 1 h, wash with medium, add fresh medium, and then add the transfection mixture containing the final recombination plasmid.
5. After 24 and 48 h, observe the expression of eGFP under fluorescence microscope, to confirm the formation of recombinant MYXV (*see* Fig. 2).
6. Harvest the cells in 500 μL medium, and release virus from infected cells by three freeze (−80 °C)-thaw (37 °C) cycles and sonication in ice-water bath for 1 min.

3.1.3 Recombinant Virus Isolation by Foci Purification (Screening for Fluorescent Reporters)

1. Seed RK13 or Vero cells in 6-well plates and allow cells to adhere overnight at 37 °C in a CO_2 incubator.
2. Next day, make ten-fold serial dilutions of the freeze-thawed infected cell lysates, and infect cells in 6-well plates with 1 mL of each dilution for 1 h at 37 °C in a CO_2 incubator.

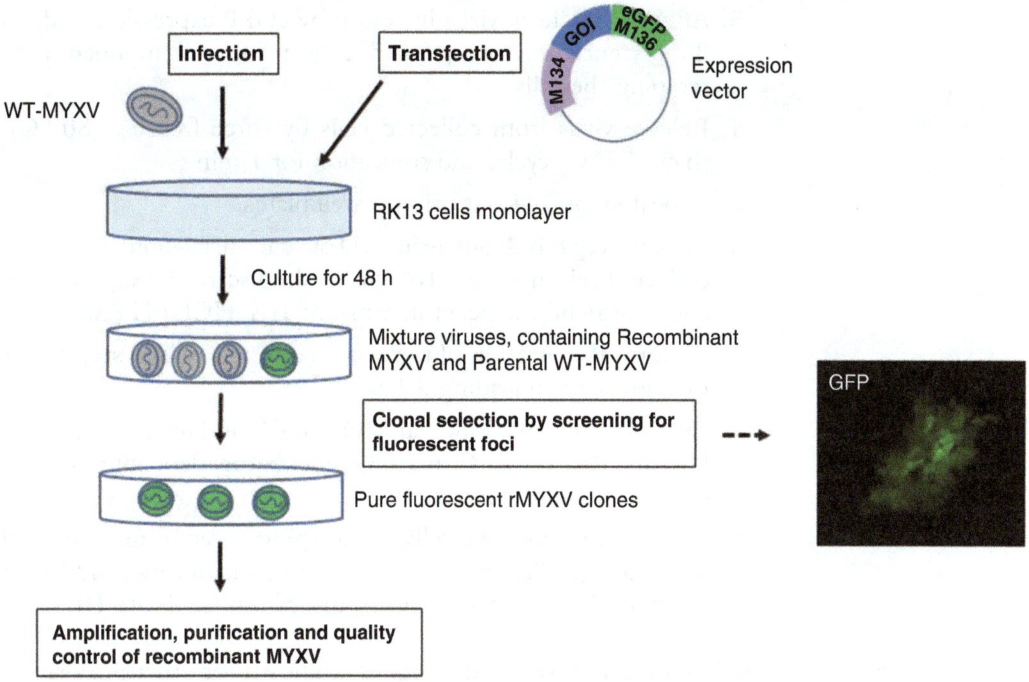

Fig. 2 Experimental procedure for constructing a recombinant Myxoma virus. *eGFP* green fluorescent protein, *GOI* gene of interest, *MYXV* myxoma virus, *RK13* rabbit kidney cells, *WT* wild type

3. Remove medium and add 2 mL of 1% low-melting agarose-containing DMEM complete medium. Allow the agarose to solidify, and store at 37 °C in the cell culture incubator.
4. After 48 h, mark the eGFP-expressing foci using a fluorescence microscope and collect the cells from the marked area using a 1 mL filter tip (confirm that the foci were picked, by checking under the fluorescence microscope).
5. Resuspend the agarose-containing cells in 500 μL medium in a 1.5 mL vial and freeze at −80 °C until use.
6. Release virus from collected cells by three freeze (−80 °C)-thaw (37 °C) cycles and sonication for 1 min, make serial dilutions, and perform multiple rounds of foci purification (**steps 1–6**) on the same cell line until pure eGFP-expressing foci (free of parental virus) are observed under microscope.
7. Select several of the recombinant MYXV clones and test for transgene insertion in the viral genome (*see* Subheading 3.2.2) and for protein expression (*see* Subheading 3.2.3).

3.1.4 Upscaling of Virus Stocks from Single Foci

1. Seed RK13 or Vero cells in 24-well plates and allow cells to adhere overnight at 37 °C in a CO_2 incubator.
2. Next day, infect with an aliquot of the freeze-thawed infected cell lysates and incubate for 24–48 h at 37 °C in a CO_2 incubator.

3. After 48 h, check virus infection by eGFP expression under a fluorescence microscope and collect the cell monolayer by scraping the cells.
4. Release virus from collected cells by three freeze (−80 °C)-thaw (37 °C) cycles and sonication for 1 min.
5. Repeat **steps 1–4** but using 6-well plates.
6. Repeat **steps 1–4** but using a 160 cm^2 dish. Spin down the collected cells from the 160 cm^2 dish, discard the supernatant, and resuspend the pellet in 1 mL of Tris–HCL pH 8.0.
7. Perform the titration of the crude collected cells in **step 6** (for titration *see* Subheading 3.1.5).
8. Infect 25160 cm^2 dishes, at MOI of 0.1, and incubate the cells for 48–72 h at 37 °C in a CO_2 incubator, depending on the progression of the infection.
9. Harvest the infected cells by scraping directly into the cell monolayer, collect the cells in a 50 mL falcon tube, and freeze at −80 °C or proceed to virus purification (*see* **Note 10**).

3.1.5 Virus Titration

MYXV does not form lytic plaques as known for other poxviruses, and plaques (representing foci of virus-infected cells) are usually detected by the expression of fluorescent markers (*see* **Note 11**).

1. Seed RK13 or Vero cells in 24-well plates and allow cells to adhere overnight at 37 °C in a CO_2 incubator.
2. Next day, thaw virus material and homogenize in cup sonicator for 30 s.
3. Make ten-fold serial dilutions (ranging from 10^{-1} to 10^{-9}) of virus material in 1 mL virus growth medium.
4. Plate on RK13 24-well plates 250 μL of virus suspension per well in duplicates, using 10^{-5} to 10^{-9} dilutions. Incubate for 1 h at 37 °C in a CO_2 incubator.
5. Replace inoculum with cell growth medium. Incubate for 48 h at 37 °C in a CO_2 incubator.
6. Count the foci under the fluorescence microscope (*see* **Note 12**).

3.2 Quality Control of Recombinant MYXV

After the recombinant MYXV (rMYXV) is generated, several quality control assays must be performed to assure that the new virus is functional: First, rMYXV should be evaluated for its replication capacity when compared to the parental virus. Second, it must be tested if the insertion of the genetic cassette is in the right locus. Third, the protein expression of the transgene should be evaluated (*see* **Note 13**). Fourth, it should be assessed in vitro for the cell killing capacity of the new constructed MYXV vector.

3.2.1 Multistep Growth Analysis of Recombinant MYXV

Multistep growth analysis allows detection for whether the new rMYXV has any defect in virus replication when compared to parental virus WT-MYXV.

1. Seed RK13 or Vero cells in 6-well plates and allow cells to adhere overnight at 37 °C in a CO_2 incubator.
2. Next day, infect cells with MYXV-GFP and rMYXV at a MOI of 0.01 for 1 h.
3. After 1 h, un-adsorbed free virus is removed, cells are washed with medium, and complete growth medium is added. Cells are incubated at 37 °C in a CO_2 incubator.
4. Harvest cells in 1 mL complete medium at multiple time points postinfection (e.g., 4, 12, 24, 48, and 72 h), and store at −80 °C until processed.
5. Release virus from the infected cells by three freeze-thaw cycles and sonication on ice for 1 min.
6. Virus titers from the lysates are determined by serial dilutions and titration on RK13 cell monolayers (*see* Subheading 3.1.5).

3.2.2 DNA Purification and Characterization of Recombinant MYXV Genome by PCR

1. Infect RK13 or Vero cell monolayer of one well in 24 well with rMYXV at multiplicity of infection of 5, and incubate for 24–48 h at 37 °C.
2. Discard medium, harvest cell monolayer in 200 μL PBS, and transfer into a 1.5 mL vial.
3. Add 20 μL of Proteinase K and proceed to DNA purification using the Mini Blood DNA Purification Kit (as described by the fabricant).
4. Prepare PCR reaction mix on ice by adding 11 μL GoTaq green Mix, 1 μL (10 μM) forward primer, 1 μL (10 μM) reverse primer, 100 ng template DNA, and distilled water for a final volume of 25 μL/reaction.
5. Perform a PCR reaction adjusted to the conditions of the primers and amplicon.
6. Use 10 μL aliquot of each sample to perform agarose gel electrophoresis to visualize amplified PCR amplicons and determine molecular weights.

3.2.3 Expression of Recombinant Protein Assessed by Western Blot

1. Seed RK13 or Vero cells in 24-well plates and incubate for 24 h at 37 °C in a CO_2 incubator.
2. Infect subconfluent (80–90%) cell monolayers with rMYXV at a MOI of 5 in 1 mL medium.
3. Incubate for 24–48 h until 90–100% of the cells are infected; check infectivity by GFP signaling under fluorescence microscope.

4. Harvest cells by scraping the cell monolayer, transfer to a 1.5 mL vial, and pellet cells by centrifugation for 1 min at maximum speed.
5. Resuspend pellet in ice-cold 50 μL of 1× SDS-loading buffer and freeze samples until use.
6. Proceed with Western blot analysis of the lysates for the recombinant protein using specific antibodies for the detection (*see* **Note 14**).

3.2.4 Expression of Recombinant Protein Assessed by Immunocytochemistry

1. Infect subconfluent (80–90%) RK13 cell monolayer in a 6-well plate with the rMYXV at a MOI of 0.01 in 1 mL of cell growth medium.
2. Incubate for 1 h at 37 °C in a CO_2 incubator, remove inoculum, and add 2 mL of fresh cell growth medium.
3. Incubate for 48 h until small foci are visible.
4. Wash cells once with PBS and fix them for 3 min with 2 mL of 2% PFA for 12 min.
5. Wash twice with PBS.
6. Permeabilize with 1 mL of 0.2% of Triton X-100. Incubate for 2 min (*see* **Note 15**).
7. Wash twice with PBS.
8. Add 2 mL of blocking buffer per well and incubate for 30 min at 37 °C.
9. Dilute your primary antibody (directed against your recombinant protein of interest) in blocking buffer.
10. Replace the blocking buffer in each well with 1 mL of the antibody solution and incubate for 30 min at 37 °C.
11. Remove primary antibody solution and wash cells six times with 2 ml of blocking buffer.
12. Dilute your secondary antibody conjugated with a fluorescent protein (directed against your primary antibody) in blocking buffer, add 1 mL of antibody solution to the wells, and incubate for 30 min at 37 °C.
13. Remove secondary antibody solution and wash cells six times with 2 mL of blocking buffer. Keep the cells with 1 mL of blocking buffer.
14. Visualize the antigen-specific staining under the fluorescence microscope (*see* **Note 16**).

3.2.5 In Vitro Cytotoxicity Assay

The evaluation of the in vitro cell killing capacity of the new constructed recombinant MYXV with respect to the parental MYXV is another useful control test. This assay is essential when the inserted transgene is intended to enhance the cell oncolytic capacity of the engineered MYXV vector, as well as to confirm that the oncolytic capacity was not affected during the generation of the new recombinant MYXV.

1. Using DMEM complete medium, seed HeLa cells (or any given tumor cell line) in 96-well plates (5000 cells/well) and allow cells to adhere overnight at 37 °C in a CO_2 incubator (*see* **Note 17**).
2. Next day, remove the media and apply to the cells (in triplicate) 30 μL of virus at a MOI of 10 resuspended in complete DMEM. Include a control of vMyx-GFP along with a mock control.
3. Incubate for 1 h at 37 °C in 5% CO_2.
4. Add 70 μL of complete DMEM (to a final volume of 100 μL).
5. After 48 or 72 h, add 20 μL of CellTiter 96® AQueous One Solution Reagent to each well (*see* **Note 18**).
6. Incubate for 2 h at 37 °C in 5% CO_2 (*see* **Note 19**).
7. Record the absorbance at 490 nm using a 96-well plate reader.
8. Analyze the readout using PRISM GraphPad or Excel.

4 Notes

1. The primers must include the recombination sites for the Gateway Recombination system (attB sequences). The sequences and guidelines for primer design are provided in the manufacturer's protocol. The forward primer (attB4r) for the heterologous recombinant protein must include a viral promoter (commonly synthetic early-late or late P11). An early-late viral promoter is commonly preferred for most transgenes since it will guarantee high levels of protein expression, whereas in the case that the heterologous protein promotes cell death it is recommended to use a late promoter that will allow the completion of the replication cycle. In case the transgene is intended to be secreted, it must include a secretion signaling peptide. The transgene sequence might be codon optimized when required.
2. One of the flanking region fragments, in this case right flank M136, must be previously fused to the reporter gene (eGFP), and then the fused cassette is inserted into the pDONR221 p3-p2. This methodology saves time, since once the two flanking entry vectors for a given locus have been constructed, it will only be necessary to engineer the middle entry vector for the GOI. Then, a last recombination step will be performed for constructing the final expression plasmid (with the whole cassette).

3. To generate mutant MYXV viruses we use the same methodology, previously described in detail by Rahman, MM [11]. To build recombination vectors alternative cloning strategies can be used, like the conventional ligation methods.
4. The transgene may be included in a vector purchased from a commercial company; in case some modifications are needed, it can be custom-made by gene synthesis, another service provided by different companies. For PCR amplification the use of a proofreading enzyme is encouraged to avoid undesired modifications.
5. Follow the instructions included in the Gateway cloning system user guide. For the BP reactions use 150 ng of the pDONR plasmid and 150 ng of the PCR product (either purified or gel-purified). For long amplicons it is advantageous to incubate the reaction overnight at RT.
6. Alternatively a colony PCR could be performed, by adding a small amount of a colony directly to the PCR reaction.
7. We recommend checking the correct insert by sequencing the plasmids using universal M13 forward and reverse primers or specific primers for the GOI.
8. Follow the instructions included in the Gateway cloning system user guide. For the LR reactions use 150–200 ng of the three pDONR plasmids and 150–200 ng of the destination vector, incubating the reaction overnight at RT.
9. The expression of the recombinant protein can be detected from the entry clones or the recombination vectors, by performing a transfection/infection assay, preferably using high MOI = 5, to guarantee a high protein level expression.
10. MYXV purification has been previously described by Smallwood et al. [12].
11. A protocol for titration of poxviruses using the tissue culture infectious dose 50 ($TCID_{50}$) method was reported before by Kremer et al. [13].
12. Equation to calculate focus-forming units: FFU = #virus foci (average 3 wells)/ (volume x dilution).
13. The expression of the recombinant protein can be performed by Western blot or immunocytochemistry. For secreted proteins and to have a more quantitative data an ELISA assay might be developed; additionally, consider ELISA assays for polypeptides which Western blot might not be sensitive enough to detect the protein in the supernatant.
14. Protocols for the development of SDS-PAGE and Western blot have been previously described by Kurien et al. [14].

15. The permeabilization step with Triton X-100 is essential when checking for intracellular proteins; for proteins exclusively in the membrane it is not required.
16. Instead of fixing the cells directly in the plate, it is possible to use slides and cover slides that would allow to use a higher magnification at the fluorescence microscope, and therefore get better resolution images.
17. The optimal cell number must be standardized for each type of cell.
18. Alternatively use CellTiter Glo or other cell proliferation assays to count the number of live cells. In some particular cases it might be more accurate to use an assay that measures cell death directly, like SYTOX™ Orange.
19. The manufacturer recommends an incubation time of 1–4 h, since optimal time varies from cell line to cell line and must be standardized prior to the experiment.

References

1. Torres-Dominguez LE, McFadden G (2019) Poxvirus oncolytic virotherapy. Expert Opin Biol Ther 19(6):561–573
2. Chan WM, Rahman MM, McFadden G (2013) Oncolytic myxoma virus: the path to clinic. Vaccine 31(39):4252–4258
3. Chan WM, McFadden G (2014) Oncolytic poxviruses. Annu Rev Virol 1(1):119–141
4. Rahman MM, McFadden G (2020) Oncolytic virotherapy with myxoma virus. J Clin Med 9 (1):171–187
5. McFadden G (2005) Poxvirus tropism. Nat Rev Microbiol 3(3):201–213
6. Thorne SH (2014) Immunotherapeutic potential of oncolytic vaccinia virus. Front Oncol 4:155
7. Haddad D (2017) Genetically engineered vaccinia viruses as agents for cancer treatment, imaging, and transgene delivery. Front Oncol 7:96
8. Kohlhapp FJ, Zloza A, Kaufman HL (2015) Talimogene laherparepvec (T-VEC) as cancer immunotherapy. Drugs Today (Barcelona, Spain: 1998) 51(9):549–558
9. Tosic V, Thomas DL, Kranz DM et al (2014) Myxoma virus expressing a fusion protein of interleukin-15 (IL15) and IL15 receptor alpha has enhanced antitumor activity. PLoS One 9(10):e109801
10. Bartee MY, Dunlap KM, Bartee E (2017) Tumor-localized secretion of soluble PD1 enhances oncolytic virotherapy. Cancer Res 77 (11):2952–2963
11. Rahman MM (2018) Methods for identifying virus-derived serpins. Methods Mol Biol (Clifton, NJ) 1826:73–86
12. Smallwood SE, Rahman MM, Smith DW, McFadden G (2010) Myxoma virus: propagation, purification, quantification, and storage. Curr Protoc Microbiol Chapter 14:Unit 14A 1
13. Kremer M, Volz A, Kreijtz JH et al (2012) Easy and efficient protocols for working with recombinant vaccinia virus MVA. Methods Mol Biol (Clifton, NJ) 890:59–92
14. Kurien BT, Scofield RH (2015) Multiple immunoblots by passive diffusion of proteins from a single SDS-PAGE gel. Methods Mol Biol (Clifton, NJ) 1312:77–86

Chapter 5

Molecular Design and Production of AAV Viral Vectors for Gene Therapy

Raela B. Ridley, Erin M. Walsh, and Cristhian J. Ildefonso

Abstract

Adeno-associated virus (AAV) is a helper-dependent single-stranded DNA parvovirus. Over the years, AAV has become the vector of choice in the gene therapy field due to its safety profile and low immunogenicity. With a carrying capacity of 4.2 kbp, these vectors have demonstrated their clinical value, especially in the field of ophthalmology. Herein we describe methods for the molecular design and packaging of AAV viral vectors. These methods apply to the design of single-stranded or self-complementary AAV vectors.

Key words AAV, Transgene, Gene therapy, PCR, Ligation, Transformation, Cloning

1 Introduction

Adeno-associated virus (AAV) is a small parvovirus measuring approximately 20 nm diameter. This helper-dependent virus was discovered in 1965 as a contaminant to viral cultures of adenovirus [1]. AAV has single-stranded DNA genome with two inverted terminal repeats (TRs). The wild-type AAV genome encodes two genes, capsid structure and viral replication regulation genes (Cap and Rep). The capsid structure of AAV is composed of the 60 protein subunits composed of three subunits (VP1, VP2, and VP3) that form an icosahedral shell. These capsid proteins determine the interaction of the viral particles with the host cell.

AAV vectors have low immunogenicity when used as viral vectors [2]. However, AAV transduction is known to cause an initial inflammatory response due to the presence of capsid proteins [3, 4]. Wild-type AAV integrates into the host genome at AAVS1, a specific site in human chromosome 19 [5, 6]. AAV cloning plasmids only retain the TR sequences of the wild-type AAV, with the gene of interest (i.e., promoter-open reading frame and polyA signal) cloned between both TRs. The AAV vector genome remains as episomal concatemers in the nucleus of the host cell (*see* Fig. 1)

Alexandra R. Lucas (ed.), *Viruses as Therapeutics: Methods and Protocols*, Methods in Molecular Biology, vol. 2225, https://doi.org/10.1007/978-1-0716-1012-1_5,

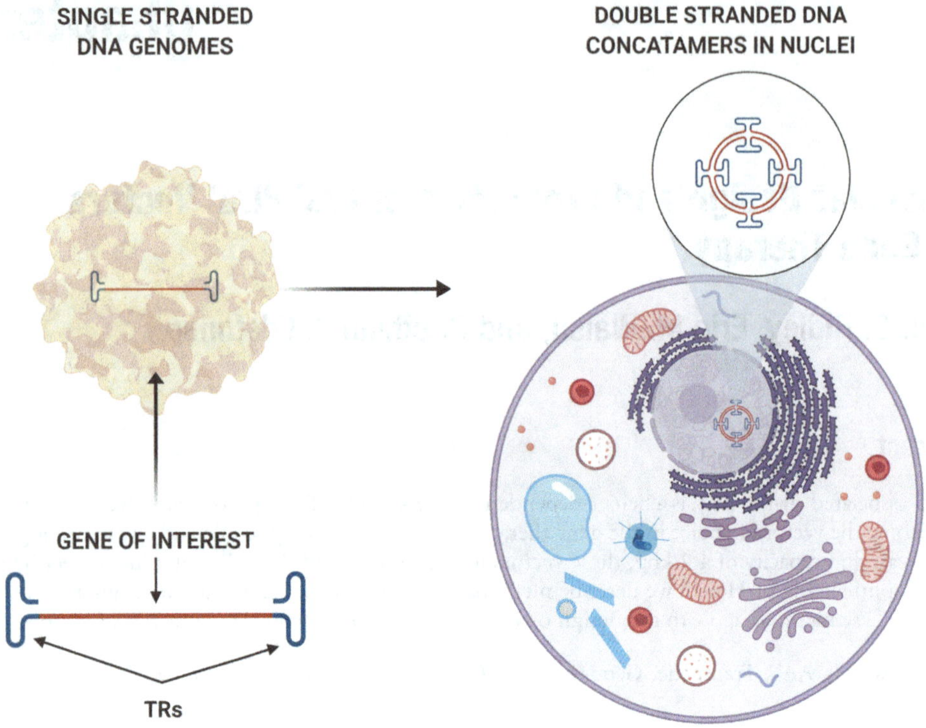

Fig. 1 Recombinant adeno-associated virus (AAV) as a gene therapy vector. A gene of interest, including its promoter and poly-A signal, is cloned between the two terminal repeat (TRs) sequences of an AAV. The TRs and the gene of interest are packaged into the AAV viral capsid using a co-transfection method in vitro. Viral particles are concentrated and purified before use. Once inside a cell, the AAV genome is delivered to the nucleus, where it becomes an episomal double-stranded DNA concatemer

[7–9]. Because of its safety profile, AAV has become the platform of choice in the field of gene therapy.

AAV vectors have several advantages over other viral vectors, such as lentivirus and adenovirus. AAV vectors infect a diverse range of cell types and have the ability to transduce postmitotic cells. AAV also exhibit long-term expression in postmitotic cells. However, AAV vectors have limitations that determine which diseases can benefit from using this platform. A significant limitation of AAV is its relatively small genetic carrying capacity. This restricts the length of the transgene that can be inserted into an AAV vector genome to about 4.5 kilobases (kb) for single-stranded AAV. For self-complementary AAV (scAAV), the size of the transgene is further limited to 2.4 kb. Considering that the median length of a human protein-coding mRNA is 2.787 kb [10], AAV can still be used in a significant number of gene transfer methods. Although it has an even more significant size disadvantage, scAAV eliminates the need for second-strand AAV DNA synthesis since the viral particles contain double-stranded genomes. The double-stranded genome of the scAAV results in faster gene expression when compared to the use of single-stranded AAV vectors.

The transduction efficiency of different AAV serotypes varies wildly based on their different capsid proteins. There are 12 wild-type AAV serotypes (AAV1–AAV12) that can be used to transduce different mammalian cell types and organs/tissues with different efficiencies [11]. Researchers have introduced amino acid-specific capsid mutations that either improve transduction efficiency and tissue penetrance or alter vector tropism [12]. Other groups have used a directed evolution approach to identify AAV variants that have higher transduction efficiency for specific cell types, but which may be poorly transduced by wild-type AAV [13].

Both of these approaches have exponentially increased the number of AAV capsid serotypes available for research. This has also improved the ability to generate gene therapy approaches with a highly specific cell-autonomous effect by combining capsid elements with cell-specific promoters. In our lab we have used AAV to express immune-modulating genes derived from another virus, from the Myxoma virus, and demonstrated efficacy in reducing uveitis in a mouse model [14]. In this chapter, we describe the methods involved in the cloning and packaging of AAV vectors. These methods can be applied to both single-stranded and self-complementary AAV production.

2 Materials

All solutions must be prepared with sterile deionized water. Reagents should be stored at room temperature unless stated otherwise. Follow current waste disposal regulations.

2.1 Polymerase Chain Reaction (PCR)

1. High-fidelity DNA polymerase.
2. 10 mM Defined deoxyribose nucleotide triphosphate (dNTP)—dNTPs.
3. Primers (10 μM each).
4. 100% DMSO.
5. DNA polymerase buffer.
6. Nuclease-free water.
7. Template DNA.
8. Polymerase chain reaction (PCR) tubes.
9. Ice.
10. Thermocycler.

2.2 Gel Purification

1. Low-melting temperature agarose.
2. Tris-borate-EDTA 1× (89 mM Tris, 89 mM borate, 2 mM EDTA) or tris-acetate-EDTA 1× (40 mM Tris, 40 mM acetate, 1 mM EDTA).
3. Ethidium bromide (10 mg/mL).
4. Restriction enzymes that can cleave the restriction sites used in cloning the cloning strategy.
5. Incubator or heat block at 37 °C.
6. Calf intestine phosphatase or shrimp alkaline phosphatase (to be used in blunt-end ligation reactions).
7. Gel electrophoresis chamber.
8. Transilluminator or gel doc.
9. Scalpel.
10. Deoxyribonucleic acid (DNA) ladder.

2.3 Ligation

1. T4 ligase (400 units/μL) concentration or bought as ready mixed.
2. 10× T4 ligase buffer.
3. Nuclease-free water.
4. Ice.
5. Incubator, water bath, or heat block at 16 °C.
6. DNA clean and concentration kit.

2.4 Transformation

1. SURE electrocompetent modified *E. coli* cells.
2. Ice.
3. Electroporator.
4. Electroporation cuvettes.
5. 1.5 mL Tubes.
6. SOC medium or Luria Broth.
7. Gel-loader tips.

2.5 Plating Cells

1. Ampicillin agar plates (100 ug/mL ampicillin).
2. Bunsen burner.
3. Spinning platform.
4. Metal spreaders.
5. Microcentrifuge.
6. Incubator at 37 °C.

2.6 Picking Colonies and DNA Isolation

1. Ampicillin (100 mg/mL).
2. Sterile terrific broth.
3. Polystyrene culture test tubes.
4. DNA mini-prep kit.

2.7 Screening Colonies

1. Ice.
2. 1% Agarose gel.
3. Restriction enzymes.
4. Incubator or heat block at 37 °C.
5. 10× Restriction enzyme buffer.
6. Nuclease-free water.
7. Gel electrophoresis chamber.
8. Transilluminator or gel doc.
9. DNA ladder.

2.8 Packaging AAV

1. Cell factory.
2. Dulbecco's modified Eagle's medium (DMEM) supplemented with 5% fetal bovine serum and 1% penicillin-streptomycin.
3. Linear polyethyleneimine (PEI, 25,000 MW).
4. DMEM without antibiotic or serum.
5. Cell culture incubator at 37 °C and with 5–7% CO_2.
6. Ultracentrifuge.
7. Fast protein liquid chromatography (FPLC) system.
8. Anion-exchange column.
9. 215 mM NaCl, pH 8.0.
10. Balanced salt solution with 0.014% Tween 20.
11. 100 K concentrator.
12. Real-time PCR thermocycler.
13. Polyacrylamide gel.
14. Silver nitrate ($AgNO_3$).
15. Luria Broth.
16. Endosafe system.

3 Methods

All methods should be conducted at room temperature unless otherwise specified.

3.1 Creating a Gene of Interest PCR Product

1. The gene of interest (GOI) can be amplified by PCR to introduce the specific restriction sites needed for downstream cloning. Briefly, a total of 10–15 nucleotides from either the 5′ or the 3′-end of the GOI should be selected considering that their melting temperature (Tm) difference is no more than 3 °C. Afterwards, the 5′ or forward primer should include the Kozak sequence to improve translation, while the 3′ or reverse primer

should include a stop codon (e.g., TGA). Finally, the desired restriction sites sequences should be added to the sequence. Care should be taken when designing primers to include extra nucleotides at the 5′-end so as to maximize the restriction digest efficiency. It is recommended to include extra TAT (forward primer) or ATA (reverse primer) bases at the 5′-end of each primer. An example is provided in Fig. 2.

2. Create a master mix in a PCR tube using the components listed in Table 1 on ice. Gently mix by flicking.
3. After creating the master mix, load it into a thermocycler. Use Table 2 as a guide for creating PCR conditions.

3.2 Gel Purification

1. Digest the pTR plasmid and PCR insert with the proper restriction enzymes (*see* **Note 1**). Digestion should be prepared with at least 1 μg of pTR plasmid to compensate for DNA loss during the next steps.
2. Incubate the digestion reaction at 37 °C or the optimal temperature of the restriction enzyme for at least 2 h. Longer incubations should be considered (*see* **Note 2**).

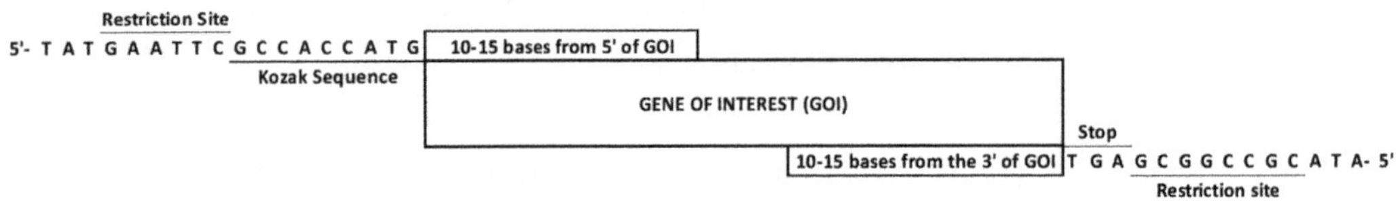

Fig. 2 Primer design to amplify GOI flanked by restriction sites. Forward and reverse primers should contain between 10 and 15 complementary bases to the GOI sequence. These should be immediately followed by either the Kozak sequence (forward primer) or a stop codon (reverse primer). Next, both primers should include the sequence of the restriction enzyme to be used during cloning. Finally, an extra three bases should be included on the 5′-end of each primer to improve restriction enzyme efficiency cleavage

Table 1
Representative PCR master mix

Reagent	Volume (μL)
5× High-fidelity DNA polymerase buffer	10.0
dNTPs (10 mM each)	1.0
10 μM of forward primer	2.5
10 μM of reverse primer	2.5
High-fidelity DNA polymerase (2 units/μL)	0.5
Template DNA	Variable (~1 ng)
Nuclease-free water to 50 μL	Up to 50
DMSO (optional)	1.5

The amount of DNA template should be between 1 and 10 ng

Table 2
Representative PCR conditions for the amplification of a GOI

Step	Temperature (°C)	Time
Initial denaturation	98	30 s
Melting	98	10 s
Annealing 35 Cycles	45–72	10 s
Extension	72	15–30 s/kb
Final extension	72	5 min
Hold	12	

3. Run the digests in a 1% agarose gel (*see* **Note 3**). Run the gel at low voltage (100–200 V) until the loading dye reaches the end of the gel.
4. Visualize the bands on the gel with a transilluminator or gel doc (*see* **Note 4**).
5. Cut and excise the bands with the expected size using a scalpel. Collect the bands in 1.5 mL tubes.
6. Purify the DNA bands using a gel purification kit per the manufacturer's instructions.
7. Measure the DNA concentration using a nanodrop or UV spectrophotometer.

3.3 Ligation

1. Thaw T4 ligase buffer and place it immediately on ice (*see* **Note 5**).
2. Determine the best molar ratio for the GOI. As a guide, a one kbp insert can be successfully ligated using a 1:4 ratio of a linearized pTR plasmid to insert.
3. Set a control ligation as well as use more nuclease-free water in place of the insert (*see* **Note 6**). An example of the recipes for each ligation is provided in Table 3.
4. Incubate for 2 h at 16 °C and then for 15 min at room temperature (*see* **Note 7**).
5. After ligation, use a DNA clean and concentration kit to purify the recombinant DNA (*see* **Note 8**).

3.4 Ligation Cleanup

1. Concentrate the ligation reaction using a DNA concentration kit as per the manufacturer's instructions (*see* **Note 9**).
2. Quantify the DNA concentration with a nanodrop or spectrophotometer.

Table 3
Ligation reaction setup

Reagent	Volumes for experimental ligation (μL)	Volumes for control ligation (μL)
Linearized pTR plasmid	1.0	1.0
Digested GOI insert	4.0	0.0
T4 ligase buffer (10×)	2.0	2.0
T4 ligase	1.0	1.0
Nuclease-free water	12.0	16.0

Examples of a 1:4 ratio ligation and a control ligation are provided. Multiple ratios should be explored to identify the optimal ratio for any specific GOI insert

3.5 Transformation

1. Pre-chill the electroporation cuvettes on ice. Label the cuvette with the sample names.
2. Place 1.5 mL tubes on ice and label them with the sample names.
3. Place a 25 μL aliquot of SURE Competent cells (*see* **Note 10**) and place on ice to thaw (about 30 s).
4. Add 1–2 μL of ligation reaction to the cells and mix the tube by flicking.
5. Pipette the cells and ligation mix into the chilled cuvette between the two metal plates without introducing bubbles.
6. Tap the cuvette gently on the bench to remove any bubble.
7. Wipe the cuvette and place it in the slot of the electroporation device with the two metal plates of the cuvette facing the electrodes.
8. Set the electroporator to 1.25 kV with a capacitance of 25 μF.
9. Slide the cuvette into place and conduct a pulse of 8 s with the described settings.
10. Remove the cuvette and add 500 μL of defined SOC medium to the cuvette and pipette up and down until the media becomes cloudy.
11. Place the cells in the pre-chilled tubes from **step 3**.
12. Incubate the cells at 37 °C while shaking at 500 rpm for an hour to allow them to recover.
13. Place six ampicillin plates (100 μg/mL) (three for experimental, three for control) into the incubator to warm at 37 °C.

3.6 Plating Cells

1. Plating bacteria should be done at the end of the day (*see* Fig. 3).
2. Turn on a Bunsen burner, flame the metal spreaders, and put them to the side.

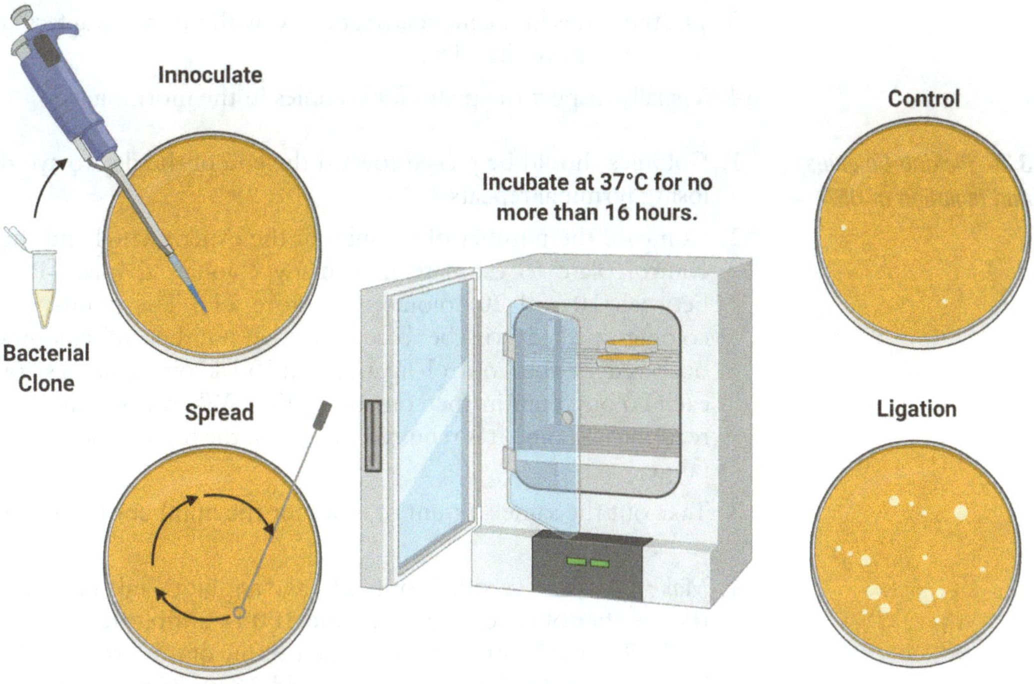

Fig. 3 Selection of bacterial colonies with recombinant DNA. Bacterial clones are resuspended in broth and added to an ampicillin plate. The bacteria are spread with a circular motion until no liquid is observed on the plate. Plates are incubated upside down in a 37 °C incubator for no longer than 16 h. The control plate could have some colonies. However, the ligation plate should have at least four times more colonies than the control. Smaller colonies should be picked over larger colonies since these are more likely to retain the TRs

3. After the cells finish shaking, take the plates out of the incubator.
4. Label the bottom of the plates.
5. Place the plate on the spinner.
6. Pipette 40 μL of cells on one plate, and 200 μL of cells on another plate.
7. Concentrate the cells by centrifuging them for 5 min at 1000 × *g*.
8. Cool off the spreader by placing it on the edge of the plate for a few seconds.
9. Spread the cells while spinning the plate on a platform. Make sure to spread the cells evenly.
10. Remove half of the media and resuspend the cells in the remaining media.
11. Add this remaining amount of cells to the final plate the same way as in **steps 4** and **5**.
12. Repeat these steps for the control of ligation cells.

13. Put the plates in the incubator at 37 °C with the labeling facing up for no more than 16 h.
14. Visually inspect the plates for colonies in the morning.

3.7 Picking Colonies and Isolation of DNA

1. Colonies should be picked toward the end of the day to avoid losing terminal repeats.
2. Compare the number of colonies in the experimental and the control plate to estimate how many colonies to pick. Pick between 10 and 20 colonies (*see* **Note 11**). The number of colonies picked will be based on the number of colonies observed in the control ligation, and pick two colonies for each colony seen in the control ligation. When doing nondirectional cloning, it is required to pick on the higher end of this range.
3. Take out the same amount of tubes for the number of colonies to be picked.
4. Make a stock of Terrific Broth (TB) and antibiotic mix based on the number of tubes you have: Add 1000× antibiotic solution to the Terrific Broth. Add the same volume of mL broth in μL. For example, for 36 mL of broth, add 36 μL of antibiotic (*see* **Note 12**). Mix gently by shaking.
5. Aliquot 3–5 mL of TB with the antibiotic solution into each tubes.
6. Only pick colonies that are not in contact with other colonies. Do this with a pipette tip.
7. Carefully touch the colony with the tip of the pipette.
8. Put the tip into the tube with the TB and eject it into the broth. This will ensure that the colony makes it into the broth.
9. Do this for the rest of the tubes and then place the tubes into the incubator shaking at 37 °C. Shake for 16 h (*see* **Note 13**).
10. Take the tubes out of the incubator and centrifuge for 15 min at 2500 × *g*.
11. After the centrifugation, decant the supernatant (*see* **Note 14**).
12. Isolate plasmid DNA using a miniprep kit following the manufacturer's instructions.
13. Determine the DNA concentration using a nanodrop or any other spectrophotometer.

3.8 Screening Clones for GOI Insert

1. Perform a restriction digest to identify clones containing your GOI. The same restriction enzymes used to linearize the pTR backbone may be used.
2. Select an enzyme buffer in which both enzymes are highly active to ensure optimal plasmid digestion.

Table 4
This table includes an example of how to set up a restriction digest recipe to screen colonies for the presence of insert

Reagent	Volume (μL)
First enzyme	0.5
Second enzyme	0.5
10× Buffer	1.0
DNA	Variable (~300 ng of DNA)
Water	Up to 10 μL

Table 5
Example of XmaI digestion to screen for terminal repeats

Reagent	Volume (μL)
XmaI	0.5
10× Buffer	1.0
DNA	Variable (~300 ng DNA)
Water	Up to 10 μL

3. Prepare a digestion reaction and maintain on ice.
4. Incubate digestions at 37° for 1 h. An example of a digest recipe is in Table 4.
5. Prepare a 1% agarose gel.
6. Load the samples in the gel and run at 200 V until the samples have reached the bottom of the gel.
7. Image the gel using a transilluminator or a gel doc system to identify clones that contain the GOI band.
8. Select clones containing the GOI.
9. Complete another digest with XmaI enzyme to check for TRs (*see* **Note 15**). Digest for 1 h. An example of such a digest is described in Table 5.
10. Prepare a 1% agarose gel and run the gel the same as in **step 6** above.
11. Image the gel as done previously to identify clones with terminal repeats. An ideal clone will have at least 70% of intact TRs.
12. Clones with TRs should be sent for Sanger sequencing to confirm the integrity of the GOI sequence.

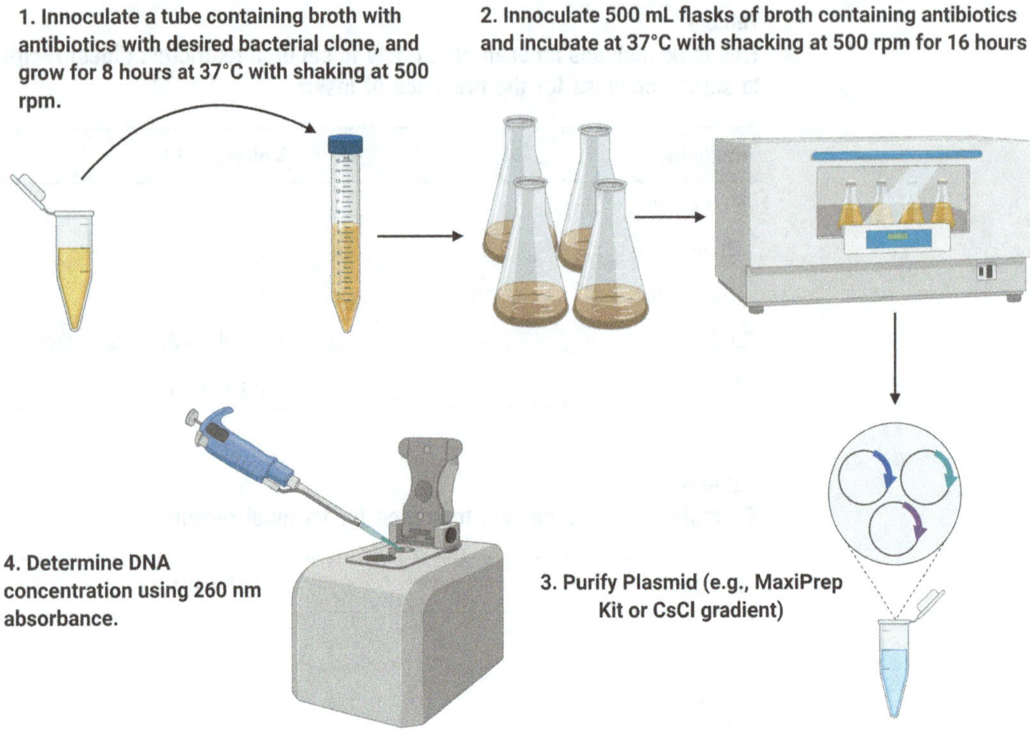

Fig. 4 Amplification of AAV plasmid for virus packaging. A bacterial clone carrying the desired AAV plasmid is used to inoculate a small culture (5–10 mL of broth containing selection antibiotic) and grown for 8 h at 37 °C with shaking. This small culture is then divided into multiple flasks containing broth with selection antibiotics to further grow the plasmid for 16 h at 37 °C with shaking. Bacteria are harvested into pellets, and plasmid DNA is isolated by either Maxi/Giga prep kit (per manufacturers' protocol) or CsCl gradient. The quality and concentration of the plasmid preparation are determined by absorbance at 260 nm with a spectrophotometer/ nanodrop

Once a clone is identified by DNA sequencing and TR screening, this DNA should be expanded by following the steps described in Fig. 4. The DNA purification can be performed using a Maxi or Giga Prep from a reputable company. A minimum of 700 μg of plasmid DNA is needed to generate AAV viral particles with the method described below.

3.9 Packaging AAV

Adeno-associated virus (AAV) vector preparations are produced with a two-plasmid co-transfection method [15]. Before packaging AAV particles, it is essential to select what serotype/capsid will be used based on the target cell [16] (*see* **Note 16**). A summary of the process described below is given in Fig. 5.

1. Prepare a cell stack with approximately 10^9 HEK 293 cells cultured in Dulbecco's modified Eagle's medium, supplemented with 5% fetal bovine serum and 1% antibiotics.
2. A polyethyleneimine (PEI) transfection is set up by mixing a 1:1 molar ratio of rAAV vector plasmid DNA and serotype/ capsid-specific rep–cap helper plasmid DNA.

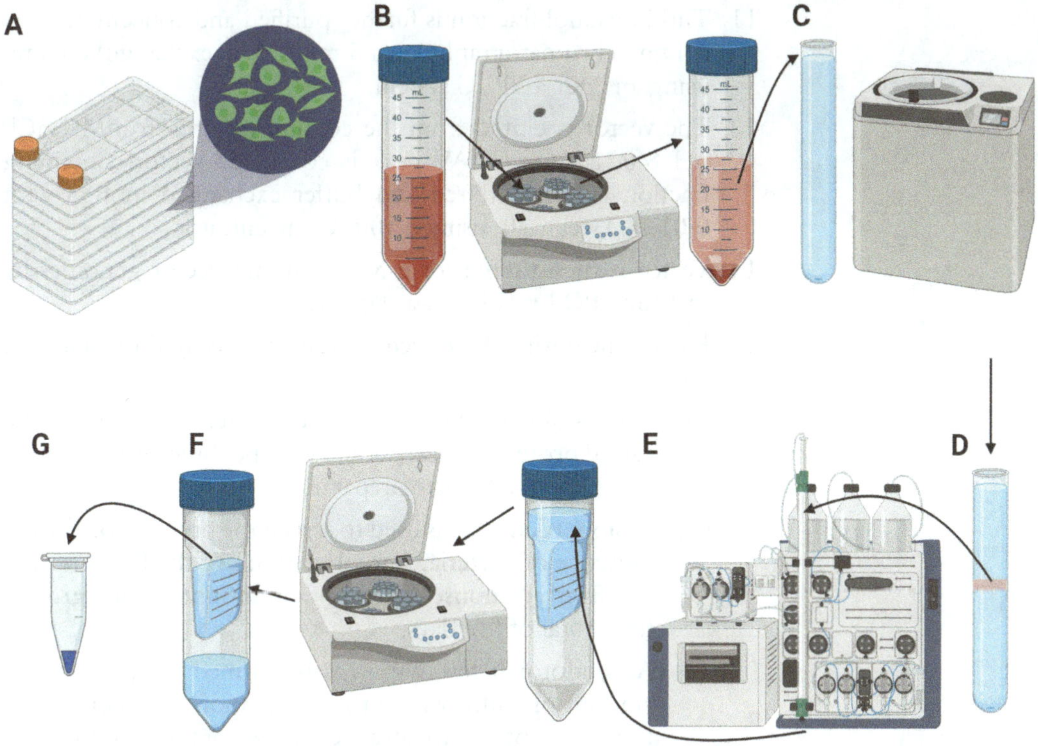

Fig. 5 Packaging and purification of AAV vectors. (**a**) The AAV plasmid is co-transfected with the Rep and Cap helper plasmids into HEK293 cells grown in a cell factory. Cells are grown for 60 h before harvesting by centrifugation. (**b**) Cells are lysed by freeze-thaw cycles and then cleared by centrifugation to remove debris. (**c**) The AAV-containing supernatant is then loaded onto an iodixanol gradient and centrifuged at ultrahigh speed. (**d**) The band containing the viral particles is then isolated. (**e**) AAV is cleaned using a chromatography column in an FPLC system. (**f**) The cleaned vector is subjected to a buffer exchange using a 100 kDa concentrator. (**g**) Vector is then aliquoted in 20–25 μL samples

3. For the DNA:PEI complexes, use a 1:3 (ug of DNA:μL of PEI) ratio.
4. This complex is added to 1.1 mL of NIL DMEM (*see* **Note 17**) (serum and antibiotic free), and then add to the cell monolayer.
5. The transfection is allowed to incubate at 37 °C, 7% CO_2, for 60 h.
6. The cells are then harvested by centrifugation and lysed by three freeze/thaw cycles.
7. Clarify crude lysate by centrifugation at 3000 × *g* for 15 min.
8. Divide the resulting vector-containing supernatant among four discontinuous iodixanol step gradients.
9. Centrifuge the gradients at 350,000 × *g* for 1 h.
10. Remove 5 mL of the 60–40% step interface from each gradient and combine it.

11. This iodixanol fraction is further purified and concentrated by column chromatography on a 5 mL (anion exchange) column using an AKTA FPLC system.
12. The vector is eluted from the column using 215 mM NaCl, pH 8.0, and the rAAV peak is collected. Vector-containing fraction is concentrated and buffer exchanged in BSS with 0.014% Tween 20, using a 100 K concentrator.
13. Vector is then titered for DNase-resistant vector genomes by real-time PCR relative to a standard.
14. Finally, the purity of the vector is validated using three standard assays:
    (a) Silver-stained SDS-PAGE is used to detect the three AAV capsid proteins. There should only be three visible protein bands in an adequate preparation.
    (b) A bioburden assay is used to screen the preparation for the presence of bacteria. A total of 10 μL of the final product to 15 mL of nonselective LB and monitored for growth for a total of five days.
    (c) An endotoxin test system is performed to quantify the amount of endotoxin in the preparation. The concentration of endotoxin for an adequate vector preparation is $\leq$5 EU/mL.
15. Once the preparation has been validated by all of the assays it can be aliquoted into tubes (25 μL) and stored at −80 °C.

4 Notes

1. When only one restriction enzyme is used to digest the pTR plasmid, treat the backbone plasmid with phosphatase before gel purification and ligation. Common types of phosphatase include calf intestine phosphatase (CIP) and shrimp alkaline phosphatase (SAP). Follow the manufacturer's directions.
2. When the pTR plasmid used as backbone liberates a $\leq$ 50 base pair insert an overnight digest followed by phosphatase (even if you digest with two enzymes) is recommended. The agarose gel containing the sample should be run entirely without letting the sample run out of the gel.
3. Use a wide gel comb for this gel. Less DNA is lost when the digest is loaded into one well. Alternatively, run the multiple samples of the digestion side by side.
4. Use low-energy UV light to visualize the band while cutting. UV exposure damages DNA.

5. Keep the ligase buffer cold to minimize ATP degradation in the buffer.
6. A control ligation is recommended to estimate how much nonrecombinant plasmid you should expect in the experimental ligation. Use the proportion of experimental to control ligation colonies to choose the number of colonies to pick.
7. Ligation can also be done at 16 °C overnight.
8. This is an especially important step when using electrocompetent cells to transform.
9. Ethanol precipitation can be used as an alternative to the DNA concentration kit.
10. Other competent *E. coli* strains can be used, such as Stbl3, Sure 2, or TOP10. Follow the manufacturer's instructions for these cells. When working with a > 8 kb plasmid, it is recommended to use XL10 cells. If more than 70% of TRs are lost using XL10 cells, retransform plasmid in Sure Competent cells and screen colonies for the presence of intact TRs.
11. A significant number of colonies in the control ligation plate indicate a high background ligation. In this case, it is recommended to start with a new ligation. When a single restriction enzyme is used, it is recommended to pick at least 20 clones to increase the chances of identifying a positive clone.
12. Bacterial cultures should not be shaken for more than 16 h to avoid the loss of TRs.
13. Prepare a master solution of LB containing the antibiotic needed at the final concentration, and then aliquot in individual tubes for inoculation.
14. Bacterial pellets can be stored in a −20 °C freezer. Freeze-thawing the pellet helps increase the yield of DNA.
15. Successful packing of AAV requires plasmids with at least 70% of TRs. TR loss can be minimized by plating bacteria and picking colonies later in the day.
16. AAV capsid should be selected based on the target cell type. Even within the same serotype, new capsid mutants are currently available that differ in the efficiency or distribution of the vector within the tissue.
17. The use of nil media improves transfection efficiency by decreasing the number of negatively charged molecules that interact with the transfection particles. Alternatively, serum-reduced media such as Opti-MEM-I can be used for transfection.

Acknowledgments

We would like to thank all the members of the Ocular Gene Therapy Core lab for the preparation of the AAV used in our studies. This work was funded by a grant from the Bright Focus Foundation (M2017165), a grant from NEI-NIH (EY026268), and an unrestricted funds grant from the Research to Prevent Blindness.

References

1. Atchison RW, Casto BC, Hammon WM (1965) Adenovirus-associated defective virus particles. Science 149:754–755. https://doi.org/10.1126/science.149.3685.754
2. Martino AT, Markusic DM (2020) Immune response mechanisms against AAV vectors in animal models. Mol Ther Methods Clin Dev 17:198–208. https://doi.org/10.1016/j.omtm.2019.12.008
3. Dauletbekov DL, Pfromm JK, Fritz AK, Fischer MD (2019) Innate immune response following AAV administration. Adv Exp Med Biol 1185:165–168. https://doi.org/10.1007/978-3-030-27378-1_27
4. Nidetz NF, McGee MC, Tse LV et al (2020) Adeno-associated viral vector-mediated immune responses: understanding barriers to gene delivery. Pharmacol Ther 207:107453. https://doi.org/10.1016/j.pharmthera.2019.107453
5. Cheung AK, Hoggan MD, Hauswirth WW, Berns KI (1980) Integration of the adeno-associated virus genome into cellular DNA in latently infected human Detroit 6 cells. J Virol 33:739–748
6. Kotin RM, Siniscalco M, Samulski RJ et al (1990) Site-specific integration by adeno-associated virus. Proc Natl Acad Sci U S A 87:2211–2215. https://doi.org/10.1073/pnas.87.6.2211
7. Afione SA, Conrad CK, Kearns WG et al (1996) In vivo model of adeno-associated virus vector persistence and rescue. J Virol 70:3235–3241
8. Flotte TR, Afione SA, Zeitlin PL (1994) Adeno-associated virus vector gene expression occurs in nondividing cells in the absence of vector DNA integration. Am J Respir Cell Mol Biol 11:517–521. https://doi.org/10.1165/ajrcmb.11.5.7946381
9. Yang J, Zhou W, Zhang Y et al (1999) Concatemerization of adeno-associated virus circular genomes occurs through intermolecular recombination. J Virol 73:9468–9477
10. Piovesan A, Caracausi M, Antonaros F et al (2016) GeneBase 1.1: a tool to summarize data from NCBI gene datasets and its application to an update of human gene statistics. Database (Oxford) 2016:baw153. https://doi.org/10.1093/database/baw153
11. Lisowski L, Tay SS, Alexander IE (2015) Adeno-associated virus serotypes for gene therapeutics. Curr Opin Pharmacol 24:59–67. https://doi.org/10.1016/j.coph.2015.07.006
12. Büning H, Srivastava A (2019) Capsid Modifications for Targeting and Improving the Efficacy of AAV Vectors. Mol Ther Methods Clin Dev 12:248–265. https://doi.org/10.1016/j.omtm.2019.01.008
13. Bartel MA, Weinstein JR, Schaffer DV (2012) Directed evolution of novel adeno-associated viruses for therapeutic gene delivery. Gene Ther 19:694–700. https://doi.org/10.1038/gt.2012.20
14. Ridley RB, Young BM, Lee J et al (2019) AAV mediated delivery of myxoma virus M013 gene protects the retina against autoimmune uveitis. J Clin Med 8:2082. https://doi.org/10.3390/jcm8122082
15. Zolotukhin S, Byrne BJ, Mason E et al (1999) Recombinant adeno-associated virus purification using novel methods improves infectious titer and yield. Gene Ther 6:973–985. https://doi.org/10.1038/sj.gt.3300938
16. Beltran WA, Boye SL, Boye SE et al (2010) rAAV2/5 gene-targeting to rods: dose-dependent efficiency and complications associated with different promoters. Gene Ther 17:1162–1174. https://doi.org/10.1038/gt.2010.56

Chapter 6

Murine Model of Thermal Burn Injury for Evaluating Protein Therapeutics Derived from Viruses

Gabriella S. Stuart, Nicola C. Jones, and Lyn M. Wise

Abstract

In vivo wound healing models are predictive preclinical tests for therapeutics that enhance skin repair or limit scarring. Large animals, such as swine, heal in a manner similar to humans, but testing is impractical and expensive. Experiments in mice are more economic, but may be less translatable as this species heals primarily through contraction, not by the processes of epithelialization and granulation tissue formation as seen in human wounds. Here, we describe a murine model of thermal burn injury that closely mimics human healing, resulting in a large, hypertrophic-like scar. This practical, reproducible model is ideal for testing promising wound-healing therapies, such as virus-derived growth factors and immune-modulatory proteins.

Key words Virus protein, Therapeutics, Skin, Thermal burn, Wound, Scar, mouse model

1 Introduction

Impaired wound healing greatly impacts the length and quality of life for a significant proportion of the world's population. This is particularly true for burn victims [1], with approximately 11 million requiring medical attention and 180,000 deaths per year. To improve outcomes for these patients, new therapies must be developed to overcome major healing impairments and excessive scarring [2]. In humans, healing of burns, and other wounds, is a continuous process consisting of overlapping phases [3, 4]. The response starts with production of cytokines, chemokines, and other mediators that bring in inflammatory cells to phagocytose necrotic and infected tissue. These cells in turn produce growth factors that initiate the migratory and proliferative responses in epithelial, endothelial, and fibroblast cells that form the new epithelium and granulation tissue. Over time, this newly formed tissue reduces in cellularity and undergoes substantial matrix remodeling to improve scar pliability. This wound-healing response is greatly impacted by confounding factors such as infection, age, diabetes, or

Alexandra R. Lucas (ed.), *Viruses as Therapeutics: Methods and Protocols*, Methods in Molecular Biology, vol. 2225, https://doi.org/10.1007/978-1-0716-1012-1_6,

obesity, which can exacerbate or prolong inflammation. Excessive inflammation impedes normal tissue cell migration and proliferation in slow-to-heal wounds, while the persistence of inflammation prevents normal tissue remodeling leading to hypertrophic and keloid scars.

Progression of the wound-healing response is tightly regulated by cytokines and growth factors [5]. Advanced understanding as to the role of these regulatory factors in successful and impaired healing has led to the development of therapies aimed at bolstering or suppressing their activity [6, 7]. While most therapies have utilized recombinant human cytokines and growth factors, another exciting source of protein therapeutics lies in viruses [8, 9]. Many large DNA viruses, such as members of the *Herpesviridae* and *Poxviridae*, encode and secrete immune-modulatory proteins. This includes orthologues of human cytokines, proteases, and growth factors, as well as binding proteins that inhibit human cytokines and chemokines [10–13]. During infection, these secreted proteins prime cells to support virus replication while suppressing the hosts' inflammatory and immune responses. In the absence of virus infection, however, recombinant viral proteins may have utility as wound-healing therapeutics. Indeed, the therapeutic administration of viral orthologues of epidermal growth factor and vascular endothelial growth factor to skin wounds has been shown to promote repair processes [14–16]. In addition, wound inflammation and scar formation were suppressed following the treatment of wounds with interleukin-10 [17], a virus-derived immune-modulating serine protease inhibitor [18], and two different chemokine-binding proteins derived from viruses [19, 20]. To explore their utility and translate virus proteins into wound-healing therapeutics, it is important to test them in a reproducible experimental model that replicates human healing as closely as possible.

The complexity of the wound-healing response, and its associated complications, is impossible to replicate in vitro. Animal models must therefore be used to identify novel therapeutic proteins and to optimize treatment regimens [21, 22]. Many studies have utilized large animals, such as swine, as their healing processes are thought to be equivalent to humans [23]. However, experimentation with these animals requires specialized housing facilities and dedicated handlers, making them difficult and expensive to perform for most researchers. Small animals, such as mice, offer a more readily available and cost-effective solution, but can be limited due to differences in how they heal [24–26]. For example, murine skin heals primarily through contraction, and not epithelialization and granulation tissue formation as observed in humans [26]. This is partly due to the panniculus carnosus, a muscle layer which allows the skin of mice to contract independently of deeper muscles. Mice do, however, mount a robust inflammatory and proliferative response to injury [27], expressing many of the same wound regulators as humans. But generally, this response does not lead to the excessive scarring observed in humans [26]. This suggests that murine wound models may be appropriate for testing viral protein therapeutics, but it is unclear as to which is the most appropriate to use. Numerous injury techniques have been tried in mice, including surgical incisions, biopsy excisions with or without splinting, and thermal burns, but with varying consistency with regard to their depth [24–26]. A standardized injury model is therefore needed for future testing of virus proteins and other new therapeutics.

In the model described here, 8 mm metal rods, heated to 100 °C for 5 min, were applied for 4 s to shaved dorsal skin. Two burn wounds were applied per mouse to each side of the midline, each representing 2% of total body surface area. To reduce animal numbers, each wound on the mouse can act as independent replicates, or as matched treatment and vehicle controls. Early debridement of the wound mimics the care of human burns, and standardizes the rate of closure. Therapies are applied to open wounds topically or by subcutaneous or intradermal injection. Occlusive dressings are recommended to aid in the retention of topical treatments at the burn site. Digital photography during the course of healing allows for morphological measurements of the wound and scar area. Wound and scar biopsies taken from euthanized mice allow for histological analysis of epithelialization, granulation tissue formation, and wound contraction, and immunohistological analysis of cellular infiltrates and regulatory factors. Using this method, the damage extends from the epidermis to the dermis, leaving the panniculus carnosus and hypodermis intact [24]. Wound closure occurs over a 15-day period, equally through tissue formation and contraction, resulting in a substantial scar by 6 weeks postburn [24]. This economical and easy-to-perform model can readily be adapted to mimic confounding factors associated with impaired healing in humans.

2 Materials

2.1 Experimental Animals

1. Obtain Animal Ethics Committee approval for all experiments to be performed.
2. Use female C57BL/6 mice, aged 8–10 weeks, weighing 18–22 g (*see* **Notes 1** and **2**).
3. Keep in standard conditions of 21 °C and a 12-h light-dark cycle with free access to food and water (*see* **Note 3**).

2.2 Equipment Required for Surgery

1. 0.5 mL 29 g × 1.27 cm insulin syringes.
2. Electric animal hair clippers.

3. Cotton gauze and cotton swabs.
4. Sterile 0.9% saline: Store at 37 °C, until required.
5. Small-animal incubator or thermal pad.
6. Sterile forceps.
7. Sterile surgical drapes with 4 × 2 cm aperture.
8. Dry heat block suitable for 1.5–2 mL centrifuge tubes. Pre-warm to 100 °C.
9. Filter-sterilized ultrapure water.
10. Sterile 8 mm diameter aluminum rods, 10 cm in length.
11. Timer and digital metronome.
12. Pliers.

2.3 Medicines Required for Surgery

1. Anesthesia solution: 25 mg/mL Ketamine, 0.2 mg/mL medetomidine, 0.0163 mg/mL atropine and sterile saline. For ten mice, combine 0.6 mL ketamine, 1 mL domitor, and 0.6 mL atropine in a disposable vacutainer tube, and mix by inversion. For each mouse, label an insulin syringe and load with 0.22 mL of the anesthesia solution.
2. Lubricating eye ointment.
3. Hair removal cream (sensitive skin).
4. Analgesic solution: 0.2 mg/mL Bupivacaine. For each mouse, label an insulin syringe and load with 0.2 mL of the analgesic solution.
5. Antibiotic solution: 10.2 mg/mL Trimethoprim. For each mouse, label an insulin syringe and load with 0.1 mL of the antibiotic solution.
6. Antiseptic solution: 10% w/v Povidone-iodine.
7. Fluid resuscitation solution: 0.9% Saline. For each mouse, label an insulin syringe and load with 0.5 mL of the fluid resuscitation solution. Store at 37 °C, until required.
8. Reversal solution: 1 mg/mL Atipamezole. For each mouse, label an insulin syringe and load with 0.1 mL of the reversal solution.

2.4 Equipment Required for Debridement

1. Rodent anesthesia machine (isoflurane).
2. Small-animal incubator or thermal pad.
3. Sterile forceps.

2.5 Preparation of Treatments

1. Intradermal (ID) virus protein treatment: Dilute in phosphate-buffered saline (PBS) to the concentration required. For each burn site, label an insulin syringe and load with 40 μL. Store on ice, until required.

2. Subcutaneous (SC) virus protein treatment: Dilute in PBS to the concentration required. For each burn site, label an insulin syringe and load with 100 μL. Store on ice, until required.
3. Topical virus protein treatment: Dilute in wound hydrogel (for example, we use SOLOSITE® Wound Gel, Smith & Nephew, Watford, UK) to the concentration required. For each burn site, apply 40 μL of protein gel. Store on ice, until required.
4. Occlusive dressing for topical treatments (we use 3M™ Tegaderm).

2.6 Equipment for Digital Analysis of Wound and Scar Area

1. Tripod camera (we use a Digital Canon EOS SLR Camera with tripod).
2. Metric ruler.
3. FIJI software.

2.7 Equipment for Animal Euthanasia and Harvesting of Tissue

1. Rodent anesthesia machine (CO_2).
2. Scissors.
3. Forceps.
4. Scalpel blades.
5. Biopsy cassettes.
6. Plastic container.
7. Fixative (*see* **Note 4**).
8. 70% Ethanol.

3 Methods

3.1 Surgical Preparation and Burn Injury

To mimic human burns, and to standardize the burn depth, we recommend the burn injury be created using heated metal rods applied to depilated skin for a set period of time. Two burns can be applied to each mouse without compromising animal welfare.

1. Induce general anesthesia in the mouse through SC administration of anesthesia solution into the inguinal region (0.22 mL).
2. To prevent corneal drying and trauma, apply lubricating ointment to open eyes.
3. Place the mouse in a prone position, and then prepare the surgical site by removing fur with clippers from behind the front legs to 3–4 cm further down the back (*see* **Note 5**).
4. Apply hair removal cream to the site with a cotton swab, and leave for a maximum of 2 min. Wet cotton gauze with pre-warmed saline and use to remove cream and remaining fur.

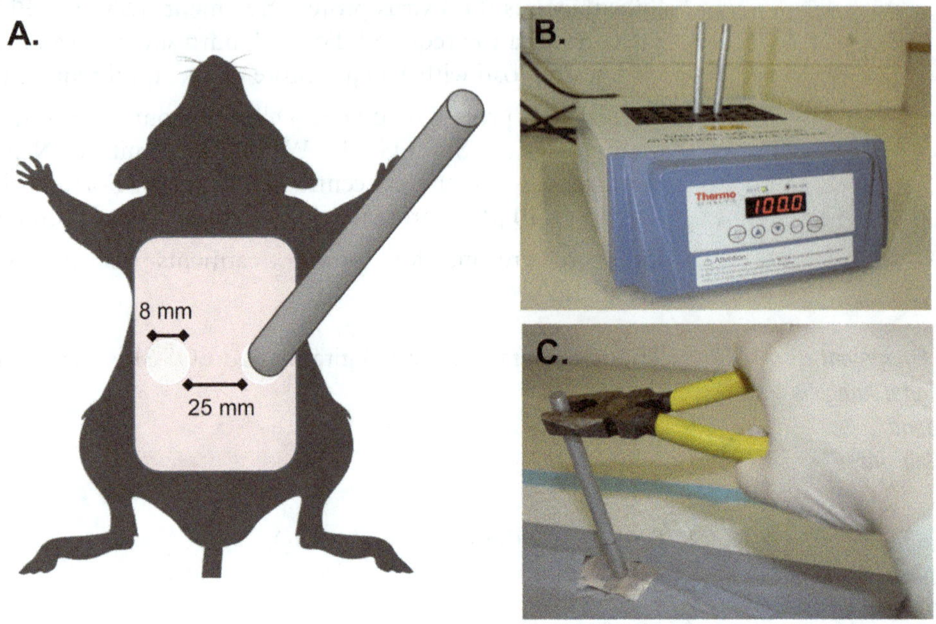

Fig. 1 A murine model of thermal burn injury. In this model, two partial-thickness burns are created on depilated skin on either side of the midline using 8 mm aluminum rods headed to 100 °C and then applied for 4 s. (**a**) Diagrammatic representation of the thermal burn procedure. (**b**) Photograph of heat block used to prepare rods. (**c**) Photograph showing application of the heated rod to murine skin using pliers and gravity

5. Administer analgesic solution to the mid-left and mid-right dorsal regions by SC injection (0.1 mL per site).
6. Administer antibiotic solution to the nape of the neck by SC injection (0.1 mL).
7. Administer pre-warmed fluid resuscitation solution to the nape of the neck by SC injection (0.5 mL).
8. Move the mouse to an animal incubator or thermal pad to keep warm. Monitor respiration for 15 min, to allow the anesthesia and analgesia to take effect.
9. Using forceps and a firm toe pinch, test for suppression of the pedal reflex.
10. Once deep pain recognition is suppressed, place the mouse in the surgical space, and cover with a surgical drape, exposing the shaved dorsal region through the aperture.
11. Wet cotton gauze with antiseptic solution and use to sterilize the surgical site; repeat twice.
12. Diagrammatic representation of the murine burn model is presented in Fig. 1a.
13. Prepare the preheated heat block by filling two wells with Milli-Q water. Immediately place aluminum rods into the wells, and

then incubate for 5 min, replenishing the water as required (*see* Fig. 1b).

14. Turn on digital metronome, ready to time the burn administration.
15. Using pliers, lift one rod and apply with only the weight of gravity to the mid-left dorsal region of the surgical site (*see* Fig. 1c). Count for 4 s using the metronome, and then lift and discard the rod.
16. Repeat the process with the other rod, applying it to the mid-right dorsal region of the surgical site.
17. Reverse the anesthesia through SC administration of reversal solution into the opposite inguinal region (0.1 mL).
18. Return the mouse into a recovery cage, warmed on a thermal pad, and observe to ensure recovery from the anesthesia and surgery. Once ambulatory, return to original housing and containment room.

3.2 Monitoring of Animal Welfare

Animals should be monitored closely for signs of pain, infection, dehydration, and weight loss post-surgery and throughout the entire experiment.

1. Diagrammatic representation of the monitoring schedule is presented in Fig. 2.
2. Until burn closure is achieved, mice are monitored twice daily for signs of pain, infection, dehydration, and weight loss (*see* **Note 6**).
3. Following burn closure mice are monitored twice a week to ensure that no complications have arisen.

3.3 Debridement and Treatment

To mimic human burn care, and to standardize wound contraction, we recommend that the burns are debrided prior to the first treatment. The eschar that forms in this model can be readily removed with little trauma to the mouse at day 3 postburn. Viral protein treatments can then be applied using topical or injectable delivery methods until closure is achieved.

1. Diagrammatic representation of the debridement and treatment schedule is presented in Fig. 2 (*see* **Note 7**).
2. Induce general anesthesia using 5% isoflurane gas (flow rate 1 L/min), and using forceps and a firm toe pinch, test for suppression of the pedal reflex.
3. While maintaining anesthesia with 1% isoflurane gas, use sterile forceps to lift the edge of the eschar and gently peel off.
4. Give topical applications directly to the debrided wound (40 μL), followed by application of an occlusive dressing. ID administration requires two injections 0.25 cm either side of

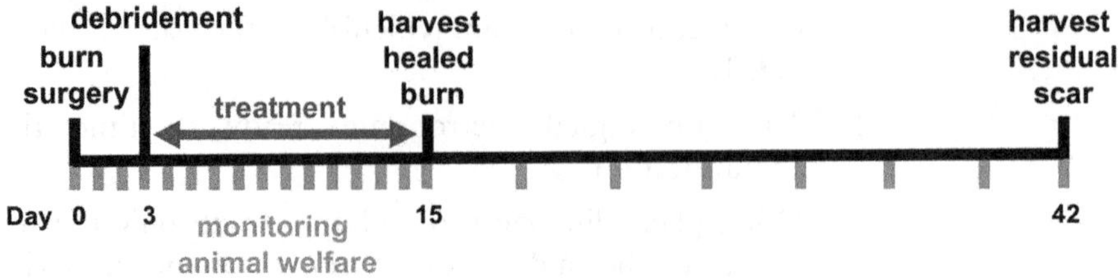

Fig. 2 Diagrammatic representation of the schedule for burn debridement, treatment, monitoring, and tissue harvesting

the debrided wound (20 μL each site). SC administration requires two injections 0.5 cm either side of the debrided wound (50 μL each site) (*see* **Note 8**).

5. Reverse the anesthesia through withdrawal from the isoflurane gas.
6. Return the mouse into a recovery cage, warmed on a thermal pad, and observe to ensure recovery from the anesthesia. Once ambulatory, return to original housing and containment room.
7. Subsequent treatments can be administered under gas anesthesia daily, or as deemed appropriate.

3.4 Digital Analysis of Wound and Scar Area

Digital photographs can be taken while animals are under general or gas anesthesia during surgery, debridement, and treatment and as desired. To assess the rate of healing photos should be taken until approximately day 15 postburn, while photos of the scar to assess its size should be taken until approximately day 42.

1. Using a digital camera photograph each side of the mouse with a ruler placed within 2 cm of the burn site (*see* Fig. 3a) (*see* **Note 9**).
2. Open digital images in the FIJI software.
3. Draw a line corresponding to 10 mm on the metric ruler. Set the scale under the analyze menu in the tool bar (*see* **Note 10**).
4. Use the freehand tracing tool to trace the perimeter of the open wound or the scar (*see* Fig. 3b).
5. Select measure area under the analyze menu.
6. After obtaining all wound areas, calculate the area of the wound at each time point as a percentage of the wound area on the day of debridement.
7. An example of burn closure data is shown in Fig. 3c.
8. An example of scar area data is shown in Fig. 3d.

3.5 Harvesting of Tissue for Histological Analysis

To assess the impact of treatments on healing and scarring processes, the burn wound or scar tissue can be harvested from euthanized mice, then fixed, and processed into paraffin wax for histochemical or immunohistochemical analyses. Histochemical

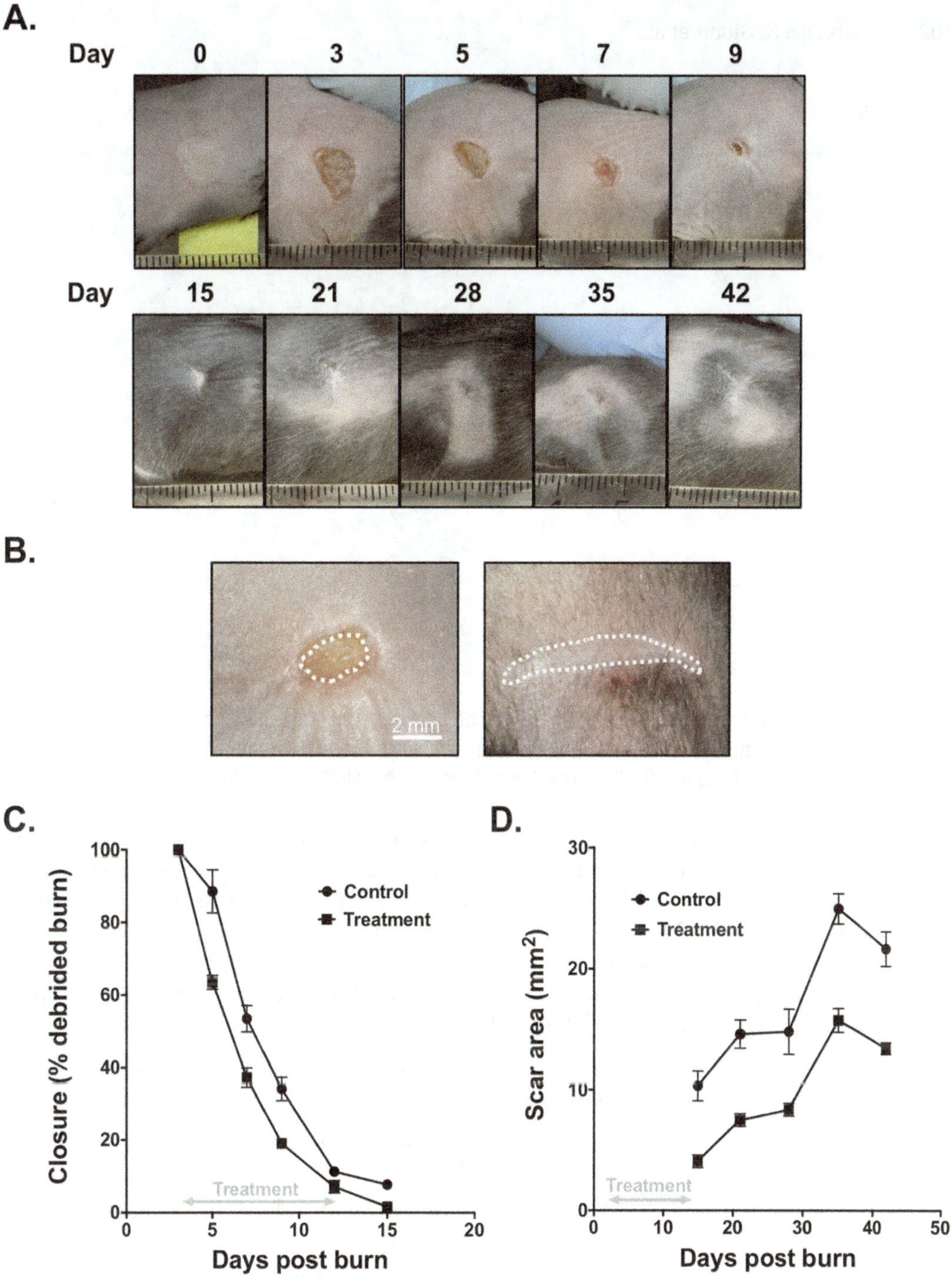

Fig. 3 Digital analysis of burn closure and scar size. (**a**) Representative images of a healing thermal burn and its residual scar. (**b**) Representative images showing the use of FIJI to trace the perimeter of a burn wound and scar as indicated by white dashed lines. Scale is indicated, and was calculated from the ruler in the original photographs as shown in (**a**). (**c**) Graph of representative burn closure data. The treatment window following debridement and terminating at wound closure is indicated. Burn area is calculated from images as shown in (**b**), with closure expressed as a percentage of the burn at debridement (day 3). (**d**) Graph of representative scar area data. The treatment window prior to closure and scar measurements is indicated. Scar area is calculated from images as shown in (**b**). Data represents the mean $\pm$ SEM, with $n = 2$ burns from each of 4 mice per treatment group

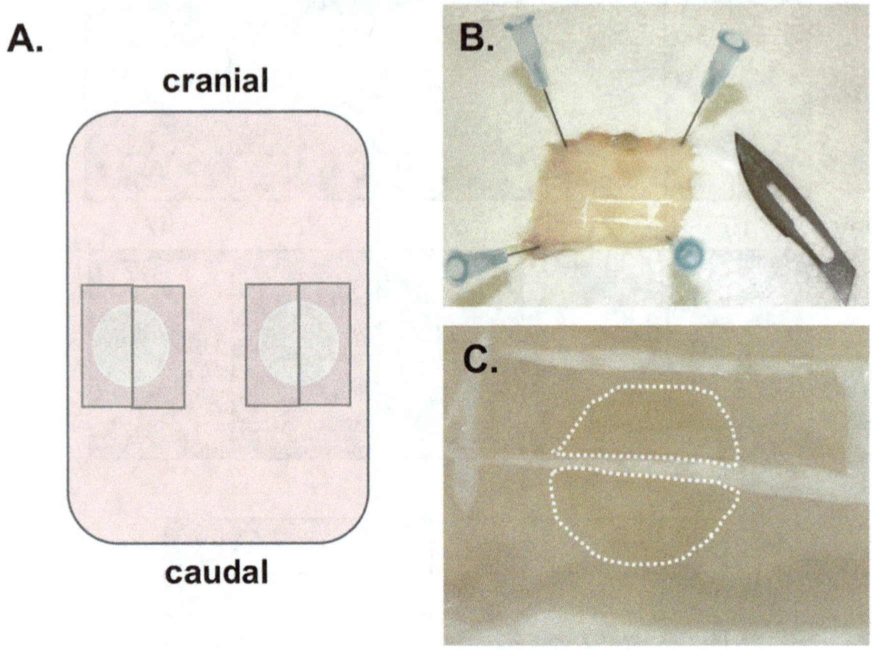

Fig. 4 Harvesting burn or scar tissue for histological analysis. (**a**) Diagrammatic representation of the harvesting procedure. (**b**) Photograph showing excised skin which includes two burn wounds. (**c**) Photograph showing a bisected day 0 burn, with the wound area indicated by white dashed lines

stains such as hematoxylin and eosin or Martius scarlet blue trichrome can be used to measure burn depth and contraction, re-epithelialization, and granulation and scar tissue formation [24]. Immunohistochemical staining can also be performed with antibodies specific for specific cells to assess their abundance and locality within the burn during the stages of inflammation, repair, and remodeling [24].

1. Select time points at which the mice are to be euthanized (*see* **Note 11**).
2. Diagrammatic representation of tissue harvesting schedule is presented in Fig. 2.
3. Euthanize each mouse with an overdose of CO_2 anesthesia, and confirm death by cervical dislocation.
4. Diagrammatic representation of tissue harvesting is presented in Fig. 4a.
5. Excise the dorsal skin surrounding the two burns using scissors, and then using a surgical scalpel excise the individual wounds or scars keeping approximately 5 mm of adjacent normal tissue (*see* Fig. 4b).
6. Bisect the wound at the widest margin along the cranial-caudal axis (*see* Fig. 4c) (*see* **Note 12**).

7. Place the biopsy in a cassette, and place in a plastic container containing fixative.
8. Store at room temperature for the time required (*see* **Note 13**).
9. Transfer the cassette to 70% ethanol, and store for 24 h or until ready to process into paraffin wax.

4 Notes

1. Thermal burns are created in mice prior to the desynchronization of hair cycling at 12 weeks of age. If experiments are initiated after this point, the site of burn injury and its rate of repair can differ between animals.
2. Male mice can be used, but they heal at a different rate so should be analyzed separately to female mice.
3. To ensure that animals are calm and compliant at the time of surgery, gently handle the mice for 7 days prior.
4. 0.5% Zinc salt solution provides effective fixation compatible with histological stain detection of immune cell markers by immunohistochemistry [24, 28].
5. Keep the surgical space clear of fur by performing the depilation on another surgical bench.
6. We observe up to 10% weight loss in the mice at 24 h post-surgery, with weights returning to that of pre-surgery by day 3. If weight loss persists and dehydration is evident, we recommend the administration of fluid resuscitation. Pain and infection are rarely observed post-surgery, unless mice are handled too roughly causing rupture of the eschar before the debridement scheduled at day 3. We therefore recommend keeping handling to a minimum until day 3 post-surgery.
7. Complete closure of the skin burn is usually achieved by day 15 post-surgery.
8. For topical administration, the two burns on each mouse can act as matched controls and treatments. Occlusive dressings may however be required to keep treatments in place and to prevent interference by the animal. The width of the dressing should be at least 2 cm greater than the burn area, so as to allow for complete attachment and to prevent it from being removed quickly by the animal. The dressing should also be removed and replaced with replicate treatments. With SC administration, we have observed that treatments, anti-inflammatories in particular, can modulate the healing response in the adjacent wound. The two burns can, however, receive the same treatment and act as independent replicates.

9. To ensure that the photographs are of a high quality that is suitable for analysis, take care to hold camera parallel to wound, and use a camera with a macro lens, tripod, and twin flash if available.
10. It is important to set the scale for each image individually as the distance between the camera and the animal may differ between photographs.
11. For analysis of burn inflammation and repair we recommend harvesting tissue at days 10–15 postburn, while for scar remodeling we recommend harvesting tissue at day 42 postburn.
12. Care should always be taken to bisect each wound in the same orientation, and at the same location, so that any histological comparisons made between treatments are fair. Burns on the mouse dorsum contract at the lateral and medial edges preferentially to the cranial and caudal edges, which means the widest part of the wound is along the cranial-caudal axis.
13. Biopsies should be stored in zinc salt fixative at room temperature for 24 h, and the solution needs to be removed and replenished for a further 24 hr.

References

1. WHO (2018) Fact sheet: burns. https://www.who.int/news-room/fact-sheets/detail/burns. Accessed 6 Mar 2018
2. Wang Y, Beekman J, Hew J et al (2018) Burn injury: challenges and advances in burn wound healing, infection, pain and scarring. Adv Drug Deliv Rev 123:3–17. https://doi.org/10.1016/j.addr.2017.09.018
3. Evers LH, Bhavsar D, Mailander P (2010) The biology of burn injury. Exp Dermatol 19 (9):777–783. https://doi.org/10.1111/j.1600-0625.2010.01105.x
4. Rose LF, Chan RK (2016) The burn wound microenvironment. Adv Wound Care 5 (3):106–118. https://doi.org/10.1089/wound.2014.0536
5. Werner S, Grose R (2003) Regulation of wound healing by growth factors and cytokines. Physiol Rev 83(3):835–870. https://doi.org/10.1152/physrev.2003.83.3.835
6. Barrientos S, Brem H, Stojadinovic O et al (2014) Clinical application of growth factors and cytokines in wound healing. Wound Repair Regen 22(5):569–578. https://doi.org/10.1111/wrr.12205
7. Shpichka A, Butnaru D, Bezrukov EA et al (2019) Skin tissue regeneration for burn injury. Stem Cell Res Ther 10(1):94. https://doi.org/10.1186/s13287-019-1203-3
8. Lucas A, McFadden G (2004) Secreted immunomodulatory viral proteins as novel biotherapeutics. J Immunol 173(8):4765–4774. https://doi.org/10.4049/jimmunol.173.8.4765
9. Shchelkunova GA, Shchelkunov SN (2016) Immunomodulating drugs based on poxviral proteins. BioDrugs 30(1):9–16. https://doi.org/10.1007/s40259-016-0158-5
10. Gonzalez-Motos V, Kropp KA, Viejo-Borbolla A (2016) Chemokine binding proteins: an immunomodulatory strategy going viral. Cytokine Growth Factor Rev 30:71–80. https://doi.org/10.1016/j.cytogfr.2016.02.007
11. Lateef Z, Wise LM (2018) Exploitation of receptor tyrosine kinases by viral-encoded growth factors. Growth Factors 36 (3–4):118–140. https://doi.org/10.1080/08977194.2018.1520229
12. Lucas A, Liu L, Dai E et al (2009) The serpin saga; development of a new class of virus derived anti-inflammatory protein immunotherapeutics. Adv Exp Med Biol 666:132–156. https://doi.org/10.1007/978-1-4419-1601-3_11
13. Ouyang P, Rakus K, van Beurden SJ et al (2014) IL-10 encoded by viruses: a remarkable example of independent acquisition of a cellular gene by viruses and its subsequent evolution

in the viral genome. J Gen Virol 95 (Pt 2):245–262. https://doi.org/10.1099/vir.0.058966-0

14. Schultz GS, White M, Mitchell R et al (1987) Epithelial wound healing enhanced by transforming growth factor-alpha and vaccinia growth factor. Science 235(4786):350–352. https://doi.org/10.1126/science.3492044
15. Wise LM, Inder MK, Real NC et al (2012) The vascular endothelial growth factor (VEGF)-E encoded by orf virus regulates keratinocyte proliferation and migration and promotes epidermal regeneration. Cell Microbiol 14 (9):1376–1390. https://doi.org/10.1111/j.1462-5822.2012.01802.x
16. Wise LM, Stuart GS, Real NC et al (2018) VEGF receptor-2 activation mediated by VEGF-E limits scar tissue formation following cutaneous injury. Adv Wound Care 7 (8):283–297. https://doi.org/10.1089/wound.2016.0721
17. Wise LM, Stuart GS, Real NC et al (2014) Orf virus IL-10 accelerates wound healing while limiting inflammation and scarring. Wound Repair Regen 22(3):356–367. https://doi.org/10.1111/wrr.12169
18. Zhang L, Yaron JR, Tafoya AM et al (2019) A virus-derived immune modulating serpin accelerates wound closure with improved collagen remodeling. J Clin Med 8(10):1626. https://doi.org/10.3390/jcm8101626
19. Ridiandries A, Bursill C, Tan J (2017) Broad-spectrum inhibition of the CC-chemokine class improves wound healing and wound angiogenesis. Int J Mol Sci 18(1):155. https://doi.org/10.3390/ijms18010155
20. Sharif S, Nakatani Y, Wise L et al (2016) A broad-spectrum chemokine-binding protein of bovine Papular stomatitis virus inhibits neutrophil and monocyte infiltration in inflammatory and wound models of mouse skin. PLoS One 11(12):e0168007–e0168007. https://doi.org/10.1371/journal.pone.0168007
21. Abdullahi A, Amini-Nik S, Jeschke MG (2014) Animal models in burn research. Cell Mol Life Sci 71(17):3241–3255. https://doi.org/10.1007/s00018-014-1612-5
22. Domergue S, Jorgensen C, Noel D (2015) Advances in research in animal models of burn-related hypertrophic scarring. J Burn Care Res 36(5):e259–e266. https://doi.org/10.1097/bcr.0000000000000167
23. Seaton M, Hocking A, Gibran NS (2015) Porcine models of cutaneous wound healing. ILAR J 56(1):127–138. https://doi.org/10.1093/ilar/ilv016
24. Lateef Z, Stuart G, Jones N et al (2019) The cutaneous inflammatory response to thermal burn injury in a murine model. Int J Mol Sci 20(3). https://doi.org/10.3390/ijms20030538
25. Medina JL, Fourcaudot AB, Sebastian EA et al (2018) Standardization of deep partial-thickness scald burns in C57BL/6 mice. Int J Burns Trauma 8(2):26–33
26. Wong VW, Sorkin M, Glotzbach JP et al (2011) Surgical approaches to create murine models of human wound healing. J Biomed Biotechnol 2011:969618. https://doi.org/10.1155/2011/969618
27. Mestas J, Hughes CC (2004) Of mice and not men: differences between mouse and human immunology. J Immunol 172(5):2731–2738
28. Mori H, Soonsawad P, Schuetter L et al (2015) Introduction of zinc-salt fixation for effective detection of immune cell-related markers by immunohistochemistry. Toxicol Pathol 43 (6):883–889. https://doi.org/10.1177/0192623315587593

Chapter 7

Deriving Immune-Modulating Peptides from Viral *Ser*ine *P*rotease *In*hibitors (*Serpins*)

Jordan R. Yaron, Liqiang Zhang, Michelle Burgin, Lauren N. Schutz, Enkidia A. Awo, Shahar Keinan, Grant McFadden, Sriram Ambadapadi, Qiuyun Guo, Hao Chen, and Alexandra R. Lucas

Abstract

Viruses have devised highly effective approaches that modulate the host immune response, blocking immune responses that are designed to eradicate viral infections. Over millions of years of evolution, virus-derived immune-modulating proteins have become extraordinarily potent, in some cases working at picomolar concentrations when expressed into surrounding tissues and effectively blocking host defenses against viral invasion and replication. The marked efficiency of these immune-modulating proteins is postulated to be due to viral engineering of host immune modulators as well as design and development of new strategies (i.e., some derived from host proteins and some entirely unique). Two key characteristics of viral immune modulators confer both adaptive advantages and desirable functions for therapeutic translation. First, many virus-derived immune modulators have evolved structures that are not readily recognized or regulated by mammalian immune pathways, ensuring little to no neutralizing antibody responses or proteasome-mediated degradation. Second, these immune modulators tend to target early steps in central immune responses, producing a powerful downstream inhibitory "domino effect" which may alter cell activation and gene expression.

We have proposed that peptide metabolites of these immune-modulating proteins can enhance and extend protein function. Active immunomodulating peptides have been derived from both mammalian and viral proteins. We previously demonstrated that peptides derived from computationally predicted cleavage sites in the reactive center loop (RCL) of a viral *ser*ine *p*roteinase *in*hibitor (*serpin*) from myxoma virus, Serp-1, can modify immune response activation. We have also demonstrated modulation of host gut microbiota produced by Serp-1 and RCL-derived peptide, S7, in a vascular inflammation model. Of interest, generation of derived peptides that maintain therapeutic function from a serpin can act by a different mechanism. Whereas Serp-1 has canonical serpin-like function to inhibit serine proteases, S7 instead targets mammalian serpins. Here we describe the derivation of active Serp- RCL peptides and their testing in inflammatory vasculitis models.

Key words Immune modulation, Inflammation, Virus, Peptides, Proteins, Serpin, Reactive center loop, Vascular, Vasculitis, Myxoma virus, Thrombotic, Thrombolytic

Alexandra R. Lucas (ed.), *Viruses as Therapeutics: Methods and Protocols*, Methods in Molecular Biology, vol. 2225, https://doi.org/10.1007/978-1-0716-1012-1_7,

1 Introduction

At the most direct level, protein activity is regulated by its synthesis and by its metabolism, the balance between active protein translation and degradation adjusting local concentrations of individual proteins. While protein activity is regulated in part through this balance of synthesis and metabolism, protein function can also be modulated through posttranslational modifications (e.g., phosphorylation) or via a network of regulators controlling gene expression, local activators, and inhibitors and through metabolites that can extend or, conversely, inhibit protein activity. Peptide metabolites derived by protein cleavage can display similar function to the original protein or can have new activities, unique functions that differ from the original protein's function [1–5]. Thus, in some cases, peptides released by protein cleavage have separate or divergent functions, which may be similar, may overlap the original protein's activity, or may differ significantly from the original parent protein (*see* Fig. 1) [1–11]. For example, the lipoprotein apolipoprotein E (ApoE) has functional domains that can moderate inflammation and atherosclerosis [3]. The N-terminus of the ApoE protein has lipid-binding and anti-atherogenic activity, while a naturally occurring 14-amino-acid C-terminal peptide fragment derived from ApoE, termed Ep1.B (239–252), reduces neointimal hyperplasia in mouse models. Similarly, the calcium regulatory protein calreticulin has domains that have independent effects on calcium binding and thrombosis [4]. In another example, the central regulator of vascular hypertension, angiotensin II, is a metabolite of angiotensinogen, a mammalian *ser*ine *p*roteinase *in*hibitor, generally termed a *serpin*. Renin cleaves the N-terminus of angiotensinogen (SERPINA8) to form the angiotensin I peptide, which is subsequently cleaved by angiotensin-converting enzyme (ACE) to form angiotensin II, the blood pressure-regulating peptide [5]. Another serpin, alpha-1 antitrypsin (SER PINA1 or AAT), is the primary endogenous neutrophil elastase inhibitor, but also has additional reported functions. AAT has been shown to act as an immune regulator by reducing inflammatory cell activation and invasion, inhibiting inflammasome-dependent maturation and release of the inflammatory cytokine interleukin (IL)-1β, and even inhibiting coagulation and apoptotic pathway enzymes [12]. AAT has shown therapeutic benefit in animal models of rheumatoid arthritis, type 1 diabetes, and transplant [1, 6, 7, 11, 13], with some functions attributed to isolated peptide domains. Serpins and peptides derived from serpins are found in a wide expanse of organisms from viruses to insects (ticks) and mammals [6, 8–19]. In viruses peptides have been identified, reported, and isolated [8–10, 18–20]. One example includes M-T5 that functions as an adaptor in E3 ubiquitin ligase

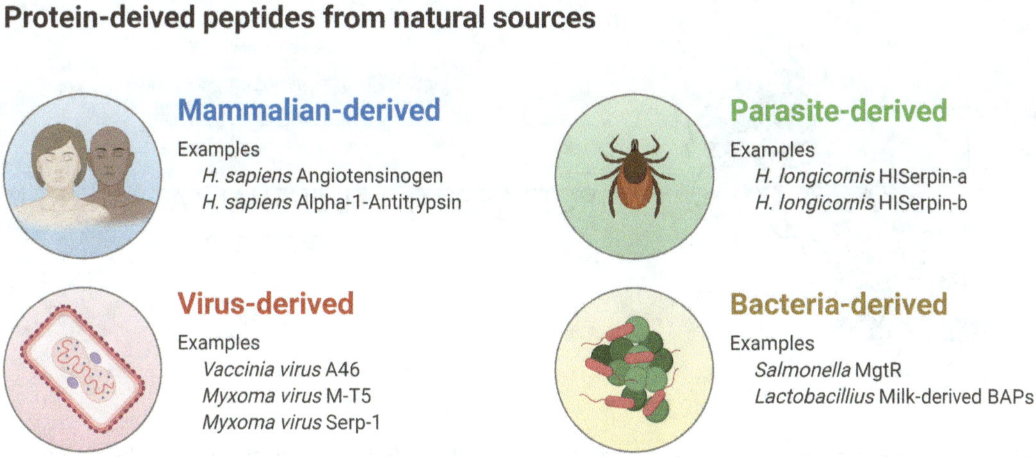

Fig. 1 Active peptides have been detected and derived from many organisms from mammals to insects and bacteria to viruses

complexes and prevents cell cycle arrest [18, 19]. This has been used to design peptides with in vitro antitumor activity [19, 20]. There is also a vaccinia protein-derived peptide that binds and inhibits toll-like receptor 4 (TLR4) [18].

There are likely to be many other protein-derived peptide metabolites, yet to be discovered, with additional functions beyond those of the parent molecules. Empirical approaches for identifying protein-derived peptides are technically challenging as protein discovery (usually by mass spectrometry) inherently induces peptides in the process of analyzing samples which may be synthetic, artifactual, or nonfunctional. Thus, discovery of novel, naturally occurring functional peptides by de novo empirical experimentation is relatively uncommon. In response to this limitation, the discovery of rationally designed and identified peptides has been improved in recent years with the development of increasingly sophisticated bioinformatic tools which process proteins in silico to provide statistically likely peptide derivatives. Here we focus on the computation prediction and testing of virus-derived serpin peptides with demonstrated preclinical anti-inflammatory function.

We have examined peptide sequences in the reactive center loop (RCL) of a serpin, Serp-1. The RCL of serpins contains a protease recognition sequence containing a P1-P1' scissile bond which acts as a bait for the target protease. In the process of cleaving the RCL, the target protease becomes covalently linked to the serpin in a transient, Michaelis complex. The high degree of metastability of the serpin results in a dramatic reconfiguration of the total protein, resulting in insertion of the RCL as the third strand of a five-strand A-beta sheet, denaturation of the target protease, and formation of an inactive suicide complex [6, 13, 17] (*see* Fig. 2). In relatively rare cases, cleavage of the RCL is completed by the target protease

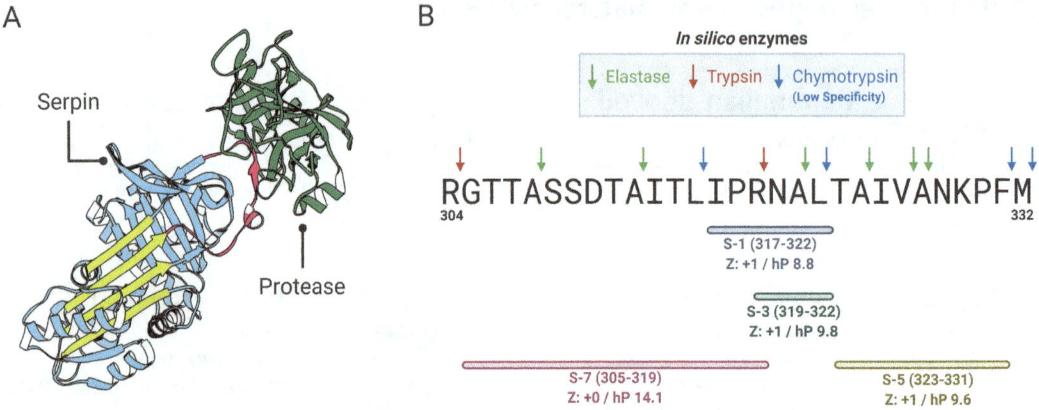

Fig. 2 (**a**). Interaction between the serpin-reactive center loop (RCL) and the targeting protease. Crystal structure used is RSCB 1OPH [24]. (**b**) Peptide identification based upon predicted metabolizing enzyme cleavage sites in the RCL. Predicted protease cleavage sites in the RCL are marked for elastase, trypsin, and chymotrypsin. Chymotrypsin cleavage was performed with low specificity parameters. The predicted peptide sequences and metabolites from the Serp-1 RCL are shown

without the formation of a suicide complex. Thus, the RCL has critical biological functions and is a natural site for potential formation of active peptide metabolites that have the potential to extend serpin function.

We used online open-source algorithms to predict natural cleavage sites targeted or cleaved by proteolytic enzymes, such as trypsin or other serine proteases. We identified a series of predicted peptides derived as likely metabolites from the reactive center loop (RCL) of Serp-1 and assessed the potential immune-modulating function of each peptide. The individual peptides from myxoma virus Serp-1 protein S1, S3, S5, and S7 (*see* Fig. 2) [8–10, 15, 16]. Serp-1 targets and inhibits thrombolytic proteases, urokinase-type and tissue-type plasminogen activators (uPA and tPA, respectively), as well as thrombotic proteases thrombin (with heparin) and factor Xa. These identified RCL peptides from Serp-1 were compared to a series of RCL peptides identified in a similar manner from the mammalian serpin, neuroserpin that inhibits only thrombolytic proteases, uPA and tPA [15, 16]. RCL peptide function is assessed by analyzing binding and inhibition of protease activity using gel shift assays and spectroscopic activity assays. We tested peptide inhibition for uPA and tPA in the thrombolytic protease cascade and thrombin in the thrombotic cascade [10, 15, 16]. RCL-derived peptide inhibition of related native mammalian serpins that regulate the coagulation cascades was also tested, based upon the capacity of the serpin RCL to insert into the A-beta sheet of the target serpin. We detected a unique capacity for serpin RCL peptides that reduced inflammation in aortic allografts and in MHV68-induced vasculitis in mice [10, 15, 16]. These active Serp-1 RCL peptides did not inhibit protease activity, but did block serpin inhibition of protease activity in vitro [8–10, 15,

16]. A similar mechanism involving aberrant insertion of an adjacent RCL into a neighboring serpin is, along with domain swapping, the presumed basis for genetic mutation-induced serpinopathies, which are characterized by inactive protein aggregates (*see* **Note 1**).

Due to the complex nature of circulating protease and serpin function, we have in most cases approached the analysis of peptide metabolites through initial testing of in vivo function in mouse models of vascular inflammation: (i) aortic allograft transplants [8, 15] and (ii) murine gamma herpesvirus 68 (MHV68)-induced inflammatory vasculitis [8–10, 16]. The aortic allograft model provides a means to test against the aggressive immune reaction that causes transplant vasculopathy, a leading cause for graft loss after the first year posttransplantation. The mouse gamma herpesvirus 68 (MHV68) infection model recapitulates important pathological mechanisms of the rare but devastating inflammatory vasculitic syndromes, specifically Takayasu's disease and giant-cell arteritis. MHV68 infection in interferon gamma receptor knockout (IFNγR$^{-/-}$) mice causes a lethal vasculitis with pulmonary hemorrhage and consolidation and severe gut dilatation. Once in vivo anti-inflammatory function is detected, we investigate predicted molecular targets in vitro.

Here, we provide a chapter partially combining methods described in prior chapters published for the assay of serpin peptides derived from the RCL [21–23], providing one source for analysis of peptides using in vivo animal model assays and in vitro protease and serpin activity assays. In this chapter we describe the methods used (1) to identify potential RCL peptide metabolites, (2) to test RCL peptide function in vitro (specifically, modulation of protease and serpin activity using protease activity assays), and (3) to test peptide-mediated immune-modulatory functions in vivo in mouse models of vascular inflammation.

2 Materials

2.1 Identification of Predicted Peptide Metabolites in the Serpin-Reactive Center Loop (RCL)

2.1.1 Online Peptide Identification Programs

We used the PeptideCutter utility hosted on the ExPASy bioinformatics resource portal (http://www.expasy.org) to predict native cleavage sites in the serpin RCL (*see* [18, 19, 21, 22], Fig. 2).

2.1.2 Calculation of Peptide Physical Characteristics

As a secondary step we calculate the physicochemical properties of individual peptide sequences using the online PepDraw utility (Wimley lab, Tulane University, New Orleans LA), which provides predicted charge and hydrophilic and hydrophobic properties.

2.2 Peptide Synthesis

The peptide synthesis center can be either university or commercially based. For our work we used the University of Florida Peptide Center for peptide synthesis. Cost will vary depending on purity, scale of synthesis, and methods of removing contaminants. Our minimum requirements are 90% or greater purity and standard desalting at a scale of 10–20 mg.

2.3 Protease and Serpin Inhibition Assays

2.3.1 Background

Activity of purified Serp-1 as well as Serp-1-derived peptides is measured as inhibitory activity for uPA and tPA assayed by commercially available kits measuring absorbance, per manufacturers' directions and as previously described [18, 19, 21–23]. Potential serpin-derived peptide inhibition of serpin activity was also assessed using these same assays by measuring a reduction of serpin-mediated inhibition of target proteases (i.e., an inhibition of serpin-mediated inhibition). For the Serp-1-derived peptides we assayed for loss of activity for the mammalian serpin plasminogen activator inhibitor-1 (PAI-1) known to bind and inhibit tPA and uPA (*see* Fig. 3). Assays are conducted per the manufacturer's specifications. Blockade of PAI-1 inhibition of tPA or uPA is measured as a reduction in tPA or uPA inhibition. Simple gel shift assays are also used to measure protein binding and potential for inhibition (*see* **Note 2**).

2.3.2 Materials

1. Gel shift assay.
   - (a) Semidry transfer apparatus.
   - (b) Gel-casting apparatus.
   - (c) Polyacrylamide gel electrophoresis (PAGE) apparatus.
   - (d) Gel-staining boxes.
   - (e) Gel-imaging apparatus.
   - (f) 30% Acrylamide/bis solution, 29:1 ratio.
   - (g) Tetramethylethylenediamine (TEMED).
   - (h) Ammonium persulfate (APS), 10% solution.
   - (i) 20% Sodium dodecyl sulfate (SDS) solution.
   - (j) Resolving gel buffer (1.5 M Tris–HCl, pH 8.8).
   - (k) Stacking gel buffer (0.5 M Tris–HCl, pH 6.8).
   - (l) Water-saturated n-butanol.
   - (m) Tris/glycine/SDS running buffer (25 mM Tris base, 192 mM glycine, 0.1% SDS, pH 8.3).
   - (n) Coomassie stain or similar protein stain (e.g., silver stain).

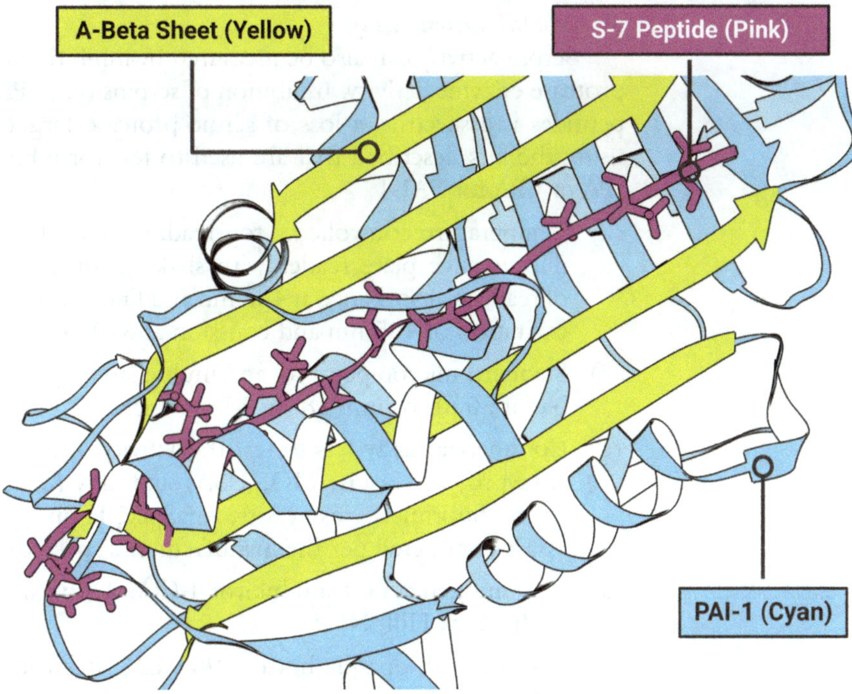

Fig. 3 Model for Serp-1 RCL S-7 peptide (pink) insertion and binding into the A-beta sheet of the mammalian serpin, PAI-1 (cyan). Crystal structure used is RSCB 3CVM [25]

(o) Non-denaturing protein-loading buffer (e.g., 62.5 mM Tris–HCl, pH 6.8, 40% glycerol, 0.01% bromophenol blue).

(p) Protein molecular weight marker.

(q) Towbin buffer (25 mM Tris, 192 mM glycine, pH 8.3).

(r) Double-distilled water.

(s) Synthesized peptides (>10 mM concentration in saline buffer).

(t) Plasminogen activator inhibitor-1(PAI-1) (we use Pepro-Tech, Rock Hill, NJ).

(u) Antibody specific to Serp-1 or other serpins to be tested.

(v) Recombinant Serp-1 (we have used good manufacturing product (GMP) quality or purified protein expressed in Chinese hamster ovary (CHO) cells with greater than 90% purity provided by Viron Therapeutics, Inc.)

(w) Serp-1 antibody (we have used Serp-1 monoclonal antibody supplied by Viron Therapeutics, Inc): Antibodies can also be developed commercially.

(x) Recombinant tPA and uPA with His-tag or similar affinity tag (e.g., GST).

(y) Anti-6×-His antibody, or appropriate anti-affinity tag antibody.

2. *ProteaseInhibition Assays*

   Serpin activity can also be measured by inhibition of serine protease enzyme activity. Inhibition of serpins (e.g., PAI-1) by peptides is assayed as a loss of serine protease target activity using the kits described that are used to test for inhibition of tPA or uPA for PAI-1.

   (a) Temperature-controlled, top-reading absorbance and fluorescence plate reader with shaking function capable of reading absorbance at 405 nm and fluorescence with an excitation of 405 nm and emission at 510 nm.

   (b) Standard micropipette set and multichannel pipette capable of 50 μL volume (8- or 12-channel).

   (c) Commercial assay kits to test the activity of uPA (we use Assaypro, St. Louis, MO, USA) and tPA (e.g., SensoLyte® fluorimetric assay kits, AnaSpec, Fremont, CA)—assays conducted per the manufacturer's specifications.

   (d) Plasminogen activator inhibitor-1 (PAI-1) (we use PeproTech, Rock Hill, NJ).

   (e) Black-walled, opaque bottom 96-well plates with round-bottom wells.

2.4 Animal Models of Vasculitis

All animal models are approved by the local institutional animal care and use committee (IACUC) and performed according to local, national, and international animal care guidelines. Equal numbers of male and female mice are used for all studies. Power calculations are performed prior to initiating studies to establish the numbers of animals required to detect statistically significant changes after treatments with immune-modulating proteins or peptides. Also see prior methodology chapter on models of vascular inflammation [21]. We focus here on the MHV68 infection model of severe inflammatory vasculitis and lung hemorrhage. We provide a brief overview of aortic transplant which is an excellent model; however, this technique requires many months of intense microsurgery training before this can be successfully performed. The MHV68 infection in interferon gamma receptor-deficient (IFNγR$^{-/-}$) mice allows for an in vivo assessment of severe vasculitis and survival with experimental treatments. This model can be used successfully by non-surgery-trained individuals.

2.4.1 Mouse Gamma Herpesvirus 68 (MHV68) Infection in IFNγR$^{-/-}$ Mice

1. MHV68 virus: Titer 6–12.5 × 10^6 PFU MHV68 in 0.1 mL of Dulbecco's minimal essential medium (DMEM).
2. IFNγR$^{-/-}$ mice (B6.129S7-Ifngr1$^{tm1Agt/J}$, C57BL/6 background 5–7 weeks of age; we source mice from JAX labs for breeding pairs, catalog #003288).

3. Purified peptides: We derived RCLpeptides from virus-derived Serp-1 and mammalian neuroserpin and analyzed these for immune-modulating functions.
4. Dulbecco's minimal essential medium (DMEM) with 10% FBS and 1% penicillin-streptomycin.
5. Sterile syringes, sterile needles.
6. Sterile cell scraper.
7. 3 T12 fibroblast (ATCC) cell culture.
8. 6-Well tissue culture plates.
9. 1% Methyl cellulose in DMEM.
10. Cell sonicator.
11. Neutral red stain (1 part 1% neutral red with 49 parts 1× PBS).

2.4.2 Aortic Allograft Transplants: C57Bl/6 Donor to BALB/C Recipient Mice

1. C57Bl/6 and BALB/c mice (commercially available or bred in-house) (*see* **Note 2**).
2. Allow mice to become acclimated on arrival for 1–2 weeks in the animal facility with adequate food and water ad libitum and standard light/dark cycles.
3. Sterile surgical instruments: Scissors, scalpels.
4. Sterile 11–0 nylon sutures for aortic anastomosis and 4–0 coated VICRYL polyglactin and sterile nylon suture for closing the abdomen.
5. Sterile clamps, vessel clips to isolate the site for aortic graft implant.
6. Surgical microscope.
7. Anesthetic: Ketamine, xylazine, or isoflurane gas: (a) For gas anesthetic, use a vaporizer with a rodent mask (we use a Surgi-Vet gas delivery system) at a dose of 1–3% to affectchange affect to effect. (b) For injectable anesthetic we use a ketamine/xylazine cocktail—120 mg/kg ketamine and 5 mg/kg xylazine at a dose of 100 μL per 25 g mouse by intraperitoneal injection.
8. Buprenorphine analgesic (0.05–0.1 mg/kg).
9. Betadine or chlorhexidine wash.
10. Heating pad for postoperative care.
11. CO_2 exposure or euthanasia by injection (e.g., euthanyl containing 240 mg sodium pentobarbital), as per veterinary staff (IACUC) recommendations for euthanasia, with secondary euthanasia by cervical dislocation.
12. If a gas anesthetic is used, the room should have a gas-scavenging system or hood.

2.4.3 Histological Analysis of Lung Consolidation, Gut Dilatation, and Vascular Inflammation in Animal Models

Tissues analyzed differ for the two models.

1. Reagents and equipment for performing routine histological staining, such as hematoxylin and eosin (H&E). These methods, including dehydration and embedding of tissues as well as cutting and staining, are described in detail in other chapters of this book.
2. Reagents for performing Verhoeff-Van Gieson (VVG) stain of elastic fibers:
   (a) 5% Alcoholic hematoxylin
   (b) 10% Aqueous ferric chloride (prepare fresh)
   (c) Weigert's iodine solution.
   (d) 5% Aqueous sodium thiosulfate
   (e) Van Gieson's counterstain (5 mL 1% aqueous acid fuchsin in 100 mL saturated aqueous picric acid).
3. Antibodies for immunohistochemical staining (IHC):
   (f) T cells: Rabbit polyclonal anti-CD3, 1:200 dilution (we use ab5690, Abcam, Cambridge, MA, USA).
   (g) Macrophage: Rabbit polyclonal anti-F4/80, 1:200 dilution (we use ab75476, Abcam, Cambridge, MA, USA).
   (h) Secondary: Goat anti-rabbit IgG, HRP conjugated, 1:500 dilution (we use 111–035-144, Jackson ImmunoResearch, West Grove, PA, USA).
4. RPMI media with 10% FBS for collecting spleens for flow cytometry.

3 Methods

3.1 Protease and Serpin Inhibition Assays

The gel shift assays and Western blots are performed as previously described [22].

3.1.1 Gel Shift Assay

3.1.2 Protease Inhibition Assay

For protease activity assays, thaw reagents on ice. Keep protease substrate on ice and protect from light.

1. *uPA and tPA Activity Assays:*
   (a) To assess the inhibitory activity of viral proteins or peptides toward uPA activity, the release of p-nitroaniline (pNA) from the pyro-Glu-Gly-Arg-pNA (e.g., chromogenic substrate S-2444) is measured in the presence of urokinase (e.g., Sigma U0633).
   (b) Mix 87 μLTris buffer (50 mmol/L Tris and 38 mmol/L NaCl at pH 8.8) with urokinase 3 units (in 3 μL) and incubate at 37 °C for 10 min.

(c) Incubate the protease mixture at 37 °C for 2 min.

(d) Add 10 μL chromogenic substrate (3 mmol/L) to the mixture and incubate at 37 °C for 5 min.

(e) Stop the reaction by adding 10 μL of 20% acetic acid.

(f) Record the absorbance (A) of the sample in a photometer at 405 nm. The absorbance is proportional to the urokinase activity and the positive controls with amounts of urokinase and substrate stated above have been found to be in the linear range in our lab.

(g) Inhibition of uPA activity by viral proteins and peptides is measured by adding the inhibitor along with the urokinase (before the introduction of the substrate to the well) and incubations are carried out as described above.

(h) Fluorometric assays for activities of tPA are carried out according to the instructions (e.g., SensoLyte kit, AnaSpec, Fremont, CA).

2. *PAI-1 SerpinActivity.*

(a) Assay for peptide-mediated blockade, e.g., inhibition of PAI-1 activity by measuring a loss of inhibitory activity for PAI-1 against the known substrates uPA and tPA.

(b) Add PAI-1 to the uPA or tPA activity assays, as described above in Subheading 3.1.2, **step 1**.

(c) Determine the optimal dose range for PAI-1 inhibition of uPA or tPA.

(d) Add peptides or serpin to be investigated at increasing concentrations in order (we used 500 ng of individual peptides in each reaction mix to detect optimal inhibitory activity for PAI-1).

(e) Measure absorbance at 405 nm to assess for return of activity of uPA or tPA in the presence of PAI-1.

3.2 Mouse Gamma Herpesvirus 68 (MHV68) Infection in Interferon Gamma Receptor Knockout (IFNγR$^{-/-}$) Mice (See Note 3)

3.2.1 MHV68 Infection in IFNγR−/− Mice, Inflammatory Vasculitis Model

IFNγR$^{-/-}$ mice are infected by intraperitoneal (i.p.) injection of 12.5 × 10^{6} PFU MHV68 given in 0.1 mL DMEM [8–10, 16]. (Note: Alternative models use intranasal infections but the model described here provides an excellent model for a severe and lethal inflammatory vasculitis with pulmonary consolidation and hemorrhage.)

3.2.2 MHV68 Virus Culture

Generate MHV68 viral stocks in NIH 3 T12 fibroblasts:

1. Culture 3 T12 fibroblasts in Dulbecco's minimal essential medium with 10% heat-inactivated FBS (DMEM) with 1% penicillin-streptomycin.
2. At 50% confluency, infect cells with MHV68, MOI of 0.1.
3. Harvest MHV68 virus with cell scraper at 7 days postinfection or earlier depending on cytopathic effects.
4. Centrifuge harvested cell lysate for 1 h at 15,000 rpm, 4 °C.
5. Discard media and resuspend in 200 μL PBS.
6. Sonicate three times at 10 s each (Sonics Vibra-Cell 505, 49% amplitude), 60 s in between each cycle.
7. Centrifuge at 3000 rpm for 10 min at 4 °C.
8. Transfer supernatant to a new tube and repeat centrifuge (3000 rpm, 10 min, 4 °C).
9. Store virus at −80 °C in 50 μL aliquots.
10. Titer MHV68 virus by plaque assay in duplicate:
11. Plate NIH 3 T12 cells at 1×10^5 per well (6-well plate) and incubate cells overnight at 37 °C.
12. Cells should be about 60% confluence the next day.
13. Remove media, rinse with PBS, and infect cells with virus dilutions for 1 h, rocking the plates back and forth, side to side every 15 min.
14. After an hour, add 4 mL of 1% methyl cellulose in DMEM with 10% FBS and 1% Pen-Strep (start with 2× DMEM with 10% FBS and 1% Pen-Strep and add equal volume of methyl cellulose in a 1:1 ratio).
15. Incubate cells for 7 days.
16. Stain with neutral red solution (1 part 1% neutral red with 49 parts 1× PBS).
17. Incubate overnight at 37 °C.
18. Aspirate methyl cellulose and count plaques over a light box.

3.3 Aortic Allograft Transplants C57Bl/6 Donor to BALB/C Recipient Mice

We refer readers to prior publications detailing the specifics of this surgical technique [8, 21] (*see* **Note 4**). A brief overview of this technique is provided below. For all surgeries, mice are anesthetized and demonstrated to be unresponsive to toe pinch before surgery. The abdominal wall is shaved and cleaned with three-stage povidone iodine, alcohol, and final iodine wash (chlorhexidine may be substituted for iodine). Harvest the donor aortic graft sections from C57Bl/6 mice immediately before transplant surgery. Euthanize the mouse at the time of aortic artery harvest. Implant the aorta from the C57BL/6 mouse in line into the BALB/c mouse aorta as a functioning section in the recipient mouse aorta. Inject the test protein or peptide into the vein (IVC or penile vein in male mice) once visible pulsation is seen after removing the clamps above and below the aortic graft implant. All procedures are performed in a sterile field. A sterile drape is placed over the sites of surgery. The aortic allograft transplant model requires special skills and microsurgery training.

3.4 Histological Analysis of Lung Consolidation, Gut Dilatation, and Vascular Inflammation in Animal Models

3.4.1 MHV68 Inflammatory Vasculitis Model: Harvest of Tissues for Histology Analysis

1. Euthanize mice at predetermined follow-up or when deemed at risk.
2. Isolate ascending aorta, aortic arch, and thoracic aorta as well as heart, lungs, spleen, and colon.
3. Divide harvested organs into two parts—one for immediate fixation in neutral-buffered formalin for later histological analysis and the second part for freezing in RNA*later* (or appropriate RNA-stabilizing solution) for later mRNA and protein analyses.
4. For spleens it is best to immediately isolate splenocytes for flow cytometry analysis (we have not included a description of FACS analysis here).
   (a) Cut spleen in half; place half in neutral-buffered formalin for histologic staining and half in sterile DMEM or RPMI containing 10% fetal bovine serum, on ice, for immediate flow cytometry (*see* **Note 5**).
5. Take blood samples by cardiac puncture at the time of euthanasia.
6. Formalin-fixed specimens are processed and embedded for routine H&E staining as well as immunohistochemical analysis as described elsewhere in this book.

3.4.2 For the Aortic Allograft Model: Harvest of Tissues for Histology Analysis

1. After euthanizing the grafted mouse, expose the transplanted aorta.
2. Isolate the donor aortic transplant site and adjacent recipient aorta (0.5–0.6 cm length).
3. Cut the harvested aortic transplant site into two 0.25–0.3 cm pieces.
4. Fix, paraffin-embed, and cut each graft site into 5 μm sections (two per aortic section, providing four sections for analysis of each transplanted mouse aorta).

3.4.3 Staining Elastic Fibers by Verhoeff-Van Gieson (VVG) Method

1. Deparaffinize and rehydrate 5 μm sections on charged glass slides into distilled water.
2. Prepare Verhoeff's working solution: Mix 20 mL of 5% alcoholic hematoxylin, 8 mL of 10% ferric chloride, and 8 mL of

Weigert's iodine solution. The solution will be jet black. Prepare freshly, immediately before use, and discard after use. Solution will not stain properly if kept for more than one working day.

3. Stain in Verhoeff's solution for 1 h at room temperature. The tissue will appear completely black.
4. Rinse slides in 2–3 changes of tap water.
5. Differentiate the stain in 2% ferric chloride at room temperature for 1–2 min.
6. Rinse the slides in several changes of tap water to stop the differentiation process. Slides can be checked microscopically to determine that elastic fibers are stained properly (black fibers on a gray background). It is preferred to underdifferentiate at this stage as stain will be slightly extracted further during counterstaining.
7. Treat slides with 5% sodium thiosulfate for 1 min at room temperature. Discard solution after each use.
8. Wash slides in 3–5 changes of tap water for 5 min at room temperature.
9. Counterstain slides in Van Gieson's solution for 3–5 min at room temperature.
10. Dehydrate slides through graded alcohol into xylene and mount with a xylene-based mounting medium.
11. Upon observation, elastic fibers will appear blue-black to black, nuclei will appear blue to black, collagen will appear red, and other tissue elements may appear yellow.

3.4.4 Routine and Immunohistochemical Staining

1. For routine first analysis, stain histology sections with hematoxylin and eosin (H&E). Methods for this are described elsewhere in this book (*see* Chapter 16).
2. Perform morphometric analysis to measure areas of inflammatory cell invasion, tissue scarring, and necrosis as well as arterial plaque, lumen size, and internal elastic lamina areas.
3. Measure aortic graft plaque by means of an image analysis program, calibrated to the microscope objective (e.g., we use cellSens Dimension, ImageJ) (*see* **Note 6**).
4. For atheroma measurement, calculate the mean total cross-sectional area of the intima, using sections with the largest detectable plaque area, for each aortic specimen (*see* **Note 7**).
5. For inflammatory cell invasion area, tissue hemorrhage and necrosis, or areas of pulmonary consolidation, measure the area of damage or hemorrhage and divide by the total area (*see* **Note 8**).

6. To assess lung damage, measure alveolar septal thickness as well as areas of hemorrhage using an image analysis program, calibrated to the microscope objective (e.g., we use cellSens Dimension, ImageJ).
7. Count numbers of invading mononuclear cells or cells staining positively on IHC-stained sections per high-powered field area at three sites in the intimal, medial, and adventitial layers—choose sites with the largest numbers of detected invading inflammatory cells or positively stained cells for all samples and sections on IHC to avoid biased readings. Methods for performing immunohistochemical staining of cells in tissue are described in other chapters of this book.
8. Calculate the mean count for each arterial layer and/or other organ specimens (*see* **Notes 6–8**).
9. Analyze statistical significance for arterial or aortic transplant plaque area, lung alveolar septal thickness, areas of hemorrhage, or cell counts per high-power field using analysis of variance (ANOVA) with secondary Fisher's analysis and/or unpaired Student's T-test depending upon the numbers of treatment groups analyzed.

4 Notes

1. Serpins are reported to form inactive aggregates via insertion of the RCL of one serpin into the A-beta sheet of adjacent serpins, the proposed basis of genetic serpinopathies. In normal serpin function the RCL is cleaved by a protease and then the cleaved RCL is inserted into the A-beta sheet dragging the protease to the opposite end of the serpin molecule forming a suicide inhibitory complex. In serpinopathies, genetic modifications lead to interactions between adjacent serpins where the RCL of one serpin becomes inserted into the A-beta sheet of an adjacent serpin, leading to inactive clumps of aggregated serpins and loss of function. We have proposed that serpin RCL-derived peptides have the capacity to inhibit other serpins, not just proteases. Testing serpin RCL-derived peptides for inhibition of mammalian serpin inhibitory actions led to the discovery of this unique serpin inhibition mediated by the RCL-derived peptide, S-7.
2. For breeding mice, we generally place two females with one male starting at 6–8 weeks of age. Providing simple paper tubes and cardboard can enhance the environment for the mice. The mice should be genotyped intermittently if bred in-house. For first-time mothers there is often a greater loss of new pups. If examining changes in the microbiota, it is best to house mice

individually where possible. Mice are coprophagic and will ingest bacteria in the pellets/feces of other mice sharing a cage.

3. For aortic allograft anastomosis, in general six sutures will provide a good anastomosis and hemostasis. For small leaks surgical glue can be applied but can cause increased local inflammatory responses.
4. We do initial follow-up at 4 weeks; once a significant change is detected then earlier follow-up times are examined for each treatment. This saves time as efficacy is determined at the later time point and earlier follow-up is then performed once a potential response to individual peptide treatments is determined. This approach also necessitates less time for follow-up; that is, it is more efficient as the shorter time follow-up can be done after the initial longer term analysis or even while completing late follow-up.
5. Immediate flow cytometry analysis of spleen cells is considered preferable for more accurate flow analysis for cellular responses; however frozen splenocytes have been studied where necessary. It should be noted that flow cytometry is considered qualitative overall due to the loss of spatial information.
6. We measure plaque area and cell counts in a minimum of three sites per section.
7. The mean plaque area, lumen area, and cell counts are calculated for each specimen analyzed and used for all subsequent analyses, to avoid selection bias.
8. Histological analysis should be performed where possible with observers blinded to the treatment groups. To avoid biased analyses, we measure histological changes initially unblinded and then as a second blinded reading by scientists who do not know the treatment used.

Acknowledgments

The authors would like to gratefully acknowledge Dr. Efrem Lim and Ms. Emily Kaelin (ASU Biodesign Institute) for their helpful contributions to the section of this chapter describing the production of MHV68 virus.

References

1. Subramaniyam D, Glader P, von Wachenfeldt K et al (2006) C-36 peptide, a degradation product of alpha1-antitrypsin, modulates human monocyte activation through LPS signaling pathways. Int J Biochem Cell Biol 38:563–575
2. Lorenz N, Loef EJ, Kelch ID et al (2016) Plasmin and regulators of plasmin activity control the migratory capacity and adhesion of human T cells and dendritic cells by regulating cleavage of the chemokine CCL21. Immunol Cell Biol 94:955–963

3. Bocksch L, Rider BJ, Stephens T et al (2007) C-terminal apolipoprotein E-derived peptide, Ep1.B, displays anti-atherogenic activity. Atherosclerosis 194:116–124
4. Dai E, Stewart M, Ritchie B et al (1997) Calreticulin, a potential vascular regulatory protein, reduces intimal hyperplasia after arterial injury. Arterioscler Thromb Vasc Biol 17:2359–2368
5. Yan Y, Zhou A, Carrell RW et al (2019) Structural basis for the specificity of renin-mediated angiotensinogen cleavage. J Biol Chem 294:2353–2364
6. Marijanovic EM, Fodor J, Riley BT et al (2019) Reactive centre loop dynamics and serpin specificity. Sci Rep 9:3870
7. Shapira E, Proscura E, Brodsky B et al (2011) Novel peptides as potential treatment of systemic lupus erythematosus. Lupus 20:463–472
8. Ambadapadi S, Munuswamy-Ramanujam G, Zheng D et al (2016) Reactive center loop (RCL) peptides derived from serpins display independent coagulation and immune modulating activities. J Biol Chem 291:2874
9. Mahon BP, Ambadapadi S, Yaron JR et al (2018) Crystal structure of cleaved Serp-1, a myxomavirus-derived immune modulating Serpin: structural design of serpin reactive center loop peptides with improved therapeutic function. Biochemistry 57:1096–1107
10. Yaron JR, Ambadapadi S, Zhang L et al (2020) Immune protection is dependent on the gut microbiome in a lethal mouse gammaherpesviral infection. Sci Rep 10:2371
11. Ma H, Lu Y, Lowe K et al (2019) Regulated hAAT expression from a novel rAAV vector and its application in the prevention of type 1 diabetes. J Clin Med 8:1321
12. Siebers K, Fink B, Zakrzewicz A et al (2018) Alpha-1 antitrypsin inhibits ATP-mediated release of interleukin-1β via CD36 and nicotinic acetylcholine receptors. Front Immunol 9:1–15
13. Carrell RW, Evans DL, Stein PE (1991) Mobile reactive centre of serpins and the control of thrombosis. Nature 353:576–578
14. Wang F, Song Z, Chen J et al (2020) The immunosuppressive functions of two novel tick serpins, HlSerpin-a and HlSerpin-b, from Haemaphysalis longicornis. Immunology 159:109–120
15. Dai E, Viswanathan K, Sun YM et al (2006) Identification of myxomaviral serpin reactive site loop sequences that regulate innate immune responses. J Biol Chem 281:8041–8050
16. Chen H, Zheng D, Abbott J et al (2013) Myxomavirus-derived serpin prolongs survival and reduces inflammation and hemorrhage in an unrelated lethal mouse viral infection. Antimicrob Agents Chemother 57:4114–4127
17. Scott BM, Sheffield WP (2020) Engineering the serpin α1 -antitrypsin: a diversity of goals and techniques. Protein Sci 29:856–871
18. Lysakova-Devine T, Keogh B, Harrington B et al (2010) Viral inhibitory peptide of TLR4, a peptide derived from vaccinia protein A46, specifically inhibits TLR4 by directly targeting MyD88 adaptor-like and TRIF-related adaptor molecule. J Immunol 185:4261–4271
19. Istivan TS, Pirogova E, Gan E et al (2011) Biological effects of a De novo designed myxoma virus peptide analogue: evaluation of cytotoxicity on tumor cells. PLoS One 6:e24809
20. Almansour NM, Pirogova E, Coloe PJ et al (2012) A bioactive peptide analogue for myxoma virus protein with a targeted cytotoxicity for human skin cancer in vitro. J Biomed Sci 19:65
21. Chen H, Ambadapadi S, Dai E et al (2018) Analysis of in vivo Serpin functions in models of inflammatory vascular disease. Methods Mol Biol 1826:157–182
22. Zhang L, Yaron JR, Ambadapadi S et al (2018) Viral serpin reactive center loop (RCL) peptides: design and testing. Methods Mol Biol 1826:133–142
23. Yaron JR, Ambadapadi S, Zhang L et al (2018) Kinetic measurement of Serpin inhibitory activity by real-time fluorogenic biochemical assay. Methods Mol Biol 1826:65–71
24. Dementiev A, Simonovic M, Volz K et al (2003) Canonical inhibitor-like interactions explain reactivity of α 1 -proteinase inhibitor Pittsburgh and antithrombin with proteinases. J Biol Chem 278:37881–37887
25. Jensen JK, Gettins PGW (2008) High-resolution structure of the stable plasminogen activator inhibitor type-1 variant 14-1B in its proteinase-cleaved form: a new tool for detailed interaction studies and modeling. Protein Sci 17:1844–1849

Chapter 8

Methods for Crystallization and Structural Determination of M-T7 Protein from Myxoma Virus

Christopher Gisriel, Petra Fromme, and Jose M. Martin-Garcia

Abstract

The myxoma virus has become of interest in human medicine in the last two decades as it has the ability to infect many types of human cancer cells and is being used as a platform to develop viro-therapeutic agents that suppress aggressive and damaging immune responses and inflammation. Furthermore, the myxoma virus encodes proteins that have strong immunosuppressive effects, and several of the myxoma virus-encoded immunomodulators are being developed to treat systemic inflammatory syndromes such as cardiovascular disease and transplant rejection. Myxoma virus encodes the M-T7 protein, the most abundantly secreted protein expressed in myxoma virus-infected cells, originally identified as a rabbit species-specific interferon-gamma (IFN-γ) receptor homolog and as a chemokine-modulating protein binding a wide range of mammalian chemokines. M-T7 is a critical virulence factor for viral pathogenesis that increases virus lethality when expressed. Although M-T7 has been extensively studied using biochemical and biophysical techniques and its interactome map is well known, its three-dimensional (3D) structure remains elusive. Obtaining the 3D structure of M-T7 would be greatly beneficial and is a crucial step toward advancing M-T7 research through understanding the molecular function and activity of M-T7 as a novel therapeutic reagent and to rationally develop this protein as a drug. This chapter provides an overview of the structural determination techniques, especially X-ray crystallography, that can be applied toward the goal of achieving the first high-resolution structure of M-T7. In addition, details of up-and-coming methods are discussed, including X-ray diffraction at X-ray free electron lasers (XFELs), nuclear magnetic resonance (NMR), cryo-electron microscopy (cryo-EM), Micro-electron diffraction (Micro-ED), and small-angle X-ray scattering (SAXS), and their potential applications to M-T7 structural biology.

Key words Myxoma virus, M-T7 protein, Structure determination, X-ray crystallography, X-ray free electron lasers (XFELs), Cryo-electron microscopy (cryo-EM), Micro-electron diffraction (Micro-ED), Small-angle X-ray scattering (SAXS)

1 Introduction

As for any protein, determining the 3D structure of the myxoma virus T7 protein (M-T7) will provide us with the molecular basis of the mechanisms that govern M-T7 activity inside and outside the cell. Furthermore, knowledge of the structure will allow identification of potential ligand-binding sites that can be used in drug

Alexandra R. Lucas (ed.), *Viruses as Therapeutics: Methods and Protocols*, Methods in Molecular Biology, vol. 2225, https://doi.org/10.1007/978-1-0716-1012-1_8,

design or in the development of protein mimics and/or other inhibitors with therapeutic applications. To date, although the M-T7 sequence shares homology with the rabbit interferon gamma receptor (IFN-γ), M-T7 also has exhibited binding and inhibition of chemokine. M-T7 blocks chemokine binding to glycosaminoglycans (GAG), not chemokine receptor binding, in a wide range of mammals from mouse, rat, and rabbits to man. Interestingly, M-T7 is also a glycoprotein and binds glycosaminoglycans. M-T7 has proven potent to immune-modulating functions, reducing inflammation in rodent and rabbit atherosclerotic and transplant vasculopathy models. Thus, this broad and potent inhibition of chemokine to GAG binding is of great potential therapeutic interest. However, there is little structural information for M-T7 and specifically there is little information for M-T7 structural binding and inhibition of chemokines, so the intent of this book chapter is to provide valuable tools and resources for understanding the relationship between sequence, structure, and function of a unique virus-derived immune-modulating protein, specifically the M-T7 protein.

To this end, it will be useful to address the known, and also the potential, characteristics of M-T7 (based on similarity) through the lens of a structural biologist for making structural predictions that will explain the capacity of M-T7 to interfere with or block chemokine binding to GAGs in the glycocalyx. M-T7 is a 37 kDa N-terminus-linked glycoprotein that presents as a homotrimer (113 kDa) in solution in vitro [1]. M-T7 exhibits ~18.5% sequence identity with the rabbit interferon-gamma receptor (IFN-γR, an IFNγ-binding protein). The IFN-γR molecular structure has been solved by X-ray crystallography to 2.2 Å resolution (PDB 3BES) [2]. Thus, unsurprisingly, M-T7 inhibits IFN-γ binding in rabbits [3]. M-T7 also acts, as noted, as a nonspecific chemokine-binding protein in humans, rabbits, rats, and mice, interacting with a conserved heparin-binding domain and interfering with GAG binding for a wide range of C, CC, and CXC class chemokines [4]. The structures of all classes of chemokines have been demonstrated to share extensive structural similarity. A review of heparin-binding domains, with a focus on structure, can be found in Muñoz, 2004 [5]. It has been shown that M-T7 binds to the RANTES (regulated on activation of normal T cell expressed and secreted) C-C chemokine. However, M-T7 binding to CC chemokines is partially inhibited in the presence of rabbit IFN-γ, implying similarity in the binding of M-T7 to RANTES and to IFN-γ [1]. A contradiction, however, is that heparin competes with M-T7 binding to RANTES [1], but not IFN-γ [6]. This has led to the hypothesis that M-T7 binds to the GAG-binding domain (GBD) of RANTES, and other classes of chemokines, although this has not yet been confirmed experimentally [7]. Various point mutants of His-tagged M-T7 have been studied that modulate its C-C chemokine anti-inflammatory activity, indicating disruption of ligand binding. Specifically, the N-terminal glycosylation may be important for interfering with chemokine to GAG binding as for the chemokine RANTES [7].

Careful characterization of the activity as well as the structure of M-T7 not only will advance our understanding of the broadly observed immune response-modulating biological mechanisms of M-T7, but may also inform rational anti-inflammatory drug design. X-ray crystallography is by far the most well-established and successful technique for accurate protein structure determination. In general, X-ray crystallography is performed by achieving a highly pure protein isolation, identifying optimal crystallization conditions, collecting X-ray diffraction data of protein crystals, calculating an electron density map from the diffraction data, and finally building a model into the map that obeys reasonable biological constraints.

Here, we present various crystallization strategies for proteins similar to M-T7 and methods for crystal characterization, describe the preparation of crystals for use in X-ray diffraction experiments and subsequent data collection, and also cover topics on data analysis and model-building strategies that can be used for M-T7 in X-ray crystallography. We close the chapter by providing brief descriptions of other common structural biology techniques that may be useful for structural characterization of M-T7 including cryo-electron microscopy (cryo-EM), micro-electron diffraction (Micro-ED), and small-angle X-ray scattering (SAXS). Our hope is to provide a wide variety of tools to guide those who would advance the structural understanding for M-T7 and other virus-derived immune response-modulating protein biologics as treatments in medicine and biology.

2 Structural Predictions

Due to the current lack of structural information for M-T7, the consideration of biophysical characteristics (i.e., molecular weight and solubility), previously predicted structural hypotheses based on biochemical analyses, identification of protein structures exhibiting sequence similarity, and analysis of such homology should be the first step in the consideration of performing structural studies on M-T7. We see this as akin to performing a literature search prior to designing experiments.

The sequence of M-T7 is available in the UniProt database (https://www.uniprot.org/) [8]. Initially, one could use the National Center for Biotechnology Information's (NCBI) Basic Local Alignment Search Tool (BLAST) (https://blast.ncbi.nlm.nih.gov/Blast.cgi) [9] to search for similar protein sequences and to then identify those similar sequences for which a molecular structure has been solved, which is helpful for hypothesizing characteristics of the protein structure in question. Alternatively, or additionally, various homology-modeling software and databases exist that provide excellent tools for preliminary structural considerations. A good example is the Swiss Model server (https://swissmodel.expasy.org/) [10] which uses an input sequence to search for protein structures within the Protein Data Bank (PDB) (https://www.rcsb.org/) [11] that exhibit sequence similarity and subsequently uses the most similar structure as a template for structural homology modeling. It is wise to create a homology model before experimental efforts are made to familiarize the experimentalist with the characteristics of the protein (i.e., sequence, molecular mass, probable conserved regions, and cofactors) and to make structural predictions. A homology model will ultimately be useful in providing phases for diffraction data (termed "molecular replacement").

2.1 Predictions Based on Homology

The structure in the PDB that exhibits the highest sequence similarity to M-T7 is a 2.2 Å resolution structure of the IFN-γ-binding protein C4R (hereafter referred to as "C4R", PDB 3BES) [2] which has ~18.5% sequence identity. A sequence-based homology model was used in a previous study to inform targets for site-directed mutagenesis [7], but an in-depth sequence analysis may provide further structural insights and expose the robustness of the homology model. On one hand, the low sequence identity calls into question the reliability of such a homology model. On the other hand, it could be that specific regions of the sequence exhibit high local sequence similarity, but that the rest of the sequence has especially low sequence similarity. Understanding the limitations of a homology model is an important aspect of this type of analysis.

An example of this type of sequence-based structural analysis can be shown with the sequence of a reductase from *Rhodococcus rhodochrous* compared to a similar reductase from *Burkholderia cepacian*, the latter of which has an associated crystal structure (PDB 2PIA) [12]. Although there is no known structure of the reductase from *R. rhodochrous*, the sequence compared to that from *B. cepacian* is ~24% similar. The low sequence similarity implies structural dissimilarity, but it is possible that certain conserved stretches of residues can provide insights into characteristics of the unknown structure.

In this example we will focus only on the 2Fe-2S cluster-binding site, although many more structure-based predictions about the reductase from *R. rhodochrous* could be made based on the two sequences. In the reductase crystal structure from *B. cepacian*, a 2Fe-2S cluster is coordinated by four cysteine residues: Cys272, Cys277, Cys280, and Cys308 (*see* Fig. 1a). All except Cys308, which aligns with a Leu residue in the *R. rhodochrous* sequence (highlighted in red in Fig. 1b), is conserved in the sequence from

A

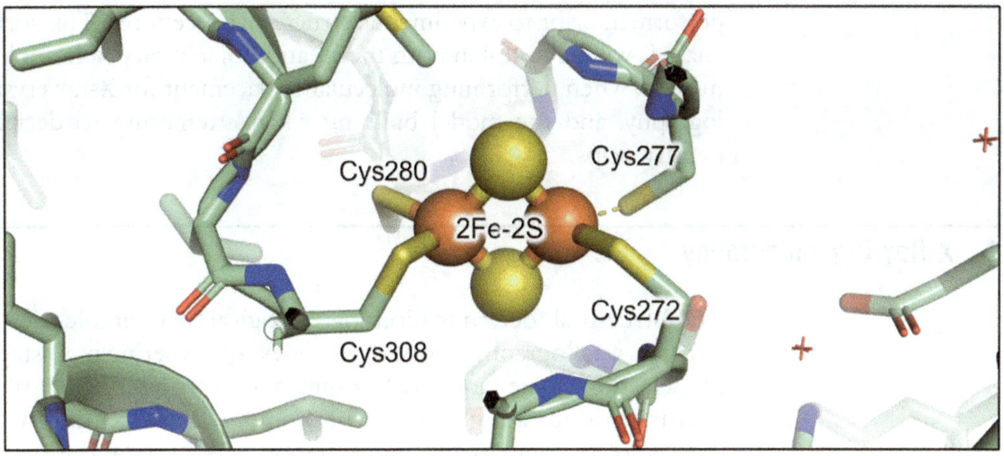

B

```
R. rhosochrous MTTVEVPHGSRSVILTVSAVVEETADTRSIVFAVPDELR-DKFAYRPGQFLTLRIPSDRT 59
B. cepacian    ---MTTPQEDGFLRLKIAS--KEKIARDIWSFELTDPQGAPLPPFEAGANLTVAVPNGS- 54
                 : .*: .  : *.:::  :*.          * : *          :. *  **: :*..

R. rhosochrous GSVARCYSLASSPFTDDAPKVTVKRTSDGYGS-NWLCDNIATGQTLEVLPPAGVFTPKSL 118
B. cepacian    ---RRTYSLCNDSQERNRYVIAVKRDSNGRGGSISFIDDTSEGDAVEVSLPRNEFPLDKR 111
                  ***...     :    ::*** *:* *.    : *: : *:::**  * . *   ..

R. rhosochrous DHDFLLFGAGSGITPVISILKSALTQGGGKVVLVYANRDEKSVIFAEELRALAEKYPTRL 178
B. cepacian    AKSFILVAGGIGITPMLSMARQLRAEGLRSFRLYYLTRDPEGTAFFDELTS--------- 162
                :.*:*...* ****::*: :.   ::*   .. * * .** :.. * :** :

R. rhosochrous TVVHWLESVQ-----GLPT--ADQLAAIAAPYESYEAFMCGPGPFMDTVHQALNTVGMPR 231
B. cepacian    --DEWRSDVKIHHDHGDPTKAFDFWSVFEKSKPAQHVYCCGPQALMDTVRDMTGH--WPS 218
                 .* ..*:      * **   *  :.:      : ..: ***  :****::  .    *

R. rhosochrous ARVHAEVFNSLSGDPFADQAPVEVSDEDAADAATVEVELDGEVHKLSWPRKQTLVDIMLA 291
B. cepacian    GTVHFESFGATNT--------------NARENTPFTVRLSRSGTSFEIPANRSILEVLRD 264
               . ** * *.: .                :* : : . *.*. .  .:. * ::::::::

R. rhosochrous KGIDVPYSCQEGECGSCACTVLEGKVEMENCDVLDPEDIEAGYILGCQARPVTDHLKIEF 351
B. cepacian    ANVRVPSSCESGTCGSCKTALCSGEADHRDMVL--RDDEKGTQIMVCVSRAKSAELVLDL 322
                .: ** **:.* ****  :: .*:.: .:  :   :* :.   *: * :*  : .* :::
```

Fig. 1 Structure predictions and sequence analysis of a reductase from two different organisms. (**a**) 2Fe-2S cluster from the *B. cepacian* structure is shown as ball representation. All cysteine residues coordinating the iron-sulfur cluster are highlighted and shown in stick representation. (**b**) Sequence alignment of the reductase sequences from *R. rhodochrous* and *B. cepacian*. The alignment was performed using Clustal Omega [126]. Real or potential 2Fe-2S-coordinating residues are highlighted as discussed in the text

R. rhodochrous (highlighted in green in Fig. 1b), implying that this cofactor is conserved in both structures. Furthermore, nearby Cys residues that are not found in the *B. cepacian* structure (highlighted in blue in Fig. 1b) may provide the missing Cys coordination.

The creation of homology models to predict secondary and tertiary structure, and sequence analysis to predict ligand binding, like the example presented here, are powerful tools that can be performed prior to experimental structural procedures. This sort of analysis will be useful in both the creation of a library of homology models, when performing molecular replacement for X-ray crystallography, and the model building when attempting to decipher density maps.

3 X-Ray Crystallography

The structural determination of biological macromolecules by X-ray crystallography requires a series of experimental stages: (1) selection of the target; (2) cloning and expression; (3) purification to produce a highly pure and homogeneous solution of the macromolecule at high concentration; (4) crystallization for obtaining large, well-diffracting crystals; (5) X-ray diffraction data collection; and (6) structure determination and refinement of the 3D model. An illustration of the overall pipeline from gene to structure is given in Fig. 2. In this chapter, we apply this workflow to purified M-T7 protein analysis by X-ray crystallography, assuming that the interested reader has a purified sample as a starting point, and discuss the principles of crystallography and X-ray diffraction as well as the computational tools that are necessary for protein structure determination.

3.1 Protein Crystallization (M-T7 Protein as a Practice Example)

Despite all the biochemical and biophysical information available [1, 7, 13–15], the 3D structure of M-T7 remains elusive. As for any protein, having structural information of M-T7 at an atomic level is crucial for understanding the mechanisms and function of this unique and highly potent immune response-modulating chemokine-binding protein on the cells. While other methods can produce important structural and dynamic data, X-ray crystallography has demonstrated for over 155 years to be, along with neutron diffraction, the only technique that can yield atomic resolution structures of proteins. Thus, X-ray crystallography becomes the most appropriate and thus the technique of choice for M-T7 structure determination.

X-ray crystallography is dependent on achieving crystals of the macromolecule of a sufficient size and order to produce high-resolution diffraction patterns. Often, creating high-quality crystals is a rate-limiting step and remains the major obstacle to obtaining crystallographic structures.

3.1.1 General Aspects on Macromolecular Crystal Growth: A Quick Overview

Macromolecular crystal growth has always been empirical in nature, demanding a lot of patience, perseverance, and intuition from the crystallographer. Crystallization of any molecule implies the arrangement of protein molecules in an ordered fashion. The crystallization process can be divided into two consecutive stages: (1) the nucleation stage, in which protein molecules associate in

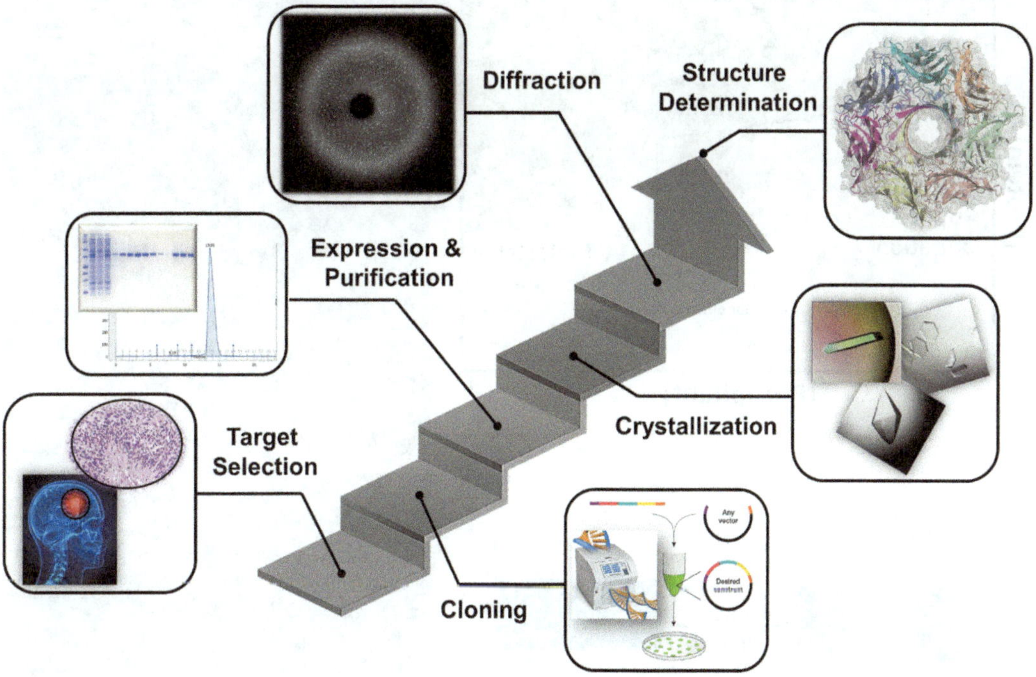

Fig. 2 Flowchart representing the main steps of the three-dimensional structure determination of biological macromolecules by X-ray crystallography. The protein target is first identified and characterized. The sequence corresponding to the gene of the selected macromolecule is cloned in an expression vector and produced classically in a bacterial organism (typically *Escherichia coli*) and then purified as indicated by size exclusion chromatography (SEC), where the protein is eluted as represented by the blue peak. Protein purity is later assessed by SDS-PAGE. The crystallization process is symbolized by the single crystals. The following X-ray diffraction experiment is represented by the picture of a diffraction pattern, which shows the diffracted reflections as small spots (white spots) over the whole detector area (black background). The final structure elucidation procedure is symbolized by the 3D crystal structure of a protein

three dimensions forming an ordered aggregate (the crystalline nuclei) that provides surfaces for further growth, and (2) the growth stage, which is governed by the diffusion and assembly of protein molecules to the surface of the crystalline nuclei making them reach a critical size. Crystal growth continues until the solution is sufficiently depleted of protein molecules.

To understand a crystallization experiment, one must examine the phase diagram of crystal growth (*see* Fig. 3a). The primary requisite for crystallization of a macromolecule is to drive the concentration of a protein solution into a supersaturated state within the metastable and/or labile regions. Depending on the level of supersaturation this zone of the diagram can be divided into three regions: very high supersaturation ("precipitation"), where molecules form amorphous aggregates; intermediate supersaturation ("labile"), where both nucleation and growth occur; and lower supersaturation ("metastable"), where only growth is supported. Both the supersaturation level and the velocity at which

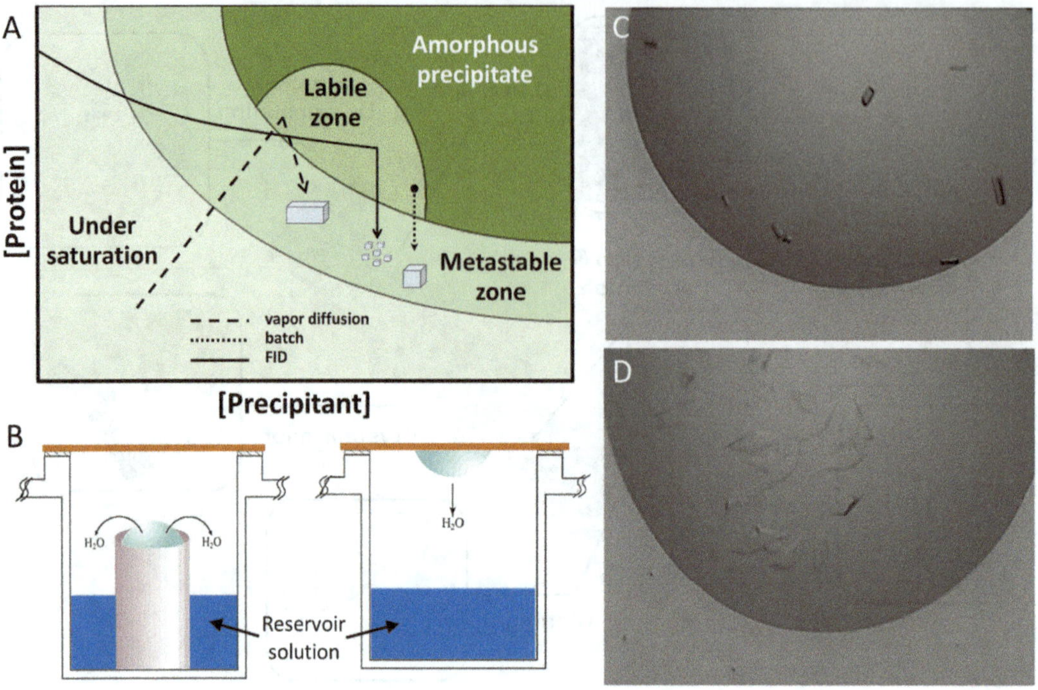

Fig. 3 The principle of protein crystallization. (**a**) Schematic of a phase diagram representing the different routes of reaching nucleation and metastable zones for the three main crystallization techniques. The diagram shows the stable states of the protein (liquid, crystalline, and precipitate) with the concentration of the protein on one axis and the concentration of the precipitant on the other. The supersaturated zone can be divided into three regions: very high supersaturation (precipitation), where molecules form amorphous aggregates; intermediate supersaturation (labile), where both growth and nucleation occur; and lower supersaturation (metastable), where only growth is supported. Here, the diagram is used to illustrate the kinetic pathway for different techniques of crystallization: vapor diffusion (dashed line), batch (dotted line), and free-interface diffusion (FID) (solid line). (**b**) Classical sitting-drop (left) and hanging-drop (right) vapor diffusion experiment setups. Equal volumes of precipitant and protein are present in the drop. Water will diffuse out and both the precipitant and protein concentration will be doubled until equilibrium is achieved between the drop and the reservoir solution. (**c**) and (**d**) Crystals of M-T7 protein obtained from initial crystallization screening using the sitting-drop method

supersaturation is achieved will determine the number, size, and order of the generated crystal(s). In an ideal crystallization process, protein concentration will be in the labile region and nuclei will form. After a few nuclei form, the protein concentration, although still supersaturated, gradually decreases, driving the system into the metastable region. As only crystal growth is now supported and not nucleation, proteins join the growing crystals until the solution is depleted of protein.

There are various techniques that can be used to drive a system into the supersaturated region, but commonly there are three main methods of crystallization: batch, vapor diffusion, and liquid-liquid diffusion [16]. They all have been extensively discussed in other publications and a vast amount of information can be found regarding the different setups and modifications to exert better control of the systems [17–20]. We focus here on the vapor diffusion method as it is the most common. Note, however, that each technique exploits the phase diagram differently to influence nucleation and concomitant crystal growth, though the final composition of the system may be the same (*see* Fig. 3a).

3.1.2 Crystallizing M-T7 Protein for the First Time: Initial Crystallization Screening

Identifying crystallization conditions for a new protein (e.g., M-T7) can be daunting; there are typically no obvious correlations between crystallization conditions and protein structure, nor are there any set rules or "magic bullets" that can guarantee the production of good crystals. For example, in most cases, the only information available for M-T7 is its amino acid sequence. In Subheading 2, we discussed the value of sequence-based homology models, but the sequence may also be exploited to assess the results obtained from the crystallization screening [21] or to select an initial crystallization condition based on sequence similarity to other proteins that have been crystallized [21]. It is therefore useful to identify physical properties of the protein in several Web-based servers to obtain critical information such as the isoelectric point, identification of flexible regions, known cofactors, and ligands that may stabilize the protein in solution. It is also possible to predict the propensity of a protein to be crystallizable based on just the amino acid sequence through several algorithms [22, 23], or to gather information from previously crystallized homologous proteins from the PDB [24].

Because there is no previous literature on M-T7 crystallization, the first step is to set up screening trials, exposing the protein to a large variety of crystallization reagents in order to identify "hits" or "leads" that point to conditions that may be conducive to crystallization. The most popular screens are sparse-matrix screens [25] available from various vendors. There are several hundred commercially available sparse-matrix screens (commonly known as crystallization kits) accounting for approximately 16,000 different formulations from numerous companies including Hampton Research, Molecular Dimensions, Jena Bioscience, Rigaku, and Qiagen. Such screens contain a number of formulations (precipitants, salts, and buffer solutions at certain pH values), based on conditions that have been known to produce crystals. Because there is substantial redundancy of conditions between and within the screens offered by different companies, it is recommended to have a full set of "kits" that cover a large range of conditions from at least one of these companies. Alternatively, one can use the screening kits with the most successful commercial conditions identified in a recent study by Fazio and co-workers [26].

Most laboratories with high-throughput capabilities tend to perform over 1,000 screens per protein. Robots make the initial crystallization screening less labor intensive and decrease sample consumption by automating microfluidic systems for many stages of the procedure: mixing the stock ingredients into a trial set of conditions, dispensing protein plus precipitant mixture at sub-μL volumes into plates, monitoring crystal growth, and scoring the outcomes with the use of image recognition software. Individual laboratories with comparatively modest resources are generally satisfied with 300–500 screening conditions per protein.

Initial crystallization screening can be performed by using any of the above mentioned crystallization techniques. However, vapor diffusion is by far the most routinely used method in most laboratories for screening purposes; one of the two configurations are possible: the sitting-drop (using either 96- or 24-well crystallization plates) and hanging-drop (using 24-well crystallization plates, Fig. 3b). This methodology creates a sealed environment in which a drop containing both the protein and precipitant solutions is equilibrated against a reservoir of higher concentration precipitant solution at a greater volume. The reservoir draws water from the protein/precipitant drop through vapor diffusion, slowly increasing the protein/precipitant solution concentration that allows the formation of the supersaturated region creating an environment ideal for nucleation of the protein (*see* Fig. 3b).

As an example, for the initial crystallization screening of M-T7, highly concentrated protein (typically of the order of 10 mg/mL) can be introduced into a high-throughput pipeline based on sub-μL scale sitting-drop vapor diffusion experiments using a fully automated robotic system.

The screening procedure purpose is to identify a starting point (initial condition(s)) for further crystal optimization. The following four outcomes are typical in such a screening experiment: (1) clear drops; (2) drops containing amorphous light or heavy precipitate; (3) drops containing phase separation (appearing oily); and (4) drops containing crystal-like material or even well-formed crystals. Figure 3c illustrates some of the outcomes identified during crystallization screening of M-T7. While the ideal result is to observe crystals, sometimes beautiful-looking crystals may diffract poorly, not diffract at all, or are discovered to be crystals of salt rather than protein. For initial screening of M-T7, only low-resolution diffraction is currently observed.

3.1.3 Improving the Diffraction Quality of M-T7 Crystals

Crystals obtained from an initial screening matrix are usually unsuitable for X-ray data collection because of insufficient size, because they grow as inseparable clusters of multiple crystals, or because they exhibit obvious defects such as cracks and fissures. Those crystals cannot provide the high-quality data that assure an accurate and precisely determined structure. Rarely, large, good-looking crystals, like those observed for M-T7, are even obtained from initial screening but very often produce low-quality diffraction data. Either way, at this stage it is necessary to improve crystal order to obtain crystals of sufficient quality for X-ray data collection.

Protein crystal optimization has been extensively described in numerous books and articles [16, 17, 19, 27–30]. The most common optimization strategy is to perform sequential experiments with incremental changes in the chemical and physical parameters around the initial conditions that have given a "hit"—varying pH, precipitant type and concentration, ionic strength, temperature, sample volume, protein concentration, etc. [29]. Although this optimization strategy has been successfully used in many cases, there is no guarantee that one will arrive at the conditions that produce high-resolution diffraction. In fact, attempts to optimize initial M-T7crystals by changing the pH and precipitant type as well as precipitant concentration have not yet produced better crystals. Other optimization approaches exist and should be considered to optimize M-T7crystals.

Another common optimization strategy is the addition of small molecules (sometimes not that small) to the precipitating cocktail. They are commonly known as additives and can play a crucial role in specific cases. When the goal of the optimization process is to improve the diffraction quality of initial crystals, as intended for M-T7crystals, the most common additives are those that either interact with the protein with consequent favorable changes in its physicochemical properties or conformation of the protein or make interactions between the macromolecules to guide the formation of the crystal lattice [16, 29, 31]. A careful and comprehensive guide of the use of additives and their effects on crystallization has been gathered recently by McPherson and Gavira [16].

Other post-crystallization methods can be used to improve X-ray diffraction of previously grown crystals. These most commonly include soaking, cross-linking, annealing, and controlled dehydration [32–36]. Soaking of protein crystals with inhibitors or ligands can be used to stabilize protein conformation and increase the crystal order. Diffusion of additives such as heavy metals or divalent metals into pre-grown crystals can also be tested to verify if they are able to reinforce protein-protein interfaces. Cross-linking to form intermolecular covalent bonds is another possible way to obtain high-resolution diffraction data. Cross-linkers, such as glutaraldehyde, have been used several times to guarantee the structural rigidity of the protein by reacting with primary amine groups (e.g., ε-amino groups of lysine residues).

The discovery in recent years that diffraction properties of flash-cooled protein crystals may improve when they are brought to room temperature and then re-flash-cooled has had a high impact on protein crystallography [33]. Lastly, crystal dehydration procedures may produce very interesting results, especially when the decrease of crystal solvent content is performed in a systematic and controlled manner. This procedure induces an internal rearrangement of the crystal that may result in improved diffraction quality [35].

When all of the above methods and approaches fail, there are two more alternatives one can attempt to improve the quality of M-T7 crystals. The first is to consider the utilization of a different crystallization method such as counterdiffusion. This method may enhance crystal quality, as it has a different effect on nucleation and crystal growth compared to vapor diffusion. Typically, crystals grown by counterdiffusion present a higher quality and are larger than those grown by vapor diffusion [13, 37, 38]. Second is the use of protein engineering strategies to optimize the probability of crystallization by protein modifications. This approach has been well described recently in a review by McPherson and Cudney [29].

In essence, crystallization is simple in principle, but optimization may be a difficult, labor-intensive process requiring plenty of patience from the experimentalist as well as a lot of protein.

3.1.4 A Practical Guide to Micro-Seeding Technique for Improving Crystal Quality

When obtaining large, well-diffracting crystals is unsuccessful, seeding is a powerful and versatile method for optimizing crystal growth conditions. Although it is not the universal solution to all optimization problems, seeding is relatively cheap, and the experiments are very straightforward to set up (see steps below). There are two methods of seeding: micro-seeding and macro-seeding. The differences between macro-seeding and micro-seeding are based on the size of the transferred seeds. In the micro-seeding technique, crystals (submicrcoscopic size) that are too small to be distinguished are used as seeds. In the macro-seeding technique, larger crystals (5–50 μm) are used as seeds. Micro-seeding is an easy method and therefore the type of seeding to try first. Macro-seeding is a much more labor-intensive method requiring tedious transfers of the parent crystal through multiple washes. Thus, micro-seeding is probably the most widely extended and commonly used of the two. In micro-seeding, a seed (a submicroscopic crystal) is transferred from one condition, where the level of supersaturation is high, to a similar condition at a lower level of supersaturation in which either the protein or the precipitant concentration is lowered in a crystallization setup. In micro-seeding, small seeds obtained by crushing large crystals or those usually present in a large number are introduced into a fresh drop of preequilibrated protein. Seeds will usually grow in conditions where nucleation will not occur. All the steps involved in micro-seeding are described below.

1. *Establishing the metastable zone for proper growth conditions*: To prevent unwanted nucleation, seeds need to be added to drops with the protein and precipitant concentration well below their supersaturation points. Thus, it is crucial that the new drops where the seeds will be added are in the supersaturated metastable zone of the solubility-phase diagram (*see* Fig. 3), in which spontaneous nucleation cannot occur, but crystal growth from seeds can occur. As well, adding seeds to an undersaturated drop will result in dissolving the crystals. For seeding to succeed, it is critical to identify the metastable zone. In other words, the optimal concentrations of protein and precipitant solution must be appropriate. Unfortunately, the user must determine these values empirically.
2. *Preparing fresh drops for seeding*: When proper growth conditions are established (**step 1**), several drops with precipitating agent of increasing concentration are set up and preequilibrated in advance, typically overnight.
3. *Preparing the seed stock*:
   (a) Pipette 5–100 μL of the precipitant solution in which parent crystals have grown into a 1.5 mL Eppendorf tube and add some seed bead (Hampton Research, Inc.) using a spatula. The precipitant solution will prevent seeds from dissolving.

   (b) Place the crystallization plate (e.g., 24-well plate) under a stereomicroscope equipped with a light polarizer. Identify the drop well containing the crystals (parent crystals) that will be used for creating the seeds.

   (c) Open the cover slide using forceps, add 5–10 μL of the precipitant solution from the reservoir, add it to the drop, then homogenize the drop several times using a pipette to finally aspirate all crystals, and transfer them to the Eppendorf tube containing the seed beads.

   (d) Parent crystals, either a large crystal or a cluster of crystals, are crushed into crystalline particles (microscopic seeds) by vortexing (we suggest ~3 s of vortexing and 10 s in an ice bath for ten rounds). Alternatively, gentle sonication for about 3 min can also be used.

   (e) If microcrystals are initially obtained, they can be used directly as seeds and **step "d"** is unnecessary.

   (f) Because the biggest disadvantage to micro-seeding is that it is difficult to control the number of seeds that are transferred, preparing a dilution series of the seed stock is highly recommended. The dilution with the optimal number of seeds has to be determined experimentally. The goal is to find a dilution that will provide just a few large crystals per drop.

4. *Transferring the seeds to the drops*: Streak seeding is probably the most common method for transferring seeds. It usually employs some kind of animal hair or whisker as the transfer tool. The surface of the seed stock in a droplet is stroked with the hair to pick up the microcrystals. Alternatively, the whisker can be dipped into a tube containing the slurry of microcrystals. The whisker containing the microcrystals or seeds adhered to it is then streaked or dipped into the preequilibrated drops whereby the seeds are transferred. The new crystals will grow along the streak line of deposited seeds after several hours or days.

3.2 X-Ray Diffraction, Data Collection and Processing

3.2.1 Introduction to X-Ray Diffraction

The discovery of X-rays in 1895 by Wilhelm Conrad Röentgen catalyzed a revolutionary change in our understanding of the physical world. The ability of X-rays to reveal the invisible has made them one of the most important research and diagnostic tools in medicine, chemistry, and physics in the last century. One of the most powerful sources of X-rays in modern research is synchrotrons, in which electrons are accelerated by radiofrequency cavities to extremely high energies (typically $\geq$3 GeV) and transferred to a storage ring in which they can be maintained at high current in stable "bunches" for hours. The electrons are then subsequently perturbed in an undulating electromagnetic field to induce X-ray emission. The first synchrotron facility was built in the early 1970s, and due to the increasing demand from both the scientific community and the private research sector, the number of synchrotron facilities has grown quickly since then. There are now over 100 beamlines dedicated to macromolecular crystallography at more than 20 synchrotrons worldwide.

3.2.2 Traditional X-Ray Crystallography

X-ray methods have played an increasingly important role in the life sciences as evidenced by medical X-ray imaging, X-ray scattering, and, highlighted here, X-ray crystallography. The latter typically makes use of X-rays from 3.5–20 keV (3.5–0.6 Å) to enable the determination of the atomic structure of matter. The most challenging biomolecules, such as large complexes and membrane proteins, are almost exclusively solved with synchrotron radiation (*see* Fig. 4a–c). In a traditional X-ray data collection experiment (*see* Fig. 4d), a single crystal of a macromolecule placed in a loop is mounted on a goniometer, a device used to position the crystal at selected orientations. The crystal is irradiated with a finely focused monochromatic beam of X-rays, a diffraction pattern is recorded, and the crystal is rotated at a well-defined angle. This process continues iteratively, collecting individual two-dimensional

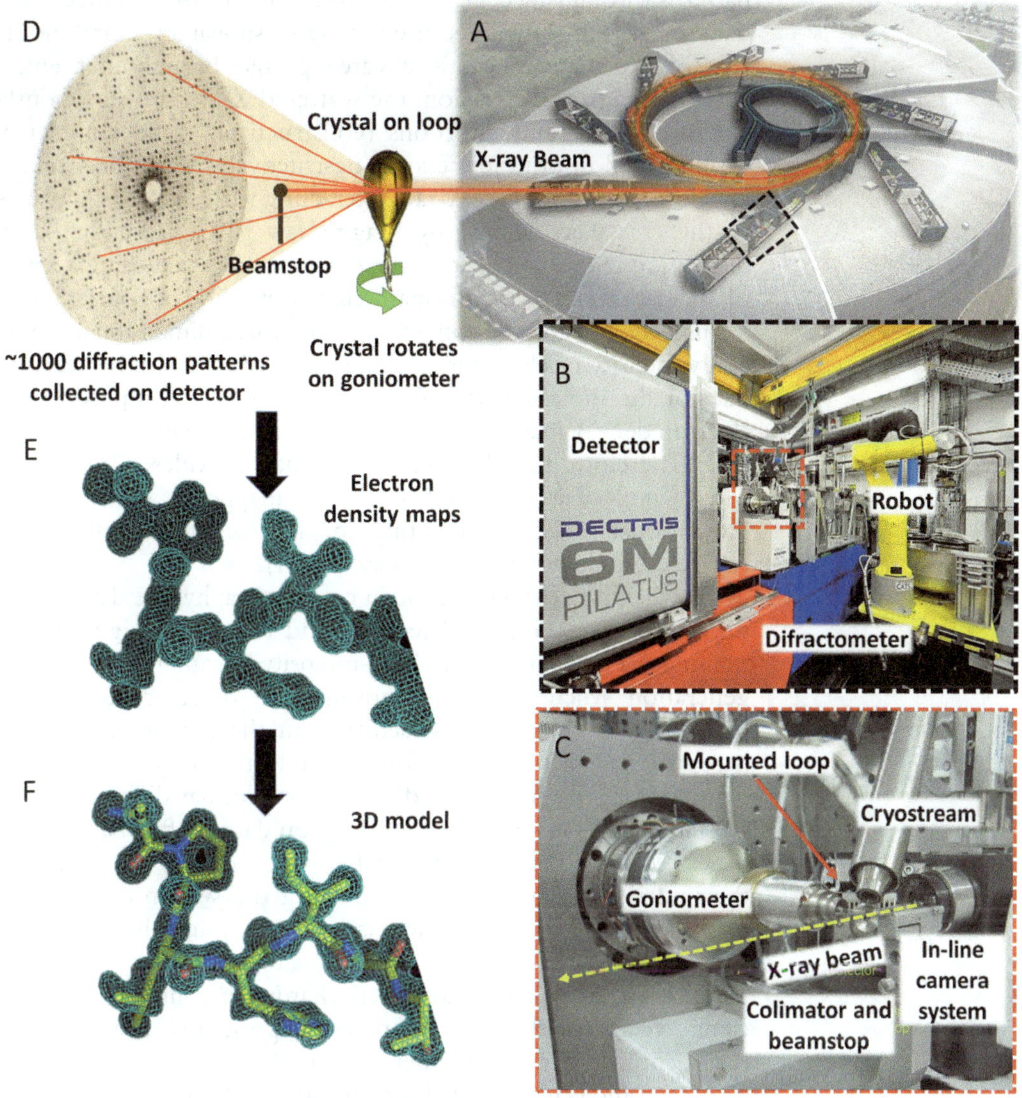

Fig. 4 X-ray protein crystallography. (**a**) Zenith view of ALBA synchrotron radiation source. The red solid line represents the X-rays. (**b**) Overview of the experimental station of XALOC beamline at ALBA highlighted by the black dashed line in (**a**). The detector and the robot for mounting crystals are shown. The XALOC experimental station is built around a diffractometer, also shown. (**c**) Closer view of the experimental station highlighted by the red dashed line in (**b**). The X-ray beam direction is indicated by the yellow dashed arrow. All device components are also shown. (**d**) At the beamline, a single crystal is placed in the way of an intense beam of monochromatic X-rays producing the regular diffraction pattern of reflections. As the crystal is gradually rotated, intensity of every spot is recorded at every orientation of the crystal. A typical full data set from a single crystal is composed of about 1000 patterns typically containing tens of thousands of reflections. (**e**) and (**f**) Data sets are processed computationally with complementary chemical information to produce the electron density maps (**e**) that will result into a model (the crystal structure) of the arrangement of atoms within the crystal after refinement steps (**f**)

patterns through a wedge of space resulting in a three-dimensional data set (*see* Fig. 4e, f). Because the highly similar structural motifs that form the individual unit cells are repeated throughout its entire volume in a periodic fashion, the scattered X-rays are enormously enhanced in certain directions (constructively), and completely extinguished in other directions (destructively). Individual reflections are the result of the X-rays interacting with electrons in planes of the crystal as defined by Bragg's law [39]. Once the X-ray diffraction data are collected, the data are processed using a series of calculations involving the mathematical method of Fourier transforms [40, 41] and are converted into a three-dimensional model of the density of electrons within a unit cell, the electron density map of the molecule of interest (*see* Fig. 4e). By inspecting the electron density map, researchers can model in the molecular structure. Therefore, the electron density map provides a framework where molecular structure can be "mapped" and a three-dimensional model representing the macromolecule of interest can be created and interpreted (*see* Fig. 4f).

In the first decades of protein crystallography, the data collection process was long and tedious, and required a high level of training. The advent of high-intensity beamlines at third-generation synchrotrons, the advances in detector technology, and the enormous progress achieved in the last decades in the hardware and software used for data collection have resulted in a dramatic increase in the speed of macromolecular diffraction data collection [42–46]. Currently, diffraction data is often successfully measured and processed by researchers who lack deep knowledge of the underlying principles by conducting the synchrotron experiments remotely from their own laboratories using their own laptops. Nevertheless, despite the availability of very powerful radiation sources, highly automatic hardware controls, very efficient detectors, and intelligent processing programs, data collection is a scientific process, not a mere technicality.

Macromolecular crystallography at synchrotron sources has proven to be the most influential method within structural biology, producing more than 148,000 structures to date.

3.2.3 X-Ray Free Electron Lasers (XFEL)

While its utility has been instrumental in advancing our knowledge of molecular structures, X-ray crystallography at synchrotrons suffers from several limitations, namely i) the need for large, well-diffracting crystals and ii) the sample suffers radiation damage that breaks bonds and obscures native structural determination. These limitations have been overcome by a new way of generating X-rays, the X-ray free electron lasers (XFELs), which have improved upon many of the properties of synchrotron radiation sources, in some cases by orders of magnitude. An XFEL instrument can be considered as a "cross" between optical lasers (light amplification by stimulated emission of radiation) and a synchrotron source. In contrast to classical synchrotron storage ring, in XFELs, a relativistic electron beam is accelerated to almost the speed of light in a linear accelerator prior to interaction with the undulator. Upon interaction with the undulator, the electrons move in curved paths via the magnets. The long length of an XFEL undulator allows the relativistic electrons to interact with their emitted radiation which causes the electrons to bunch with spacing equal to the wavelength of the emitted X-rays. As the electrons bunch progresses, their emission becomes more coherent and allows for a stronger interaction between the two. This results in a beam with highly coherent pulses with femtosecond duration. The coherence causes radiative emission proportional to the number of electrons squared in contrast to a synchrotron (incoherent), in which the radiative emission scales proportionally to the number of electrons due to cancelation effects from the out-of-phase-generated electromagnetic fields [47]. Because the number of electrons in a bunch is on the order of a billion, there is a massive increase in the number of photons per second (also known as flux) compared to what can be achieved even at the best third-generation synchrotron sources. Furthermore, the high longitudinal coherence (and thus ultrashort duration) of the pulses pushes the peak brilliance orders of magnitude higher than that achievable at a synchrotron.

The first XFEL that began operation was the Linac Coherent Light Source (LCLS) at Stanford in 2009. Four more instruments followed: the SPring-8 Angstrom Compact Free-Electron Laser (SACLA) in Harima, Japan; the European-XFEL (EuXFEL) in Hamburg, Germany; the Pohang Accelerator Laboratory-XFEL (PAL-XFEL) in Pohang, South Korea; and the Swiss-Free Electron Laser (SwissFEL), at Paul Scherrer Institute, in Villigen, Switzer land, with a total of five currently operational worldwide. The unique properties of XFELs (extremely high brightness and ultrashort pulses) facilitated the appearance of a new way of doing X-ray crystallography, the serial femtosecond crystallography (SFX) technique. In a typical SFX experiment, thousands of nano/micrometer-size crystals are delivered to the FEL X-ray beam in a serial fashion at room temperature, resulting in data sets comprised of individual snapshots, each capturing Bragg diffraction of single crystals prior to their subsequent destruction [48]. By employing strategies to initiate enzymatic activity and collecting diffraction data thereafter, SFX enables structure determination beyond just static structures at XFELs, allowing time-resolved studies to understand conformational dynamics, enzymatic reactions, and resolution of intermediate states in reactions over timescales of 100 fs to minutes. The first SFX experiment was conducted at the LCLS in 2009 using nanocrystals of photosystem I enzyme [49]. Since then, over 200 crystallographic structures determined by SFX have been deposited in the PDB. The small crystals used in SFX—those in the nano- or micrometer range—are also ideal for time-resolved studies, as a greater percentage of molecules in the crystals can be activated homogeneously by light triggering or by rapid mixing. Thus, SFX has enabled (1) rapid diffusive saturation in mix-and-inject experiments of biochemical reactions [50–53] and (2) full optical saturation of the sample by a pump laser in studies of light-driven reactions [54–64]. It is conceivable that a method of creating small crystals of M-T7 in high quantities could be used in XFEL studies in the future.

3.3 X-Ray Data Collection and Data Processing

3.3.1 X-Ray Data Collection

When obtaining crystals of suitable size (greater than 100 μm) of the protein of interest (e.g., crystals of M-T7), the next step is to check their X-ray diffraction quality. Diffraction trials can be performed on an "in-house" X-ray source or at synchrotron radiation sources. While the former are relatively weak X-ray sources suited for an estimate of diffraction quality, the latter have the advantage of more intense X-ray beams and higher quality optics, which allows much shorter exposure times and a higher signal-to-noise ratio of the diffraction pattern, leading to higher resolution diffraction. At synchrotrons, one also has the choice of measuring the crystals mounted in capillaries at room temperature or mounted in cryo-loops in a stream of liquid nitrogen at 100 K. Complete data sets are often collected with the protein crystal in a stream of cold nitrogen gas at a temperature in the range of 100–120 K. The main reason for cooling protein crystals is that at room temperature they are more susceptible to radiation damage from X-rays, which degrades the crystal, causing lower quality diffraction. Cooling the crystals slows the destructive process significantly. The method and its development are extensively discussed by Garman and co-workers [65–68].

X-ray diffraction images are, nowadays, collected on CCD detectors in "in-house" X-ray sources or using pixel array detectors (PADs) (e.g., PILATUS and EIGER) at synchrotron radiation sources. Advances in detector technology (improved signal-to-noise features and faster readout times) have significantly improved data collection at synchrotron sources in the past two decades since the first photographic films (1980s) and image plates (1990s) were developed [69, 70]. PAD detectors at synchrotron beamlines offer the further advantage of readout times in the order of milliseconds (or even faster) rather than seconds (maximum readout of CCD detectors), which can appreciably decrease the time for collection of a complete data set, especially when one considers that the exposure times at many synchrotron beamlines are less than 1 min. Before the first diffraction pattern is taken, the distance from the crystal to the detector is calculated and adjusted to allow for collection of diffracted spots, usually up to a maximum of 1.5–3.0 Å resolution. The resolution of spots collected on the detector increases as the diffracted angle increases. Hence, the highest resolution will be at the edge of the detector, and if one determines the diffracted angle required, the distance of the detector from the crystal can be adjusted accordingly.

After the first pattern is collected, one must confirm that the diffraction extends to sufficient resolution to make structure determination to near-atomic detail feasible. Essentially, this involves being able to visually detect well-ordered arrays of spots to the edge of the diffraction image. In general, spots close to or beyond 3 Å are desired and sufficient to be able to detect the amino acid side chains in the electron density map. Reflections become weaker at high resolution, limited by the internal order of the individual molecular subunits in the crystal lattice. By collecting a second diffraction pattern, one can determine the unit cell dimensions (given as three lengths: a, b, and c; and three angles: α, β, γ), the crystal system, and the space group.

Once the crystal quality is confirmed, a full data set may be collected. The number of diffraction patterns required to build a complete data set, as well as the strategy for data collection that results in the determination of a crystal structure will depend on several aspects: (1) The crystallographic symmetry present in the crystal and space group: For crystals with high symmetry (e.g., a cubic crystal), one needs only to collect diffraction data through as little as 35°. Conversely, in a lower symmetry crystal system, such as monoclinic, data might need to be collected through 180°. (2) The non-crystallographic symmetry (NCS): the number of equivalent molecules in the unit cell related by symmetry operations. A particle such as a virus, composed of many identical subunits, has a high level of NCS. In such cases, high-quality structures can be obtained from data sets that are far from 100% complete because of compensation by averaging. At the other extreme, a monomeric protein might exhibit no NCS and therefore a more complete data set will be required. (3) The availability of molecular replacement: If a sufficiently similar structure has already been solved, it can be used as a starting model and any gaps in the new data set may be "filled in." If gaps still exist, it will be necessary to collect at least one more data set, this time from a heavy atom derivative or derivatives, the data sets being referred to as "native" and "derivative," respectively. (4) The upper resolution limit required: This will usually be determined by the quality of the crystal itself and in practice most data sets are collected to the upper limit of useful diffraction. However, the amount of diffraction data required increases exponentially with higher resolution.

As mentioned, the initial diffraction analysis will give an indication of the spot spacing on the detector at the desired "crystal-to-detector" distance. This can be used to determine the oscillation range, or φ range, of each exposure. The crystal is rotated on a goniometer and exposed to the X-ray beam. For large unit cells, generally upwards of 300 Å, as little as 0.25° might be possible, whereas for small unit cells, up to 2° might be collected while still avoiding spot overlap. In an ideal situation, a complete data set will be collected from a single crystal. This greatly facilitates the scaling process (see below), but in practice it is often not possible because of radiation damage to crystals. Such damage is caused by heating and generation of free radicals; hence, radiation damage can continue for some time after the X-ray exposure is complete.

3.3.2 X-Ray Data Processing

Once a full data set is collected from our macromolecule crystal, the processing of the diffraction data is carried out by well-established algorithms that are available in many software packages and program suites; these process data and calculate an electron density map for the macromolecule.

1. The first step in the processing involves the determination of the crystal system and of the unit cell dimensions as accurately as possible. In addition, at this stage we determine the orientation of the crystal in the beam. When the cell and orientation are known, indexing can be carried out [71]. In this step, each spot on the diffraction pattern is assigned an index, quoted as three integers: h, k, and l. Computer programs for autoindexing do this by calculating a prediction of what the diffraction image will look like from the cell dimensions and orientation, and then attempting to fit the real image with the predicted one. This is an interactive process, so that the user can check for accuracy based on software-reported statistics, which helps to avoid possible errors as a result of mis-indexing.
2. The next stage of the data processing is the measurement of the spot intensities. Protein crystals diffract weakly because they are composed mainly of light atoms and they have large unit cells. Assuming that the order of the crystal is consistent, the larger the volume of the protein crystal, the stronger its diffraction. Intensities of diffracted spots vary as a result of both the amplitude of the diffracted waves and their phase relation. These factors cannot be deconvoluted at this stage; accordingly, the accuracy of the measurement of the intensities is of paramount importance. There are various programs to perform autoindexing and intensity measurements including XDS [72], MOSFL M [73, 74], HKL3000 [75], and DENZO [76]. A scale factor must be allocated so that the intensities of all the images in the data set can be related. The first image is usually allocated a scale factor of 1 and all the subsequent images will be scaled to this.
3. If data have been collected from more than one crystal, data subsets from each crystal will have to be scaled separately and then these batches scaled together in blocks, the best batch

then being given a scale factor of 1. Careful monitoring of the scaling statistics allows spots, or even whole images, to be rejected or reprocessed to preserve the quality of the overall data set. AIMLESS [77], SCALA [78], and SCALEPACK [76] are widely used programs for this step. The output from scaling is a computer file that contains the index of each spot and its measured intensity. This file must be sorted so that the spots are listed in numerical order according to index. Numerous programs exist for this purpose, a commonly used one being SortMTZ of the CCP4 suite [79].

Following scaling, the next step is to determine the amplitude of the reflections. The intensity of any diffracted spot is the result of the diffracted wave incident on the detector at that point. Therefore, the intensity will be determined by squaring the amplitude of those waves and by the phase difference between them, expressed as an angle. A phase difference of zero results in constructive interference, whereas a phase difference of 180° results in completely destructive interference. The determination of the amplitudes is mathematically simpler than that of the phases. Various computer programs exist to calculate amplitudes, for example Truncate of the CCP4 suite [79]. Usually, these take the square root of the intensity, with negative values being set to zero, and outputting amplitudes accordingly.

4 A Practical Guide for Preparation for an X-Ray Diffraction Experiment

The most common method for X-ray data collection is the goniometer method (or rotation method) at cryogenic temperatures. An important aspect to consider when working with proteincrystals is that the molecules within a crystal lattice are loosely packed with large solvent-filled holes and channels, which normally occupy 40–60% of the crystal volume, sometimes even higher. This is an advantage for the reaction of the protein with small reagent molecules as they can diffuse through these channels and reach reactive sites on all protein molecules in the crystal. However, the high solvent content usually causes problems when handling the crystals because loss of solvent destabilizes the crystals making them fragile. Therefore, protein crystals should always be kept in their mother liquor (i.e., the well contents referred to in our description of vapor diffusion crystallization), in the saturated vapor of their mother liquor, or at a sufficiently low temperature to prevent evaporation of the solvent. Below we carefully describe the steps from crystal harvesting, cryo-cooling, and crystal mounting to data collection and data processing using the rotation method of a hypothetical single M-T7 crystal.

1. *Before you start*: The assembly of a proper and extensive kit for cryo-crystallography is required to successfully flash-cool the crystals. Therefore, before harvesting the crystals, one should gather all necessary materials (e.g., pins, vials, tongs, racks, desktop dewars, covers for dewars, magnets, wands, canes, pucks, and dewars for storage and shipping) as well as prepare all solutions (e.g., cryoprotectant solution) to reduce the time between removing crystals from their mother liquor and flash-cooling. It is good practice to limit the exposure of crystals to air, avoiding their damage and degradation during the cryo-cooling process.

   For X-ray measurements under cryogenic conditions, a cryo-loop is used to scoop the crystals up from the crystallization droplet; the loop is typically pre-mounted on a magnetic base that, later on, can be easily placed on a goniometer. It is important to note that cooling the crystals from room temperature to cryogenic temperature must occur suddenly because the water in the mother liquor and in the crystal must freeze to a vitreous structure. The water should not crystallize because ice formation damages the protein crystal structure by water expansion in its transition to crystalline ice. To prevent ice formation in and around the crystal, it is carefully transferred to a solution that serves as a cryoprotectant at concentrations ranging from 15% to 30%. Glycerol, 2-methyl-2,4-pentanediol (MPD), ethylene glycol, sugars, and low-molecular-weight polyethylene glycols (PEGs) are the most popular cryoprotectants. A high salt concentration in the original mother liquor can cause problems because of salt precipitation in the cryoprotectant-containing solution causing crystal cracking. In such cases, the salt must then be replaced by a more soluble salt or by additional cryoprotectant. An alternative to mixing in a cryoprotectant is to simply coat the crystal with an oil. Excess water is removed from the crystal surface while in the oil. Many different oils and oil mixtures have been reported to work [80].

2. *Crystal harvesting:* Crystal harvesting must be done using a stereomicroscope preferably equipped with a polarizer and variable zoom for rapid identification of the crystals. Get the cryoprotectant solution (a few microliters) ready by mixing the precipitant solution with a cryoprotectant. Place the crystallization plate (e.g., 24-well plate) on the microscope, focus the microscope on a well containing crystals, and adjust the polarizer so that the birefringent crystal has a good contrast with the background. Open an individual well by removing the coverslip and place it under the microscope. It is recommendable to inject 1–2 μL of the precipitant solution into the well to reduce dehydration during crystal harvesting. Identify the crystal and pick it up from the droplet by scooping it using a mounted

cryo-loop with the inner diameter matching the size of the crystal.

3. *Cryo-cooling*: Once the crystal is in the cryo-loop, it can be carefully submerged into the droplet containing the cryoprotectant solution that was already prepared in **step 2**. Another easy transfer technique is to simply swish the crystal through the cryoprotectant solution. In either case, the crystal must spend enough time in the cryoprotectant solution to replace the liquid it was suspended in, typically a few seconds.
4. *Crystal storage*: The cryoprotected crystal is then mounted directly on a goniometer for data collection on an X-ray home source. However, the most common practice nowadays is to screen or measure crystals at synchrotrons. For that, the looped crystals are plugged into liquid nitrogen and then loaded into a cryo-vial or "puck" holder that is submerged in liquid nitrogen for longer term storage of the crystals. Pucks containing flash-cooled crystals are typically stored in a cryogenic storage dewar. This is often the preferred device for shipping crystals to synchrotron beamlines for data collection. For a more comprehensive practical guide on techniques and tools for cooling and mounting proteincrystals at cryogenic temperatures *see* [81].
5. *Data collection:* Once the crystal is mounted and aligned within the direct path of the X-ray beam, the detector distance is set to provide the desired resolution limit and two diffraction images are then collected with a conservative exposure time and an oscillation angle determined by the unit cell size (if known) or a small rotation (e.g., $0.25 \pm 0.5°$). The crystal is rotated by 90° between the two images to provide two orthogonal samples of the reciprocal lattice. Figure 5 illustrates the two diffraction patterns collected at 0° and 90° from an M-T7 crystal. These initial images are then examined carefully to determine the effective resolution limit; to check for the presence of disorder, twinning, or multiple crystals; and to obtain an approximate estimate of the mosaic spread. If the quality of diffraction is acceptable, the images are autoindexed to determine the cell dimensions and possible space groups. Finally, the data collection strategy [total rotation angle and oscillation angle(s)] and exposure time are chosen. If necessary, the detector distance is reset to provide the required resolution limit and data collection is started. Data collection parameters are typically in the order of 0.2–1-s exposure per image, 200–400 mm detector distance, and 0.1–1° oscillation angle.
6. *Data processing*: The set of diffraction patterns that contain all the X-ray reflections is next analyzed. The intensity and position of each reflection are measured, and the unit cell

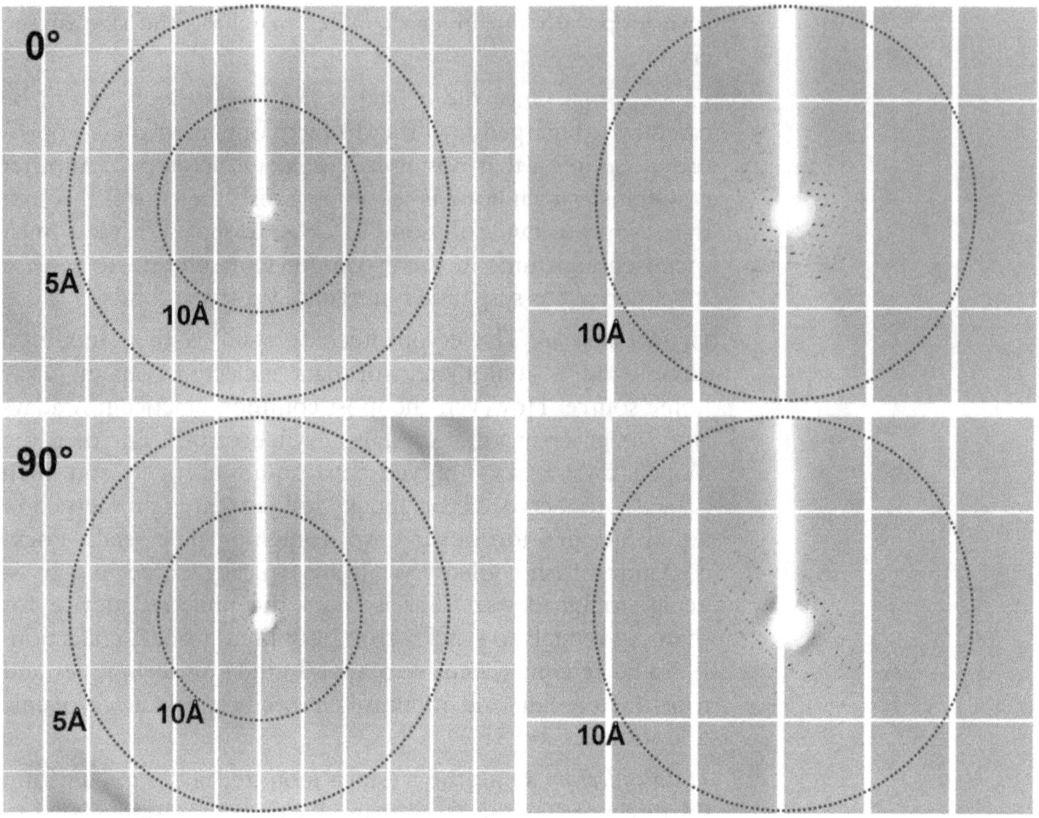

Fig. 5 X-ray diffraction patterns collected on a PILATUS detector at 0° and 90° from an M-T7 crystal. Bragg reflections are shown as black sharp spots. Right panels show a zoom in of the diffraction shown in the left panels. Resolution rings at 5 and 10 Å are also shown. The experiment was conducted at the microfocus beamline at GM/CA sector 23-ID-D at the Advanced Photon Source

dimensions and basic symmetry operation are deduced in a process known as "indexing." Indexed images are then integrated and merged into a single output file that contains the amplitudes of the diffracted waves, positions of individual diffracting planes in the form of *hkl* integers, and experimental error. Finally, a thorough statistical analysis reduces all the data into a single file (an MTZ file) containing the scaled reflection intensity data that is used for determining the electron density map and constructing a three-dimensional model of the protein of interest.

5 Model Building, Refinement, and Validation

Recent developments in X-ray data collection protocols, low-noise detectors, sample-handling robotics, synchrotron beamlines, and associated software also significantly contributed to progress in

X-ray crystallography. However, perhaps the most dramatic changes that have led to faster structure determination are in structure solution, refinement, and validation software. In this section a brief overview of the concepts on model building, refinement of the electron density maps, and structure validation will be given.

5.1 Model Building and Refinement

From an initial, non-refined model of a protein structure (for instance, obtained with isomorphous replacement or molecular replacement), important biological information can already be derived. However, for more reliable information and finer details, the structure must be refined. Refinement is the process of adjusting the model to find a closer agreement between the calculated and observed structure factors. The refinement process typically involves manual corrections of the model to improve its agreement with the electron density maps by adding/subtracting contents from the 3D model (e.g., ligands, incorrectly assigned residues, post-translationally modified residues, and solvent molecules), as well as subjecting the maps to alternating rounds of automated optimization of all refined parameters (e.g., according to least-squares or maximum-likelihood algorithms).

When an electron density map (*see* Fig. 4e) appears by inspection to have sufficient quality to permit reliable location of amino acids then model building can commence. In the model building process, we are checking (1) that the density can be traced almost continuously from one end of the protein molecule to the other. Small breaks in this continuity are often encountered due to localized protein regions of disorder or high flexibility; but this should not be a problem for model building, as long as there is no ambiguity in the route density that the protein chain takes; (2) that the quality and resolution permit reasonably reliable identification of the classes of amino acid side chain; and (3) that there are no errors in the sequence and that all residues match the electron density. Model building then requires the use of a computer graphics program to display the maps and, with the aid of the protein sequence, the insertion of each residue. COOT [82] is by far the most used and best program for model building.

The model parameters that are optimized by refinement software include, for each atom, the spatial coordinates and the atomic displacement parameter (ADP, also referred to as the temperature factor or B-factor which reflects "mobility" of an atom). The B-factor model used is usually isotropic, at least for structures solved at a resolution of 1.4 Å or lower (i.e., it assumes that the displacements have the same amplitude in all directions). Structures refined at higher resolution often use the more correct anisotropic model of atomic displacements. Note that structures are refined according to biologically understood parameters. For example, a C-C bond is typically ~1.54 Å, the residues in an alpha-helix exhibit a full turn every ~5.4 Å, and hydrophobic residues are unlikely to be in close proximity. Such characteristics are probabilistic, not absolute. Accordingly, restraints are often used by refinement software in model refinement to favor more probable atom placement; however, higher resolution structures typically decrease their dependence upon such parameters since the density map is considered more reliable than generalized protein structure characteristics.

The widely used refinement software pipelines REFMAC [83] and Phenix.refine [84] include multiple useful and adjustable options, such as the accounting for NCS (*see* Subheading 3.3.1), adjustment of stereochemical weights, automatic detection and refinement of crystal twinning, and more. For refinement of structures at high resolution (equal or higher than 1.2 Å), the software of choice is SHELXL [85], which has evolved from the philosophy of small-molecule refinement to encompass macromolecules as well.

Electron density maps can be further improved during refinement by applying solvent flattening in which the regions containing disordered solvent (~50% of the crystal) can be represented by constant electron density, thus reducing phase errors [86, 87]. One can also use molecular averaging, also known as NCS, if more than one identical molecule or subunit is contained within the asymmetric unit [84]. The geometrical operation that connects the subunits, the NCS, may be used to average the electron density of the subunits in a series of cycles in which the phases are continually improved.

5.1.1 A Practical Guide to Phasing and Structure Determination Methods for Novel Proteins

In some cases, there is no data available other than a protein sequence and predicted secondary structure, but it is far more often the case that at least partial homology exists with other proteins with known structure. We have discussed strategies for creating a homology model prior to data collection in Subheading 2. Because refinement requires an input model, one could use a full homology model for this purpose. Be warned, however, that low-confidence features should be removed from the input model to avoid phase bias. For example, if a protein of unknown structure shares 50% sequence identity with a protein of known structure, but the homology is clustered within the C-terminal 50% of the sequence, then one should exclude the N-terminal 50% of the homology model since it exhibits very low confidence. Furthermore, the user should consider the limitations of their diffraction resolution and trimming details of the structure that cannot be observed at the acquired resolution. For example, if a data set is collected at less than ~5-Å resolution, we suggest trimming the loops, and perhaps transforming the model to a poly-Ala chain for the input model. To organize this kind of thinking, we suggest the following workflow for creation of an input model that does not result in model-biased features.

1. Perform a thorough sequence alignment analysis:
   (a) The sequence analysis should inform about not only which regions of the protein exhibit low confidence in the model, but also the confidence in cofactor coordination. We suggest annotating a text document for this purpose.
2. Remove regions of the homology model that exhibit low confidence based on the alignment.
   (a) If there is a stretch of residues that exhibits especially low sequence identity, delete it from the model. One may also delete cofactors associated with this region.
   (b) Make sure to also remove any water molecules in the structure unless there is extremely high confidence (i.e., a water molecule binds a highly conserved ligand).
3. Remove structural details that are below the resolution limitations.
   (a) If the diffraction resolution is lower than ~5 Å, we suggest removing all but well-conserved secondary structure, and consider changing every existing residue to Ala.
   (b) If the diffraction resolution is between 3.5 and 5 Å, we suggest removing ligand substituents and many amino acid side chains. Around 4 Å resolution, large side chains of Tyr, Phe, and Trp are visible and can be used later to anchor the structure, filling in the residues that were previously trimmed from the initial model.
   (c) If the diffraction resolution is between 2.7 and 3.5 Å, we suggest maintaining much of the homology model, but removing only the most flexible regions.
   (d) If the diffraction resolution is higher than ~2.7 Å, using a full homology model including waters if they are present should be fine. However, we reiterate that if there are especially flexible regions where the density map is poor, remove that part of the model and build it in iteratively during refinement.

After each round of refinement, aspects of the density should become clearer and can be built into by the user. The user may find that regions which were deleted should be added back to the density. Conversely, the user may discover unexpected density into which they should build the polypeptide or cofactors. Such is the essence of novel protein structure discovery.

5.1.2 Structure Validation and Analysis, Deposition, and Publication

The ability of X-ray crystallography to provide accurate information about macromolecular structures, including molecular details of the interactions between proteins and small molecules, is unsurpassed. Consequently, and as an example, crystallography is crucial for lead generation paradigms within the pharmaceutical industry as it provides a starting point for computational structure-based drug design (SBDD). SBDD requires diffraction-quality crystals of the drug target. This can be a formidable challenge in itself, but once the structure has been determined it needs to be accurately refined, because it will serve as a template for in silico SBDD. High-resolution diffraction data produce high-quality electron density maps that allow relatively easy protein model building. However, the resolution and R-factors describe the overall quality of the structure, not the quality of local regions. Even high-resolution maps can have regions that are difficult to interpret. These regions may correspond to the macromolecule itself or may be a small molecule that was introduced, intentionally or not, during protein production, crystallization, and/or crystal harvesting. An unintentionally introduced small molecule complicates map interpretation but may shed light on the location of an active site and/or other regions of the macromolecule with high propensity for ligand binding.

Before a spate of incorrect structures was identified in the early 1990s [88–93], validation software did not exist. Subsequently, tools based on model geometry analysis were created to allow the quality of the structures to be easily inspected and potentially erroneous residues identified. While lower resolution data typically holds fast to such geometric constraints, higher resolution data may accept improbable protein geometries (within reason) if it fits the model well to the density. The stringency of these constraints may be adjusted by the user as they see fit during modeling, but ultimately overall protein geometry statistics are normalized in PDB validation (*see* the next paragraph). Since there are many sources of error, either experimental or incorrect interpretation, it is essential to validate structural and chemical geometries of both proteins and any ligands or solvent molecules that have been added to the final model during refinement process. These methods help detect inconsistencies in the final model based on the information that was not used during the refinement process, pointing to regions in need of reinspection.

1. *Validation methods*: The structure validation criterion includes the assessment of various structural features including bond lengths, bond angles, planarities, chiralities, hydrogen bonding parameters, and packing as well as backbone or side-chain torsion angles. These are all represented as root-mean-square deviation (RMSD) values comparing the observed values to probable geometry, and as a frequency of outliers for both geometry and conformation. Programs such as PROCHECK [94] and WHATCHECK [89, 95–97] provided crystallographers with an array of data that could indicate the overall quality of the structure, as well as an indication of the quality of the fit and geometry of individual atoms. The

Ramachandran plot [98] is considered one of the most informative checks of the overall quality of a protein structure because the dihedral angles of a protein measured by the plot are typically not regulated during refinement. Protein structure validation was significantly advanced by MolProbity [99, 100], which supplements geometric and dihedral analysis with an all-atom contact analysis of hydrogen atoms within the protein structure, which has rapidly become an essential analysis. Indeed, the interactive molecular building program COOT [82] can now present MolProbity results and display the steric clashes instantly.

When only atomic coordinates were available for most structures deposited in the PDB, it was possible to detect the existence of some problems with the underlying diffraction data, but almost impossible to pinpoint them precisely. Now that structure-factor deposition is mandatory for new deposited structures, much more information is available, enabling the use of tools for assessing the global quality of crystal structures and the local quality of fit to the electron density, as well as allowing data quality analysis in which the presence of experimental problems or artifacts (twinning, translational non-crystallographic symmetry (tNCS), anisotropic diffraction, data incompleteness, and potential outliers) can be assessed. Assessment of structure-factor can be done in SFCH ECK [101] and phenix.xtriage [84].

The Worldwide PDB (wwPDB) organization, working closely with the structural biology community (including journals and funding agencies), has recently established a validation pipeline that is part of the wwPDB OneDep deposition and validation system. The wwPDB Validation Service webserver (https://validate.wwpdb.org) can be used to perform checks prior to deposition [102–105].

2. *Structural analysis:* Although the structure of M-T7 is currently unknown, structural analysis will be needed once the model of M-T7 has been refined and one is confident in its quality and chemical correctness. An initial question will follow: *Is the biological unit seen within the crystalstructure?* This need not be within the asymmetric unit, since oligomerization may be arranged about a crystallographic symmetry axis. Moreover, the structure should explain the protein's observed behavior and should be consistent with prior biochemical knowledge. M-T7 has the ability to bind to different classes of chemokines through its heparin (or GAG)-binding domain [14], so this protein is a potential target for drug discovery studies with potential therapeutic applications. Therefore, M-T7 can also be analyzed for potential ligand-binding sites. Once the binding sites are determined, properties like surface

shape [106], electrostatics, and atomic composition may be investigated and probed by computational or experimental approaches to identify lead compounds. However, even small errors either in the details of the original model or in its interpretation may reduce the success rate of lead identification or slow down the process of structure-based drug design. It is difficult to overstate the importance of structure quality, because a high-quality X-ray structure of a macromolecule is only a starting point.

3. *Structure deposition and publication*: Once the crystallographic structure of M-T7 is solved, it will be deposited in the PDB. This is necessary for publication as most journals require a PDB ID to be included in any manuscript under review that describes a new crystallographic structure. The wwPDB is an invaluable resource that contains over 167,000 structures (as of July 2020). As recently published by Adams and his colleagues, the wwPDB is accepting model files in mmCIF format only for deposition as of July, 2019 (doi: https://doi.org/10.1107/S2059798319004522). To generate the mmCIF file, a two-stage process is currently recommended in phenix.refine [84]: In the first step, by default, phenix.refine will output model files in mmCIF and PDB format. If you turned off the option for mmCIF output, run a final cycle of refinement to write the mmCIF files for model and data. As the second step, the mmCIF file and a sequence file can then be processed with mmtbx.prepare_pdb_deposition program to create a mmCIF file with the sequence. This program requires the full sequence of the macromolecule to be provided.

6 The Future of Structural Biology and Impacts on the Understanding of M-T7

The function of M-T7 is only partially defined and this is mainly due to the lack of structural information. X-ray crystallography is, a priori, the primary methodology for determining the three-dimensional structure of M-T7. We have provided an overview of how traditional X-ray crystallography tools can be utilized for the crystallization, and structure determination of M-T7. However, other structural biology techniques are useful as in combination with X-ray crystallography they can provide complementary structural data that is useful in understanding the function and mechanisms of action of M-T7.

6.1 X-Ray Crystallography Advances

Traditional X-ray crystallography necessitates large, well-diffracting crystals for success, which is not always achievable and is one of the major bottlenecks of this technique. However, remarkable progress in methods for recombinant protein production [107, 108] and crystallization has accelerated the number of crystal structures deposited in the PDB in the past decades. Methodologies and strategies that make protein production more cost effective and robust include ligation-independent cloning, recombinantprotein expression, and purification on metal affinity resins [109, 110]. Automated approaches to protein crystallization, both for initial crystallization screening and hit optimization, are now routinely used by structural biology laboratories. The mining of crystallization data has resulted in the creation of more efficient crystallization cocktail screens [111] and novel approaches for sample preparation have been developed. These include in situ proteolysis [112] and reductive methylation [113] for improving the probability of crystallization of macromolecules, as well as dynamic light scattering [114] for choosing the best protein constructs, buffers, and ligands for crystallization. Recent developments in X-ray data collection protocols, low-noise detectors, sample-handling robotics, synchrotron beamlines, and associated software also significantly contributed to progress in X-ray crystallography.

Besides traditional X-ray crystallography, other forms of X-ray experiments can also provide essential structural information at different stages of structure determination. As described above, SFX uses highly focused XFELs to collect single-shot diffraction data on streams of nanocrystals, which has been shown to be useful for the investigation of structures where well-diffracting large crystals cannot be grown [49]. Over the past 10 years, SFX has cultivated new methods for nanocrystallogenesis, sample delivery, and data processing (*see* review [115]). Current and next-generation synchrotrons with much higher flux density, with beamlines using sophisticated focusing optics, submicron beam diameters, and fast low-noise photon-counting detectors, have allowed the implementation of the serial approach in these X-ray facilities [116, 117]. The small (i.e., nano- or micrometer) crystals used in SFX are also ideal for time-resolved studies as a greater percentage of molecules in the crystals can activate rapid mixing. Thus, SFX has enabled rapid diffusive saturation of an antibiotic in mix-and-inject experiments to study *"on the fly"* an enzymatic reaction inside crystals for the first time [52]. All these advancements and innovations can be implemented directly into the structural studies of M-T7. For instance, if the recently obtained large crystals of M-T7 are difficult to optimize, micro-crystallization techniques can be used to grow high-quality microcrystals for serial crystallography experiments to collect high-quality data at XFELs of synchrotrons sources. In addition, once the first crystal structure of M-T7 is solved, time-resolved experiments at XFELs can be carried out with inhibitors and native substrates to study the dynamic and functionality of this protein to a much higher level of detail that can allow us to advance our understanding of the biology of M-T7.

6.2 Cryo-EM and Micro-Electron Diffraction (Micro-ED)

Cryo-electron microscopy (cryo-EM) uses a transmission electron microscope (TEM) to image particles frozen in vitreous ice. Whereas X-ray crystallography produces electron density maps, cryo-EM produces electrostatic potential maps. Electron density maps are the result of X-rays interacting with electrons in the crystal, but electrostatic potential maps are derived from both the electron beam interacting with the electrons of the sample and the electron beam interacting with the nuclei of the sample; thus electrostatic potential maps contain additional charge information compared to electron density maps. Unlike X-ray crystallography, cryo-EM studies are limited to proteins of ~100 kDa and above, though improving technologies are decreasing this lower size limit. Since the M-T7 homotrimer is estimated to be 113 kDa, it lies near the lower limit for this method, making particle identification challenging. Still, recent advances in cryo-EM have led to an avalanche of structures deposited into the PDB using this method (more than 4000 structures to date) and it may well be suitable for a homogenous isolation of M-T7. Notably, most of the recent cryo-EM structures are solved to a resolution of ~3.0 Å, implying a practical (not theoretical) resolution limit whereas X-ray crystallography can often produce structures of much higher resolution.

In cryo-EM data collection, a series of images are collected, each containing multiple orientations of a single kind of particle—a protein for example. Individual images of the particle are extracted and used to produce a three-dimensional reconstruction of the particle. Resolution is defined by how well the Fourier space frequencies of half data sets correlate, referred to as a Fourier shell correlation. Models are refined into the resultant electrostatic potential maps similarly to X-ray-derived electron density maps but using different atomic scattering factors.

The bottlenecks of cryo-EM typically involve sample preparation. For freezing in vitreous ice, the sample solution is applied onto an EM grid containing many holes. The sample is wicked away with filter paper, referred to as blotting, leaving a very thin layer of the protein solution in each hole, and it is then plunged into liquid ethane until imaging. Blotting causes the sample in each hole to be predominantly composed of an air-water interface in which proteins are embedded. In many cases, this causes proteins to aggregate just before plunge-freezing. Thus, buffer conditions, grid treatments, and blot parameters must be optimized to prepare a sample that exhibits well-separated and homogenous particles that are suitable for single-particle analysis.

Micro-ED is an emerging technique that spans cryo-EM and X-ray crystallography, as it combines cryo-EM sample preparation and instrumentation with X-ray crystallography data collection and data analysis. Micro-ED was first introduced in 2013 by Gonen and co-workers and has experienced a rapid growth with more than 40 proteins and small molecules determined to date [118]. Micro-

ED has overcome the typical bottlenecks encountered in traditional X-ray crystallography at synchrotrons. First, there is no need for the production of large, well-diffracting crystals. Micro-ED uses electrons to measure samples, which have a lot less energy than X-rays; thus, the use of nanocrystals composed of few mosaic blocks, which are often easier to produce than large crystals, is required. Second, radiation damage is outrun. Unlike X-rays, electrons deposit considerably less damaging energy into a crystal so that electron-diffraction data can be collected at an ultralow-dose rate (~0.01 $e^{-}/Å^{2}/s$) allowing for many exposures from a single nanocrystal. Thus, when traditional X-ray crystallography at synchrotrons fails, the Micro-ED technique could become key to determine the first crystallographic structure of M-T7.

A comprehensive guide to Micro-ED sample preparation, data collection, and processing has been published previously [119, 120]. A successful Micro-ED experiment requires the initial identification of suitable nanocrystals (in the range of 0.1–0.4 μm), which are typically identified by negative-stain EM. Once suitable nanocrystals are identified, samples are prepared by vitrification methods on carbon-coated grids analogous to those used for other cryo-EM techniques [120]. Carbon grids containing the crystals are transferred into a standard TEM instrument which must be set up appropriately for Micro-ED data collection mode [120]. CMOS-based TVIPS F416 detector is ideal for Micro-ED data collection as it has high sensitivity, high dynamic range, and no "bleeding" into neighboring pixels. In a typical Micro-ED experiment, the stage is smoothly and unidirectionally rotated (up to a 140° wedge) at a constant rate while the detector constantly collects the diffraction data from a single nanocrystal. Each diffraction data set, consisting of 70–150 images, typically takes ~5–10 min to complete. Also, because the continuous rotation mode data collection allows for mostly fully measured intensities, data processing (integration, scaling, merging, phasing, and refinement) is carried out using the conventional software for X-ray crystallography described in previous sections.

6.3 Small-Angle X-Ray Scattering (SAXS)

SAXS is a powerful technique used to obtain valuable structural information (size and shape) of macromolecules at low resolution. It can also provide information about structural changes that occur at different experimental conditions [121]. SAXS experiments are performed at dedicated beamlines at synchrotron light sources [122]. In a typical SAXS experiment, a monochromatic X-ray beam illuminates a solution containing the macromolecule(s) of interest, and the intensity of the scattered X-rays is recorded on a detector and quickly processed by using specialized software. The recorded scattering pattern is reduced to a radially averaged one-dimensional scattering profile, which results in low structural resolution (>10 Å) but can still provide important structural information. Because the orientations of biomolecules in solution are random, single orientation cannot be isolated, so the information present is considered the global average of all orientations. Thus, only large-scale observations can be determined, including the overall shape of the biomolecule of interest.

SAXS becomes even more powerful and can provide more useful information when used in combination with high-resolution structural information determined from other techniques such as X-ray crystallography and NMR [123–125]. Once the first crystallographic structure of M-T7 is determined and more is determined about its multimeric state and dynamical features, this technique might be found of potential interest for the structural characterization of this protein in order to gain more structural insights that can help us understand M-T7 biology.

References

1. Lalani AS, Graham K, Mossman K et al (1997) The purified myxoma virus gamma interferon receptor homolog M-T7 interacts with the heparin-binding domains of chemokines. J Virol 71(6):4356–4363
2. Nuara AA, Walter LJ, Logsdon NJ et al (2008) Structure and mechanism of IFN-gamma antagonism by an orthopoxvirus IFN-gamma-binding protein. Proc Natl Acad Sci U S A 105(6):1861–1866
3. Upton C, Mossman K, McFadden G (1992) Encoding of a homolog of the IFN-gamma receptor by myxoma virus. Science 258 (5086):1369–1372
4. Lalani AS, McFadden G (1997) Secreted poxvirus chemokine binding proteins. J Leukoc Biol 62(5):570–576
5. Munoz EM, Linhardt RJ (2004) Heparin-binding domains in vascular biology. Arterioscler Thromb Vasc Biol 24(9):1549–1557
6. Seet BT, Barrett J, Robichaud J et al (2001) Glycosaminoglycan binding properties of the myxoma virus CC-chemokine inhibitor, M-T1. J Biol Chem 276(32):30504–30513
7. Bartee MY, Chen H, Dai E et al (2014) Defining the anti-inflammatory activity of a potent myxomaviral chemokine modulating protein, M-T7, through site directed mutagenesis. Cytokine 65(1):79–87
8. UniProt C (2008) The universal protein resource (UniProt). Nucleic Acids Res 36 (Database issue):D190–D195
9. Altschul SF, Gish W, Miller W et al (1990) Basic local alignment search tool. J Mol Biol 215(3):403–410
10. Schwede T, Kopp J, Guex N et al (2003) SWISS-MODEL: an automated protein homology-modeling server. Nucleic Acids Res 31(13):3381–3385
11. Berman HM, Westbrook J, Feng Z et al (2000) The Protein Data Bank. Nucleic Acids Res 28(1):235–242
12. Correll CC, Batie CJ, Ballou DP et al (1992) Phthalate dioxygenase reductase: a modular structure for electron transfer from pyridine nucleotides to [2Fe-2S]. Science 258 (5088):1604–1610
13. Bedard EL, Kim P, Jiang J et al (2003) Chemokine-binding viral protein M-T7 prevents chronic rejection in rat renal allografts. Transplantation 76(1):249–252
14. Liu L, Lalani A, Dai E et al (2000) The viral anti-inflammatory chemokine-binding protein M-T7 reduces intimal hyperplasia after vascular injury. J Clin Invest 105 (11):1613–1621
15. Lucas AR, Verma RK, Dai E et al (2014) Myxoma virus anti-inflammatory chemokine binding protein reduces the increased plaque growth induced by chronic Porphyromonas gingivalis oral infection after balloon angioplasty aortic injury in mice. PLoS One 9 (10):e111353
16. McPherson A, Gavira JA (2014) Introduction to protein crystallization. Acta Crystallogr F Struct Biol Commun 70(Pt 1):2–20
17. Bergfors TM (2009) Protein crystallization
18. Giegé R, Ducruix A (1992) Crystallization of nucleic acids and proteins : a practical approach. Practical approach series. p. xxiv, 331

19. McPherson A (1999) Crystallization of biological macromolecules. p ix, 586
20. Rupp B (2010) Biomolecular crystallography: principles, practice, and application to structural biology. p. xxi, 809
21. Chayen NE (2004) Turning protein crystallisation from an art into a science. Curr Opin Struct Biol 14(5):577–583
22. Jahandideh S, Jaroszewski L, Godzik A (2014) Improving the chances of successful protein structure determination with a random forest classifier. Acta Crystallogr D Biol Crystallogr 70(Pt 3):627–635
23. Kurgan L, Razib AA, Aghakhani S et al (2009) CRYSTALP2: sequence-based protein crystallization propensity prediction. BMC Struct Biol 9:50
24. Berman H, Henrick K, Nakamura H (2003) Announcing the worldwide Protein Data Bank. Nat Struct Biol 10(12):980
25. Jancarik J, Kim S-H (1991) Sparse matrix sampling: a screening method for crystallization of proteins. J Appl Crystallogr 24 (4):409–411
26. Fazio VJ, Peat TS, Newman J (2014) A drunken search in crystallization space. Acta Crystallogr F Struct Biol Commun. 70 (Pt 10):1303–1311
27. Luft JR, Newman J, Snell EH (2014) Crystallization screening: the influence of history on current practice. Acta Crystallogr F Struct Biol Commun 70(Pt 7):835–853
28. Luft JR, Wolfley JR, Said MI et al (2007) Efficient optimization of crystallization conditions by manipulation of drop volume ratio and temperature. Protein Sci 16(4):715–722
29. McPherson A, Cudney B (2014) Optimization of crystallization conditions for biological macromolecules. Acta Crystallogr F Struct Biol Commun. 70 (Pt 11):1445-1467
30. Russo Krauss I, Merlino A, Vergara A et al (2013) An overview of biological macromolecule crystallization. Int J Mol Sci 14 (6):11643–11691
31. Larson SB, Day JS, Nguyen C et al (2008) Progress in the development of an alternative approach to macromolecular crystallization. Cryst Growth Des 8(8):3038–3052
32. Harp JM, Hanson BL, Timm DE et al (1999) Macromolecular crystal annealing: evaluation of techniques and variables. Acta Crystallogr D Biol Crystallogr 55(Pt 7):1329–1334
33. Kriminski S, Caylor CL, Nonato MC et al (2002) Flash-cooling and annealing of protein crystals. Acta Crystallogr D Biol Crystallogr 58(Pt 3):459–471
34. Newman J (2006) A review of techniques for maximizing diffraction from a protein crystal in stilla. Acta Crystallogr D Biol Crystallogr 62(Pt 1):27–31
35. Russo Krauss I, Sica F, Mattia CA et al (2012) Increasing the X-ray diffraction power of protein crystals by dehydration: the case of bovine serum albumin and a survey of literature data. Int J Mol Sci 13(3):3782–3800
36. Wine Y, Cohen-Hadar N, Freeman A et al (2007) Elucidation of the mechanism and end products of glutaraldehyde crosslinking reaction by X-ray structure analysis. Biotechnol Bioeng 98(3):711–718
37. Lopez-Jaramillo FJ, Otálora F, Gavira JA (2003) Protein crystal quality in diffusive environments and its evaluation. J Cryst Growth 247(1):177–184
38. Otalora F, Gavira JA, Ng JD et al (2009) Counterdiffusion methods applied to protein crystallization. Prog Biophys Mol Biol 101 (1 3):26 37
39. Drenth J, Mesters J, MyiLibrary, et al. (2007) Principles of protein X-ray crystallography. p. 1 online resource (xiv, 332)
40. Bantz D, Zwick M (1974) The use of symmetry with the fast Fourier algorithm. Acta Crystallogr A 30:257–260
41. Ten L (1973) Crystallographic fast Fourier transforms. Acta Crystallogr A 29 (2):183–191
42. Abola E, Kuhn P, Earnest T et al (2000) Automation of X-ray crystallography. Nat Struct Biol 7(Suppl):973–977
43. Cassetta A, Deacon A, Ealick S, et al. (1999) Development of instrumentation and methods for MAD and structural genomics at the SRS, ESRF, CHESS and Elettra facilities
44. Hendrickson WA (2000) Synchrotron crystallography. Trends Biochem Sci 25 (12):637–643
45. Leslie AG, Powell HR, Winter G et al (2002) Automation of the collection and processing of X-ray diffraction data -- a generic approach. Acta Crystallogr D Biol Crystallogr 58 (Pt 11):1924–1928
46. Mitchell E, Kuhn P, Garman E (1999) Demystifying the synchrotron trip: a first time user's guide. Structure 7(5):R111–R121
47. Pellegrini C, Marinelli A, Reiche S (2016) The physics of X-ray free-electron lasers. Rev Modern Phys 88(1):015006
48. Neutze R, Wouts R, van der Spoel D et al (2000) Potential for biomolecular imaging with femtosecond X-ray pulses. Nature 406 (6797):752–757

49. Chapman HN, Fromme P, Barty A et al (2011) Femtosecond X-ray protein nanocrystallography. Nature 470(7332):73–77
50. Ishigami I, Lewis-Ballester A, Echelmeier A et al (2019) Snapshot of an oxygen intermediate in the catalytic reaction of cytochrome c oxidase. Proc Natl Acad Sci U S A 116 (9):3572–3577
51. Kupitz C, Olmos JL Jr, Holl M et al (2017) Structural enzymology using X-ray free electron lasers. Struct Dyn 4(4):044003
52. Olmos JL Jr, Pandey S, Martin-Garcia JM et al (2018) Enzyme intermediates captured "on the fly" by mix-and-inject serial crystallography. BMC Biol 16(1):59
53. Stagno JR, Liu Y, Bhandari YR et al (2017) Structures of riboswitch RNA reaction states by mix-and-inject XFEL serial crystallography. Nature 541(7636):242–246
54. Aquila A, Hunter MS, Doak RB et al (2012) Time-resolved protein nanocrystallography using an X-ray free-electron laser. Opt Express 20(3):2706–2716
55. Barends TR, Foucar L, Ardevol A et al (2015) Direct observation of ultrafast collective motions in CO myoglobin upon ligand dissociation. Science 350(6259):445–450
56. Kern J, Alonso-Mori R, Hellmich J et al (2012) Room temperature femtosecond X-ray diffraction of photosystem II microcrystals. Proc Natl Acad Sci U S A 109 (25):9721–9726
57. Kern J, Chatterjee R, Young ID et al (2018) Structures of the intermediates of Kok's photosynthetic water oxidation clock. Nature 563 (7731):421–425
58. Kupitz C, Basu S, Grotjohann I et al (2014) Serial time-resolved crystallography of photosystem II using a femtosecond X-ray laser. Nature 513(7517):261–265
59. Nango E, Royant A, Kubo M et al (2016) A three-dimensional movie of structural changes in bacteriorhodopsin. Science 354 (6319):1552–1557
60. Nogly P, Weinert T, James D et al (2018) Retinal isomerization in bacteriorhodopsin captured by a femtosecond X-ray laser. Science 361(6398):eaat0094
61. Pande K, Hutchison CD, Groenhof G et al (2016) Femtosecond structural dynamics drives the trans/cis isomerization in photoactive yellow protein. Science 352 (6286):725–729
62. Suga M, Akita F, Sugahara M et al (2017) Light-induced structural changes and the site of O=O bond formation in PSII caught by XFEL. Nature 543(7643):131–135
63. Tenboer J, Basu S, Zatsepin N et al (2014) Time-resolved serial crystallography captures high-resolution intermediates of photoactive yellow protein. Science 346 (6214):1242–1246
64. Young ID, Ibrahim M, Chatterjee R et al (2016) Structure of photosystem II and substrate binding at room temperature. Nature 540(7633):453–457
65. Garman E (1999) Cool data: quantity AND quality. Acta Crystallogr D Biol Crystallogr 55(Pt 10):1641–1653
66. Garman E (2003) 'Cool' crystals: macromolecular cryocrystallography and radiation damage. Curr Opin Struct Biol 13 (5):545–551
67. Garman EF, Owen RL (2006) Cryocooling and radiation damage in macromolecular crystallography. Acta Crystallogr D Biol Crystallogr 62(Pt 1):32–47
68. Garman EF, Schneider TR (1997) Macromolecular cryocrystallography. J Appl Crystallogr 30(3):211–237
69. Dauter Z, Wlodawer A (2016) Progress in protein crystallography. Protein Pept Lett 23 (3):201–210
70. Owen RL, Juanhuix J, Fuchs M (2016) Current advances in synchrotron radiation instrumentation for MX experiments. Arch Biochem Biophys 602:21–31
71. Kabsch W (1988) Automatic indexing of rotation diffraction patterns. J Appl Crystallogr 21(1):67–72
72. Kabsch W (2010) Xds. Acta Crystallogr D Biol Crystallogr 66(Pt 2):125–132
73. Battye TG, Kontogiannis L, Johnson O et al (2011) iMOSFLM: a new graphical interface for diffraction-image processing with MOSF LM. Acta Crystallogr D Biol Crystallogr 67 (Pt 4):271–281
74. Leslie AGW, Powell HR (2007) Processing diffraction data with mosflm. In Evolving Methods for Macromolecular Crystallography. Springer Netherlands, Dordrecht
75. Minor W, Cymborowski M, Otwinowski Z et al (2006) HKL-3000: the integration of data reduction and structure solution--from diffraction images to an initial model in minutes. Acta Crystallogr D Biol Crystallogr 62 (Pt 8):859–866
76. Otwinowski Z, Minor W (1997) Processing of X-ray diffraction data collected in oscillation mode. Methods Enzymol 276:307–326
77. Evans PR (2011) An introduction to data reduction: space-group determination, scaling and intensity statistics. Acta Crystallogr D Biol Crystallogr 67(Pt 4):282–292

78. Evans P (2006) Scaling and assessment of data quality. Acta Crystallogr D Biol Crystallogr 62(Pt 1):72–82
79. Collaborative Computational Project, N (1994) The CCP4 suite: programs for protein crystallography. Acta Crystallogr D Biol Crystallogr 50(Pt 5):760–763
80. Riboldi-Tunnicliffe A, Hilgenfeld R (1999) Cryocrystallography with oil - an old idea revived. J Appl Crystallogr 32(5):1003–1005
81. Pflugrath JW (2004) Macromolecular cryo-crystallography--methods for cooling and mounting protein crystals at cryogenic temperatures. Methods 34(3):415–423
82. Emsley P, Lohkamp B, Scott WG et al (2010) Features and development of Coot. Acta Crystallogr Sect D 66(4):486–501
83. Murshudov GN, Vagin AA, Dodson EJ (1997) Refinement of macromolecular structures by the maximum-likelihood method. Acta Crystallogr D Biol Crystallogr 53 (Pt 3):240–255
84. Adams PD, Afonine PV, Bunkoczi G et al (2010) PHENIX: a comprehensive python-based system for macromolecular structure solution. Acta Crystallogr D Biol Crystallogr 66(Pt 2):213–221
85. Sheldrick GM (2015) Crystal structure refinement with SHELXL. Acta Crystallogr C Struct Chem 71(Pt 1):3–8
86. DiMaio F, Terwilliger TC, Read RJ et al (2011) Improved molecular replacement by density- and energy-guided protein structure optimization. Nature 473(7348):540–543
87. Read RJ, Zhou A, Stein PE (2011) Solving serpin crystal structures. Methods Enzymol 501:49–61
88. Bränd'en C-I, Jones TA (1990) Between objectivity and subjectivity. Nature 343 (6260):687–689
89. Hooft RWW, Vriend G, Sander C et al (1996) Errors in protein structures. Nature 381 (6580):272–272
90. Kleywegt GJ, Jones TA (1995) Where freedom is given, liberties are taken. Structure 3 (6):535–540
91. Kleywegt GJ, Jones TA (1996) Phi/psi-chology: Ramachandran revisited. Structure 4(12):1395–1400
92. Kleywegt GJ, Jones TA (1997) Model building and refinement practice. Methods Enzymol 277:208–230
93. Morris AL, MacArthur MW, Hutchinson EG et al (1992) Stereochemical quality of protein structure coordinates. Proteins 12 (4):345–364
94. Laskowski RA, MacArthur MW, Moss DS et al (1993) PROCHECK: a program to check the stereochemical quality of protein structures. J Appl Crystallogr 26(2):283–291
95. Hooft RW, Sander C, Vriend G (1996) Positioning hydrogen atoms by optimizing hydrogen-bond networks in protein structures. Proteins 26(4):363–376
96. Hooft RWW, Sander C, Scharf M et al (1996) The PDBFINDER database: a summary of PDB, DSSP and HSSP information with added value. Bioinformatics 12(6):525–529
97. Hooft RWW, Sander C, Vriend G (1996) Verification of protein structures: side-chain planarity. J Appl Crystallogr 29(6):714–716
98. Ramachandran GN, Ramakrishnan C, Sasisekharan V (1963) Stereochemistry of polypeptide chain configurations. J Mol Biol 7:95–99
99. Chen VB, Arendall WB 3rd, Headd JJ et al (2010) MolProbity: all-atom structure validation for macromolecular crystallography. Acta Crystallogr D Biol Crystallogr 66 (Pt 1):12–21
100. Williams CJ, Headd JJ, Moriarty NW et al (2018) MolProbity: more and better reference data for improved all-atom structure validation. Protein Sci 27(1):293–315
101. Vaguine AA, Richelle J, Wodak SJ (1999) SF CHECK: a unified set of procedures for evaluating the quality of macromolecular structure-factor data and their agreement with the atomic model. Acta Crystallogr D Biol Crystallogr 55(Pt 1):191–205
102. Gore S, Sanz Garcia E, Hendrickx PMS et al (2017) Validation of structures in the Protein Data Bank. Structure 25(12):1916–1927
103. Gore S, Velankar S, Kleywegt GJ (2012) Implementing an X-ray validation pipeline for the Protein Data Bank. Acta Crystallogr D Biol Crystallogr 68(Pt 4):478–483
104. Read RJ, Adams PD, Arendall WB 3rd et al (2011) A new generation of crystallographic validation tools for the Protein Data Bank. Structure 19(10):1395–1412
105. Smart OS, Horsky V, Gore S et al (2018) Worldwide Protein Data Bank validation information: usage and trends. Acta Crystallogr D Struct Biol 74(Pt 3):237–244
106. Nicholls A, McGaughey GB, Sheridan RP et al (2010) Molecular shape and medicinal chemistry: a perspective. J Med Chem 53 (10):3862–3886
107. Almo SC, Garforth SJ, Hillerich BS et al (2013) Protein production from the structural genomics perspective: achievements and future needs. Curr Opin Struct Biol 23 (3):335–344

108. Almo SC, Love JD (2014) Better and faster: improvements and optimization for mammalian recombinant protein production. Curr Opin Struct Biol 26:39–43
109. Grabowski M, Joachimiak A, Otwinowski Z et al (2007) Structural genomics: keeping up with expanding knowledge of the protein universe. Curr Opin Struct Biol 17(3):347–353
110. Joachimiak A (2009) High-throughput crystallography for structural genomics. Curr Opin Struct Biol 19(5):573–584
111. Newman J, Egan D, Walter TS et al (2005) Towards rationalization of crystallization screening for small- to medium-sized academic laboratories: the PACT/JCSG+ strategy. Acta Crystallogr D Biol Crystallogr 61 (Pt 10):1426–1431
112. Dong A, Xu X, Edwards AM et al (2007) In situ proteolysis for protein crystallization and structure determination. Nat Methods 4 (12):1019–1021
113. Kim Y, Quartey P, Li H et al (2008) Large-scale evaluation of protein reductive methylation for improving protein crystallization. Nat Methods 5(10):853–854
114. Wilson WW (2003) Light scattering as a diagnostic for protein crystal growth--a practical approach. J Struct Biol 142(1):56–65
115. Martin-Garcia JM, Conrad CE, Coe J et al (2016) Serial femtosecond crystallography: a revolution in structural biology. Arch Biochem Biophys 602:32–47
116. Martin-Garcia JM, Conrad CE, Nelson G et al (2017) Serial millisecond crystallography of membrane and soluble protein microcrystals using synchrotron radiation. IUCrJ 4 (Pt 4):439–454
117. Martin-Garcia JM, Zhu L, Mendez D et al (2019) High-viscosity injector-based pink-beam serial crystallography of microcrystals at a synchrotron radiation source. IUCrJ 6 (Pt 3):412–425
118. Nannenga BL, Gonen T (2019) The cryo-EM method microcrystal electron diffraction (MicroED). Nat Methods 16(5):369–379
119. Hattne J, Reyes FE, Nannenga BL et al (2015) MicroED data collection and processing. Acta Crystallogr A Found Adv 71 (Pt 4):353–360
120. Shi D, Nannenga BL, de la Cruz MJ et al (2016) The collection of MicroED data for macromolecular crystallography. Nat Protoc 11(5):895–904
121. Rambo RP, Tainer JA (2013) Super-resolution in solution X-ray scattering and its applications to structural systems biology. Annu Rev Biophys 42:415–441
122. Blanchet CE, Spilotros A, Schwemmer F et al (2015) Versatile sample environments and automation for biological solution X-ray scattering experiments at the P12 beamline (PETRA III, DESY). J Appl Crystallogr 48 (Pt 2):431–443
123. Boura E, Rozycki B, Chung HS et al (2012) Solution structure of the ESCRT-I and -II supercomplex: implications for membrane budding and scission. Structure 20 (5):874–886
124. Boura E, Rozycki B, Herrick DZ et al (2011) Solution structure of the ESCRT-I complex by small-angle X-ray scattering, EPR, and FRET spectroscopy. Proc Natl Acad Sci U S A 108(23):9437–9442
125. Chalupska D, Eisenreichova A, Rozycki B et al (2017) Structural analysis of phosphatidylinositol 4-kinase IIIbeta (PI4KB) - 14-3-3 protein complex reveals internal flexibility and explains 14-3-3 mediated protection from degradation in vitro. J Struct Biol 200 (1):36–44
126. Sievers F, Higgins DG (2018) Clustal omega for making accurate alignments of many protein sequences. Protein Sci 27(1):135–145

Chapter 9

Microscopic Analysis of Viral Infection

Honor L. Glenn, Ana Lemos de Matos, Nancy Villa, and Grant McFadden

Abstract

Viruses engineered to express fluorescent proteins can be used with live-cell imaging techniques to monitor the progression of infection in real time. Here we describe a set of methods to track infection spreading from one cell population to another as well as to visualize transfer of virions between cells. This approach is extended to multiplexing with physiological readouts of cell death, which can be correlated with single-cell resolution to viral infection.

Key words Viruses, Live-cell imaging, Myxoma virus, Apoptosis, Infection, Cell death, Fluorescence, Confocal

1 Introduction

The progression of viral infection can be visually monitored by use of fluorescence or confocal microscopy [1]. This approach provides information that is not available through end-point assays such as flow cytometry or biochemical methods. There are two significant benefits of microscopy for investigation of viral infection mechanisms. First, it allows analysis of spatial relationships between cells during infection, which provides information on cell-to-cell transmission of infection as well as tracking of virions during initial stages of infection and transfer between cells. Second, live-cell imaging provides dynamic information on both viral life cycle and cellular physiology during infection. This type of data is particularly useful because a single population can be evaluated over time. This continuous temporal readout eliminates the variability inherent in piecing together a time line from individual time points collected from different samples.

Myxoma virus (MYXV) is a pox virus, which in nature is only infectious to rabbits. However, this virus has also been demonstrated to be a potent oncolytic that can selectively target many human cancers [2]. We describe here a set of methods for imaging the process of MYXV infection using genetically modified virus

Alexandra R. Lucas (ed.), *Viruses as Therapeutics: Methods and Protocols*, Methods in Molecular Biology, vol. 2225, https://doi.org/10.1007/978-1-0716-1012-1_9,

expressing the fluorescent proteins, GFP and TdTomato, under the control of early/late and/or late gene promoters (*see* Chapter 4, Torres and Rahman) [3]. Infected cells express these reporters indicating the synthesis of viral proteins (Fig. 1). MYXV infection can also be dynamically monitored by the use of MYXV engineered to express the green fluorescent protein, mVenus, linked to a structural viral protein [4]. This approach allows direct visualization of individual virions as well as an indicator for active infection as target cells begin to synthesize the fluorescent protein. It should be noted that MYXV, with a maximum diameter of about 200 nm, is below the resolution limit of conventional lens-based microscopy. Viruses of this size can be resolved using various super-resolution approaches, some of which are compatible with live-cell imaging [5, 6]. However, virus expressing mVenus can be easily detected, though not resolved, with high-numerical-aperture immersion lenses using wide-field fluorescence or confocal microscopy (Fig. 2). The methods described can be used to study mechanisms of transfer of MYXV from one cell population to another.

A wide range of live-cell, fluorescent probes are commercially available, which can be multiplexed with reporters for viral infection. This strategy allows time-dependent analysis of cellular response to virus on the single-cell level. In selecting physiological probes for live-cell imaging, it is important to determine that the probes can provide dynamic response for the duration of the assay. Many live-cell probes only provide an instantaneous response and others are cytotoxic after 1–2 h. A particularly useful assay is tracking cell death in response to viral infection. Here we describe a method for tracking cell death that is agnostic to the mechanism of death as well as a method for identifying apoptotic cells.

2 Materials

1. Jurkat cells, non-adherent, human T-cell line derived from T-cell leukemia.
2. A549 cells, adherent human epithelial cell line derived from lung adenocarcinoma.
3. RPMI complete culture medium (cRPMI): RPMI supplemented with 10% fetal bovine serum, 100 U/mL penicillin, 100 μg/mL streptomycin.
4. DMEM complete culture medium (cDMEM): Dulbecco's modified Eagle's medium supplemented with 10% fetal bovine serum, 100 U/mL penicillin, 100 μg/mL streptomycin.
5. Myxoma virus expressing green fluorescent protein (GFP) and TdTomato under early/late and late viral promoters, respectively (vMyx-GFP-TdTomato) [3].

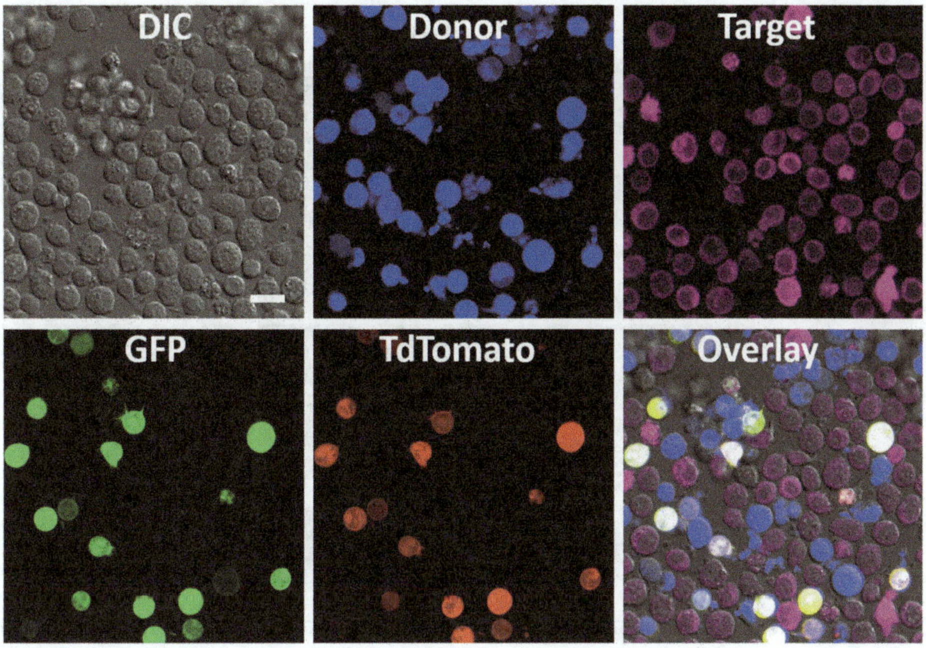

Fig. 1 vMyx-GFP-TdTomato-infected donor, Jurkat cells co-cultured with target cells. Top left (DIC): differential interference contrast image; top center (donor) (blue): cells labeled with CellTrace™ Violet; top right (target) (magenta): cells labeled with CellTrace™ Far Red; bottom left (GFP) (green): GFP expression driven by viral early/late promoter; bottom center (TdTomato) (red): TdTomato expression driven by late promotor; bottom right (overlay): overlay of all five channels. Scale bar = 20 μm

6. Myxoma virus expressing mVenus-tagged structural protein M093L (vMyx-mVenus) [4].
7. Myxoma virus expressing GFP under early/late viral promoter (vMyx-GFP) [7].
8. Myxoma virus expressing Td-Tomato under early/late viral promoter (vMyx-TdTomato) [8].
9. 8-Well chambered cover glasses (we use Ibidi brand).
10. Hanks' balanced salt solution (HBSS).
11. Hoechst stock solution: 1 mg/mL Hoechst in dH_2O. Store at −20 °C and protect from light.
12. CellTrace™ Violet stock solution: Prepare 5 mM CellTrace™ Violet in DMSO. Store at −20 °C and protect from light (CellTrace is a trademark of Thermo Fisher Scientific).
13. CellTrace™ Far Red stock solution: 1 mM CellTrace™ Far Red in DMSO. Store at −20 °C and protect from light.
14. Propidium iodide stock solution: 1 mg/mL Propidium iodide in dH_2O. Store at 4 °C, and do not freeze. Protect from light.

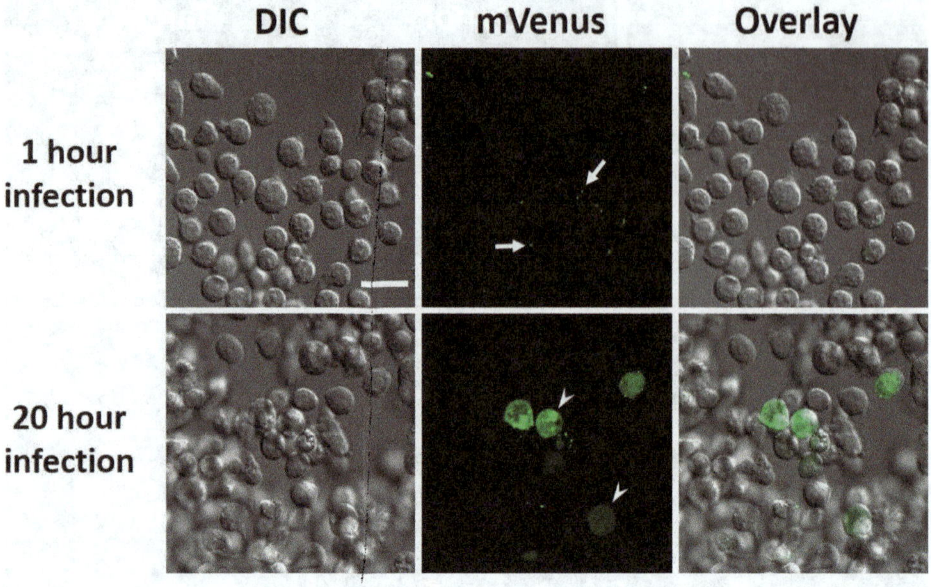

Fig. 2 Progress of infection of vMyx-mVenus-infected Jurkat cells. At early infection (top row) (1-h infection) vMyx-mVenus virions are visible as small green puncta (arrows). As infection progresses (bottom row) (20-h infection), cells begin to express viral proteins and diffuse cytoplasmic mVenus signal becomes visible (arrowheads). Left panels (DIC): differential interference contrast image; middle panels (mVenus) (green): mVenus fluorescent protein fused to MYXV structural protein; right panels (overlay): overlay both channels. Scale bar = 20 μm

15. Staurosporine stock solution: 500 mM Staurosporine in DMSO. Store at −20 °C.
16. CellEvent™ Caspase-3/7 Green Detection Reagent (Thermo Fisher Scientific): Supplied as 2 mM stock solution in DMSO. Store at −20 °C and protect from light.
17. Bench-top centrifuge for pelleting cells.
18. Automated wide-field fluorescent or confocal microscope with CO_2 incubator, capable of detection in standard 2-[4-(aminoiminomethyl)phenyl]-1H-indole-6-carboximidamide hydrochloride (DAPI) (blue emission), fluorescein isothiocyanate (FITC) (green emission), tetramethylrhodamine-isothiocyanate (TRITC) (red emission), and cyanine5 (Cy5) (far-red emission) channels, as well as differential interference contrast or phase contrast (we use a Nikon C2 laser scanning confocal). The microscope should also have a lens with numerical of 0.4–0.8 and one with numerical aperture of 1.2 or greater.

3 Methods

3.1 Imaging Fluorescent Protein Reporters for Viral Infection

3.1.1 Infection of Adherent Cells

1. Seed A549 cells 12–24 h prior to infection in two, 8-well chambered cover glasses in cDMEM such that cells are approximately 80–90% confluent at the time of infection. Use 290 μL medium per well in 8-well chambers. Prepare two identical samples, one for time-lapse imaging and one unimaged control (*see* **Note 1**).
2. Infect cells with vMyx-GFP-TdTomato [3]. Infect for 1 h at 37 °C at a multiplicity of infection of 10 in cDMEM, to allow virus adsorption. Keep at least one well on each substrate as an uninfected control. In addition, have a paired, uninfected control for each experimental condition.
3. Prepare imaging medium: 5 mL cDMEM with 1 μg/mL Hoechst, blue fluorescing DNA label.
4. Aspirate medium from infected cells and rinse with prewarmed HBSS.
5. Add 290 μL of imaging medium with Hoechst to each well. Incubate for 15 min at 37 °C.

3.1.2 Infection of Suspension Cells

1. For two 8-well chambered cover glasses, pellet 2.5×10^7 Jurkat cells for infection with MYXV (*see* **Notes 1** and **2**). Resuspend cells in 400 μL cRPMI containing vMyx-GFP-TdTomato at a multiplicity of infection of 10. Incubate cells with virus for 1 h at 37 °C to allow virus adsorption.
2. During incubation, prepare imaging medium: 6 mL cRPMI with 1 μg/mL Hoechst.
3. Pellet cells, rinse with prewarmed HBSS, and then resuspend in 400 μL imaging medium.
4. Count cells and calculate the volume of cells and culture medium necessary to suspend 3×10^5 cells in a total volume of 290 μL per well in 8-well cover glass.
5. Add imaging medium with Hoechst to wells and then add the infected cells and uninfected control cells. Incubate for 15 min at 37 °C (*see* **Note 3**).

3.1.3 Live-Cell Imaging with Fluorescent Protein Reporters

1. Secure the chambered cover glass in slide holder on confocal or wide-field fluorescence microscope with pre-warmed, humidified incubation chamber supplied with 5% CO_2, or standard conditions for the cell type under study. Keep the unimaged control samples in the normal cell culture incubator.
2. Use an air lens with numerical aperture of 0.4–0.8 (*see* **Note 4**). Image using standard excitation and emission channels for DAPI (405/450), GFP (488/510), and TdTomato (561/580). Also, image in a transmitted light channel with

differential interference contrast (DIC) or phase contrast. Use the Hoechst signal to focus on the cells (*see* **Note 5**). Excitation light intensity must be kept at a minimum to prevent phototoxicity. Excess light, particularly at low wavelengths, compromises cell health and can cause perturbation of the biological phenomena under observation (*see* **Note 6**).

3. Image each well on automated microscope every 10–60 min for 24–48 h (*see* **Notes** 7 and **8**).
4. After imaging, remove the sample and take end-point images of each well in the unimaged control using the same settings applied to the imaged sample.

3.1.4 Image Analysis of Fluorescent Protein Reporters

1. In order to confirm that the imaging has not affected cell health or progression of infection, compare cell numbers, morphology, and fluorescent protein expression levels in the sample that was imaged throughout infection to the unimaged control kept in the incubator and imaged only at the conclusion of the experiment (*see* **Note 1**).
2. To determine an overall value for infection, determine the ratio of the total signal in the green (GFP) or red (TdTomato) channel to the blue (Hoechst) channel. This normalizes the viral expression to the cell number.
3. Alternately, the proportion of GFP/TdTomato-expressing cells can be determined. Threshold each channel at the signal intensity that is unambiguously positive to generate a binary image. Use the same technique to create a binary image of the nuclei in the Hoechst channel. Then green or red cells can be counted manually or by automated analysis and expressed as a proportion of nuclei, which represent total cells (*see* **Note 9**).

3.2 Imaging Viral Particles with Fluorescent Protein Label

1. Infect Jurkat (suspension) or A549 (adherent) target cells according to the Subheadings 3.1.1 or 3.1.2 with vMyx-mVenus [4] at a multiplicity of infection of 10 (*see* **Note 10**).
2. Secure the chambered cover glass in slide holder on confocal or wide-field fluorescence microscope with pre-warmed, humidified incubation chamber supplied with 5% CO_2, or standard conditions for the cell type under study. Keep the unimaged control samples in the normal cell culture incubator.
3. Select an oil immersion lens with numerical aperture of 1.2 or greater (*see* **Note 11**).
4. Use the standard excitation and emission channels for DAPI (405/450) and GFP (488/510) to image the Hoechst and mVenus, respectively, as well as a transmitted light channel with differential interference or phase contrast.
5. Image every 1–3 min for 1–24 h (*see* **Notes 12** and **13**).

3.3 Imaging Cell-to-Cell Spread of Viral Infection

3.3.1 Imaging Spread of Viral Infection from Infected Donor Cells to Target Cells

1. For two 8-well chambered cover glasses, infect 1.5×10^7 Jurkat cells with vMyx-GFP-TdTomato at a multiplicity of infection of 10 according to Subheading 3.1.2 above (*see* **Note 10**).
2. Transfer cells into a 75 cm^2 (T75) flask containing 15 mL of cRPMI. Return to incubator and allow infection to progress for 16–24 h (*see* **Note 14**).
3. Prepare CellTrace™ staining solutions: Dilute CellTrace™ Violet stock solution to 20 μM in 500 μL HBSS, and CellTrace™ Far Red stock solution to 4 μM in 500 μL HBSS.
4. After incubation period, pellet infected Jurkat cells (donor cells) and 7.5×10^6 uninfected Jurkat cells (target cells). Resuspend the donor cells in 500 μL CellTrace™ Violet labeling solution, and the target cells in 500 μL CellTrace™ Far Red labeling solution in 1.5 mL microfuge tubes. Incubate at 37 °C for 30 min with gentle mixing every 10–15 min.
5. Pellet both samples in microfuge at ~400 × *g* for 3 min. Wash cells once in 500 μL HBSS, and then resuspend in a final volume of 400 μL of cRPMI.
6. Count cells and calculate the necessary volumes for 2×10^5 donor cells and 1×10^5 target cells for each well. To each well in the 8-well chambered cover glass, add cRPMI, then the donor cells, and target cells to a total volume of 290 μL (*see* **Note 15** for sample calculation).
7. Image cells as described above (see Subheading 3.1.3), and include a channel appropriate for Cy5 with excitation at 640 nm and emission at 660 nm to detect the CellTrace Far Red™ signal.
8. The rate of transfer of infection can be tracked by the increase in green signal (GFP) and the subsequent increase in red signal (TdTomato) that are coincident with the far-red signal in the target cell population.

3.3.2 Imaging Cell-to-Cell Transfer of Bound Virions

1. Prepare CellTrace™ staining solutions: Dilute CellTrace™ Violet stock solution to 20 μM in 500 μL HBSS, and CellTrace™ Far Red stock solution to 4 μM in 500 μL HBSS.
2. For two 8-well chambered cover glasses, pellet 1.5×10^7 Jurkat cells for the donor cell population and 7.5×10^6 Jurkat cells for the target cell population.
3. Resuspend the donor cells in 500 μL CellTrace Violet™ labeling solution, and the target cells in 500 μL CellTrace Far Red™ labeling solution in 1.5 mL microfuge tube. Incubate at 37 °C for 30 min with gentle mixing every 10–15 min.
4. Pellet both samples in microfuge at ~400 × *g* for 3 min. Wash cells once in 500 μL HBSS, and then resuspend in a final

volume of 400 μL of cRPMI. Return target cells to cell culture incubator.

5. Add vMyx-mVenus at a multiplicity of infection of 10 to the donor cells, mix gently, and incubate for 60 min at 37 °C.
6. Pellet donor cells in microfuge at ~400 × *g* for 3 min. Rinse with 500 μL HBSS. Then resuspend in 400 μL cRPMI.
7. Count cells and calculate the necessary volumes for 2 × 10^5 donor cells and 1 × 10^5 target cells for each well and sufficient medium for a total volume of 290 μL. To each well in the 8-well cover glass, add cRPMI, and then the donor cells.
8. Set up imaging as described above (*see* Subheading 3.2), using standard excitation and emission channels at 405/450 for CellTrace Violet™-labeled donor cells, 488/510 for vMyx-mVenus, and 640/660 for CellTrace Far Red™-labeled target cells, and a transmitted light channel with differential interference contrast (DIC) or phase contrast.
9. Focus on donor cells using 405/450 (blue emission) channel. Carefully add the amount of target cells calculated in **step 6** and begin imaging as soon as possible.
10. Image every 1–3 min for 1–24 h (*see* **Notes 12** and **13**).
11. A viral transfer event is identified when a target (far red) cell without detectible virus and a donor (blue) cell with a viral particle (green) come in contact and the viral particle is seen at the contact point and subsequently on the donor cell.

3.4 Imaging Physiological Probes During Viral Infection

3.4.1 Prepare Adherent Cells for Cell Death Imaging with Propidium Iodide

1. Seed A549 cells 12–24 h prior to infection in two, 8-well chambered cover glasses such that cells are at approximately 80–90% confluence at the time of infection. Use 290 μL cDMEM per well in 8-well chambers. Prepare two identical samples, one for time-lapse imaging and one unimaged control (*see* **Note 1**).
2. Infect cells with vMyx-GFP, a MYXV-expressing GFP under early/late viral promoter [7]. Infect for 1 h at 37 °C at a multiplicity of infection of 10 in cDMEM. Keep at least one well on each substrate as an uninfected control. In addition, have a paired, uninfected control for each experimental condition.
3. Prepare cell death imaging medium: 5 mL cDMEM with 1 μg/mL Hoechst and 1 μg/mL propidium iodide (*see* **Note 16**).
4. Aspirate medium and rinse cells with prewarmed HBSS.
5. Add 290 μL of cell death imaging medium (**step 3**) to each well. Incubate for 15 min at 37 °C.

3.4.2 Prepare Suspension Cells for Cell Death Imaging with Propidium Iodide

1. For two 8-well chambered cover glasses, pellet 2.5×10^7 Jurkat cells for infection with MYXV (*see* **Notes 1** and **2**). Resuspend cells in 400 μL cRPMI containing vMyx-GFP at a multiplicity of infection of 10. Incubate cells with virus for 1 h at 37 °C.
2. During incubation, prepare 6 mL cell death imaging medium: cRPMI with 1 μg/mL Hoechst and 1 μg/mL propidium iodide (*see* **Note 16**).
3. Pellet cells, rinse with prewarmed HBSS, and then resuspend in 400 μL cell death imaging medium. Prepare uninfected control cells in parallel. Prepare at least one uninfected control well in each chambered cover glass as well as a paired, uninfected control for each experimental treatment.
4. Count infected and control cells and calculate the volume of cells and culture medium necessary to suspend 3×10^5 cells in a total volume of 290 μL per well in 8-well chambered cover glass. Prepare two identical samples, one for time-lapse imaging and one unimaged control (*see* **Note 1**).
6. Add cell death imaging medium to wells and then add infected cells and uninfected control cells. Incubate for 15 min at 37 °C (*see* **Note 3**).

3.4.3 Imaging Cell Death Using Propidium Iodide

1. Secure the chambered cover glass in slide holder on confocal or wide-field fluorescence microscope with pre-warmed, humidified incubation chamber supplied with 5% CO_2, or standard conditions for the cell type under study. Keep the unimaged control samples in the normal cell culture incubator.
2. Use an air lens with numerical aperture of 0.4–0.8 (*see* **Note 4**). Image using standard excitation and emission channels for DAPI (405/450), FITC (488/510), and TRITC (561/580) to image Hoechst, GFP, and propidium iodide, respectively. Also, image in a transmitted light channel with differential interference contrast (DIC) or phase contrast. Use the Hoechst signal to focus on the cells (*see* **Note 5**). Excitation light intensity must be kept at a minimum to prevent phototoxicity. Excess light, particularly at low wavelengths, compromises cell health and can cause perturbation of the biological phenomena under observation (*see* **Note 6**).
3. Image each well on automated microscope every 10–60 min for 24–48 h (*see* **Notes** 7 and **8**).
4. After imaging, remove the sample and take end-point images of each well in the unimaged control using the same settings applied to the imaged sample.
5. Cell death can be expressed as the ratio of red, propidium iodide-positive nuclei to total, blue Hoechst-labeled nuclei. Infected cells are identified by expression of GFP.

3.4.4 Prepare Adherent Cells for Apoptosis Imaging with CellEvent™

1. Seed A549 cells 12–24 h prior to infection in two chambered cover glasses such that cells are at approximately 80–90% confluence at the time of infection. Use 290 μL cDMEM per well in 8-well chambers. Prepare two identical samples, one for time-lapse imaging and one unimaged control (*see* **Note 1**).
2. Infect cells with vMyx-TdTomato [8]. Infect for 1 h at 37 °C at a multiplicity of infection of 10 in 290 μL cDMEM. Keep at least one well on each substrate as an uninfected control, as well as paired, uninfected controls for each experimental condition. In addition, reserve an uninfected well for a positive control for apoptosis.
3. Prepare apoptosis imaging medium: 6 mL cDMEM with 1 μg/mL Hoechst and 5 μM CellEvent™ Caspase-3/7 Green Detection Reagent (*see* **Note 17**). Transfer 500 μL of the imaging medium to a separate 1.5 mL tube and add 1 μL of 500 μM staurosporine stock solution to make the 1 μM positive control working solution.
4. Aspirate medium and rinse cells with prewarmed HBSS.
5. Add 290 μL of the staurosporine-containing medium to the apoptosis positive control well and 290 μL apoptosis imaging medium to the other wells. Incubate for 30 min at 37 °C.

3.4.5 Prepare Suspension Cells for Apoptosis Imaging with CellEvent™

1. For two 8-well chambered cover glasses, infect 2.5×10^7 Jurkat cells with a multiplicity of infection of 10 (*see* **Notes 1** and **2**). Pellet cells and resuspend them in 400 μL cell culture medium containing vMyx-TdTomato [8]. Incubate cells with virus for 1 h at 37 °C to allow virus adsorption.
2. Prepare apoptosis imaging medium as follows: 6 mL cRPMI with 1 μg/mL Hoechst and 5 μM CellEvent™ Caspase-3/7 Green Detection Reagent (*see* **Note 17**). Transfer 500 μL of the imaging medium to a separate 1.5 mL tube and add 1 μL of 500 μM of stock staurosporine solution to make the 1 μM positive control working solution.
3. Pellet cells, rinse with prewarmed HBSS, and then resuspend in 400 μL apoptosis imaging medium. Prepare uninfected control cells in parallel. Seed at least one uninfected control well in each chambered cover glass as well as a paired, uninfected control for each experimental treatment.
4. Count infected and control cells and calculate the volume of cells and apoptosis imaging medium necessary to suspend 3×10^5 cells in a total volume of 290 μL per well in 8-well chambered cover glass (*see* **Note 15** for sample calculation). Prepare two identical samples, one for time-lapse imaging and one unimaged control (*see* **Note 1**).

5. Add staurosporine-containing medium to one well and apoptosis imaging medium (**step 2**) to the rest of the wells. Add the calculated volume of infected and control cells to wells, being sure to seed the staurosporine, positive control well with uninfected cells (*see* **Note 3**). Incubate for 30 min at 37 °C.

3.4.6 Imaging Apoptosis Using CellEvent™

1. Secure the chambered cover glass in slide holder on confocal or wide-field fluorescence microscope with pre-warmed, humidified incubation chamber supplied with 5% CO_2, or standard conditions for the cell type under study. Keep the unimaged control samples in the normal cell culture incubator.
2. Use an air lens with numerical aperture of 0.4–0.8 (*see* **Note 4**). Image using standard excitation and emission channels for DAPI (405/450), FITC (488/510), and TRITC (561/580) to image Hoechst, TdTomato, and CellEvent™, respectively. Also, image in a transmitted light channel with differential interference contrast (DIC) or phase contrast. Use the Hoechst signal to focus on the cells (*see* **Note 5**). Excitation light intensity must be kept at a minimum to prevent phototoxicity. Excess light, particularly at low wavelengths, compromises cell health and can cause perturbation of the biological phenomena under observation (*see* **Note 6**).
3. Image each well on automated microscope every 10–60 min for 24–48 h (*see* **Note 7**).
4. After imaging, remove the sample and take end-point images of each well in the unimaged control using the same settings applied to the imaged sample.
5. Apoptosis can be expressed as the ratio of green CellEvent-positive nuclei to total, blue Hoechst-labeled nuclei. Infected cells are identified by expression of TdTomato.

4 Notes

1. In live-cell imaging it is critical to ensure that cell health is not compromised during image acquisition and that the phenomena under study are also not perturbed. This can be of particular concern during time-lapse imaging when cells receive doses of high-intensity light over an extended period of time, such as the full viral life cycle. In time-lapse experiments, samples should be prepared in duplicate and one set should be kept under normal cell culture conditions during imaging (unimaged control). This control should then be imaged at a single time point, immediately after the time-lapse. The final state of the imaged and control sets should then be compared with regard to cell morphology, as well as all experimental

outcomes. Any variations should be investigated as possible artifacts due to imaging.

2. The recommended imaging substrates, 8-well chambered cover glasses, have a cell growth surface area of approximately 0.9 cm^2 per well; the suggested seeding densities for suspension cells are optimized for Jurkat cells. Suspension cells should settle in the wells such that they are loosely packed, with the majority of cells in contact with other cells, but not so crowded that they pile up. For cells or vessels of different sizes, optimization of cell numbers is recommended. Care must be taken when handling the chambers that no medium contacts the lid. This can cause fluid in the wells to wick up the walls of the chambers leading to sample drying and cross-contamination between wells.
3. To seed suspension cells in wells, add culture medium to the well first. Add the cells by dipping the pipette tip into the medium taking care not to contact the bottom of the well. Dispense cells slowly while moving the tip to disperse cells throughout the well while avoiding the edges. This approach is particularly important when co-culturing different cell types or treatments. Do not mix or shake the sample as this can cause the cells to collect densely in the center of the well.
4. When selecting lenses, it is important to consider the resolution appropriate for the experimental question. For experiments tracking the spread of infection by expression of fluorescent proteins in the cytoplasm, high resolution is not required since the smallest structure of interest is the cell nucleus, which is commonly used to normalize the probe expression to the cell number/DNA content. For this purpose 10X or 20X lenses with numerical apertures of 0.4–0.8 are adequate. These lower power lenses have the added advantage of providing a much larger field of view and therefore provide analysis of more cells per image. Conversely, when directly imaging fluorescent protein-tagged viral particles, high-powered, oil immersion lenses with numerical apertures above 1.2 are recommended since the virions are below the limit of resolution and are not easily detected with lower resolving power. It is important to recognize that the image of a single virion will appear as a diffraction-limited spot, larger than the actual particle. Thus, virions cannot be unambiguously counted, as multiple particles in close proximity cannot be distinguished.
5. When imaging suspension cells, even using the method described in **Note 3**, commonly cells will be quite dense in the center of the well and sparse near the edges. Usually, there will be a band surrounding the central density where cells are

dispersed well for imaging. At the ideal density, cells will be in close proximity and direct contact, but not at all piled on top of each other.

6. One of the most critical concerns with live-cell imaging is the danger of phototoxicity. One result of excitation of intracellular fluorophores is the generation of reactive oxygen species (ROS), which causes stress and damage to cells [9]. This mechanism may be particularly disruptive in the context of viral infection since this process can also trigger ROS production [10, 11]. Therefore, it is important to limit light intensity and duration during imaging, particularly for shorter wavelength light.
7. If an automatic microscope or imaging incubator is not available, cells may be kept in a cell culture incubator and infection progression can be tracked by removing the sample to image at desired time points. Periodic excursions from the incubator should not affect the progress of infection or cell viability as long as the imaging sessions are 15 min or less. However, final time-point images should be carefully compared to the control, imaged only at the end of the experiment, to ensure that the imaged sample has not been compromised. For myxoma infection, cells should ideally be imaged at least every 4 h.
8. When imaging suspension cells on an automated microscope, the speed of the stage when moving between positions can disrupt the experiment by moving the cells out of view or disturbing cell-cell interactions that may be experimentally relevant. If cells appear to be moving during imaging, the default stage speed should be decreased.
9. For automated image analysis approaches that require identification of individual cells, accuracy of results is highly dependent on cell density. Cells that are in contact, but loosely packed, such that the areas of contact are small relative to the cell circumference, are readily distinguished by automated methods.
10. In order to maximize infection, cells do not need to be washed after exposure to virus. This step is only required for experiments in which unbound virus would confound results, for example, when tracking transfer of bound virus from one cell population to another.
11. Polymer cover glass substrates are not compatible with all types of immersion oil. We have found that type A immersion oil works well with these vessels.
12. The frequency and duration of imaging should be adjusted to address the experimental question. Viral dynamics at the cell surface can be rapid and require image frequency of about a minute. However, these processes may be transient and conclude in a few minutes or hours. It is good practice to do

preliminary experiments to narrow down the time frame of interest in order to improve the efficiency of data acquisition as well as limit unnecessary cell stress from prolonged light exposure.

13. It should be noted when tracking individual virions that not all particles are infective. Cells supporting productive infection with vMyx-mVenus will begin to emit diffuse, cytoplasmic green fluorescence within a few hours of infection as they initiate expression of viral proteins (Fig. 2).
14. We find that a 24-h incubation produces a robust infection in donor Jurkat cells, but extending the incubation to 36–48 h increases the amount of virus and subsequent infection rate of target cells.
15. Sample calculation for seeding donor and target cells:
    Donor cell concentration = 2.5×10^7 cells/mL

    $$(2 \times 10^5 \text{cells/well})/(2.5 \times 10^7 \text{cells/mL}) = 8 \times 10^{-3} \text{mL} = 8 \mu\text{L/well}$$

    Target cell concentration = 1.75×10^7 cells/mL

    $$(1 \times 10^5 \text{cells/well})/(1.75 \times 10^7 \text{cells/mL}) = 5.7 \times 10^{-3} \text{mL} = 5.7 \mu\text{L/well}$$

    Volume of medium = 290 μL – 8 μL – 5.7 μL = 276.3 μL
16. Propidium iodide is a probe for late stages of cell death resulting from any mechanism and provides long-term, dynamic response [12]. This fluorophore is impermeant to live cells, but rapidly enters dead cells, which have lost membrane integrity. Intracellularly, it rapidly binds to DNA and emits bright, red fluorescence. It is compatible with Hoechst which binds DNA of both live and dead cells.
17. CellEvent™ (Thermo Fisher Scientific) is a membrane-permeable probe that contains the recognition amino acid sequence for caspase 3/7. The intact molecule is nonfluorescent, but cleavage by caspase 3/7 releases a green fluorescent product. The optimal working concentration for CellEvent™ varies for different cell types and conditions, so a range of concentrations should be tested in preliminary experiments. A shelf-stable version of this reagent is also available in convenient dropper format (CellEvent™ Caspase-3/7 Green ReadyProbes™ Reagent, Thermo Fisher Scientific).

References

1. Witte R, Andriasyan V, Georgi F et al (2018) Concepts in light microscopy of viruses. Viruses 10(4):202
2. Rahman MM, McFadden G (2020) Oncolytic virotherapy with myxoma virus. J Clin Med 9 (1):16
3. Bartee E, Mohamed MR, Lopez MC et al (2009) The addition of tumor necrosis factor plus beta interferon induces a novel synergistic antiviral state against poxviruses in primary human fibroblasts. J Virol 83(2):498–511
4. Chan WM, Bartee EC, Moreb JS et al (2013) Myxoma and vaccinia viruses bind differentially to human leukocytes. J Virol 87 (8):4445–4460

5. Godin AG, Lounis BCognet L (2014) Super-resolution microscopy approaches for live cell imaging. Biophys J 107(8):1777–1784
6. Sigal YM, Zhou RB, Zhuang XW (2018) Visualizing and discovering cellular structures with super-resolution microscopy. Science 361 (6405):880–887
7. Barrett JW, Sypula J, Wang F et al (2007) M135R is a novel cell surface virulence factor of myxoma virus. J Virol 81(1):106–114
8. Liu J, Wennier S, Reinhard M et al (2009) Myxoma virus expressing interleukin-15 fails to cause lethal Myxomatosis in European rabbits. J Virol 83(11):5933–5938
9. Icha J, Weber M, Waters JC et al (2017) Phototoxicity in live fluorescence microscopy, and how to avoid it. BioEssays 39(8):15
10. Rahman MM, McFadden G (2011) Myxoma virus lacking the pyrin-like protein M013 is sensed in human myeloid cells by both NLRP3 and multiple toll-like receptors, which independently activate the inflammasome and NF-kappa B innate response pathways. J Virol 85(23):12505–12517
11. Lin X, Wang RF, Zou W et al (2016) The influenza virus H5N1 infection can induce ROS production for viral replication and host cell death in A549 cells modulated by human Cu/Zn superoxide dismutase (SOD1) overexpression. Viruses 8(1):13
12. Zhao H, Oczos J, Janowski P et al (2010) Rationale for the real-time and dynamic cell death assays using Propidium iodide. Cytometry A 77A(4):399–405

Chapter 10

Metabolomics Analysis of Viral Therapeutics

Haiwei Gu, Xiaojian Shi, Paniz Jasbi, and Jeffrey Patterson

Abstract

Virotherapy, enabled by recent advances in the transdisciplinary field of biotechnology, has emerged as a powerful tool for use in anticancer treatment, gene therapy, immunotherapy, etc. Examining the effects of viruses and virus-derived immune-modulating therapeutics is of great fundamental and clinical interest. Here we describe a sample preparation protocol for metabolite extraction from virus-infected tissue, in addition to liquid chromatography-mass spectrometry conditions essential for subsequent analysis. This metabolomics approach delivers highly sensitive and specific metabolite information on various biospecimens. Such an approach may be adopted to monitor biological changes in over 30 relevant metabolic pathways in response to viral infection and also viral therapeutics.

Key words LC-MS/MS, Metabolomics, Sample preparation, Viral vector, Virotherapy

1 Introduction

There is great interest in the use of viruses and virus-derived components as therapeutics. Viral vectors also hold enormous potential to realize the goals of gene therapy [1] and, in the past decade, more than 1700 gene therapy trials have been enabled by the use of viral therapeutics [2]. In addition to their use as a gene therapy delivery vehicle, viral vectors may also be used as an alternative to antibiotics [3] and, recently, have proven useful in DNA vaccination [4]. Utilization of adeno-associated viruses (AAV) for the deliverance of engineered DNA to target cells [5] has led to the development of fit-for-purpose AAV vectors [6]. Further popularizing such therapeutic concepts is the advent of bacteriophage lambda vectors, or RNA-binding proteins of phage origin [7]. These phage vectors have become an attractive vaccine platform given their inexpensive production, biological stability, and suitability to a wide range of applications with minimal development needed between uses [8]. Additionally, virus-derived immune-modulating proteins may also be used for therapeutic purposes. For example, rabies-derived glycoproteins have been

Alexandra R. Lucas (ed.), *Viruses as Therapeutics: Methods and Protocols*, Methods in Molecular Biology, vol. 2225, https://doi.org/10.1007/978-1-0716-1012-1_10,

shown to facilitate drug delivery to the central nervous system (CNS) [9], while virus-derived anti-inflammatory proteins may improve the human immune response to immune-mediated diseases and in various cancers through inhibition of protease, chemokine, cytokine, and apoptotic cascades [10].

Since metabolites are sensitive to subtle differences and changes in pathological status and immune response status, metabolomics, the comprehensive study of small-molecular-weight metabolites and their dynamic changes in biological systems [11–15], provides advanced methods to identify changing metabolite levels, and has resulted in the rapid discovery of disease biomarkers during the past decade (*see* Fig. 1) [16–21]. Mass spectrometry (MS)-based metabolic profiling has proven to be a promising tool for analyzing metabolic alterations due to various diseases and, therefore, can provide sensitive and valuable diagnostic information [18, 22, 23], pathogenesis identification [24, 25], and potential therapeutic targets for clinical treatments [26] and disease monitoring [27]. Indeed, previous studies have applied metabolomics to various aspects of viral therapeutics such as the elucidation of metabolites involved in virus infection and pathogenesis [28] and accurate differentiation of vaccination status [29], in addition to characterizing the metabolic profiles of patients with hepatitis B virus-related

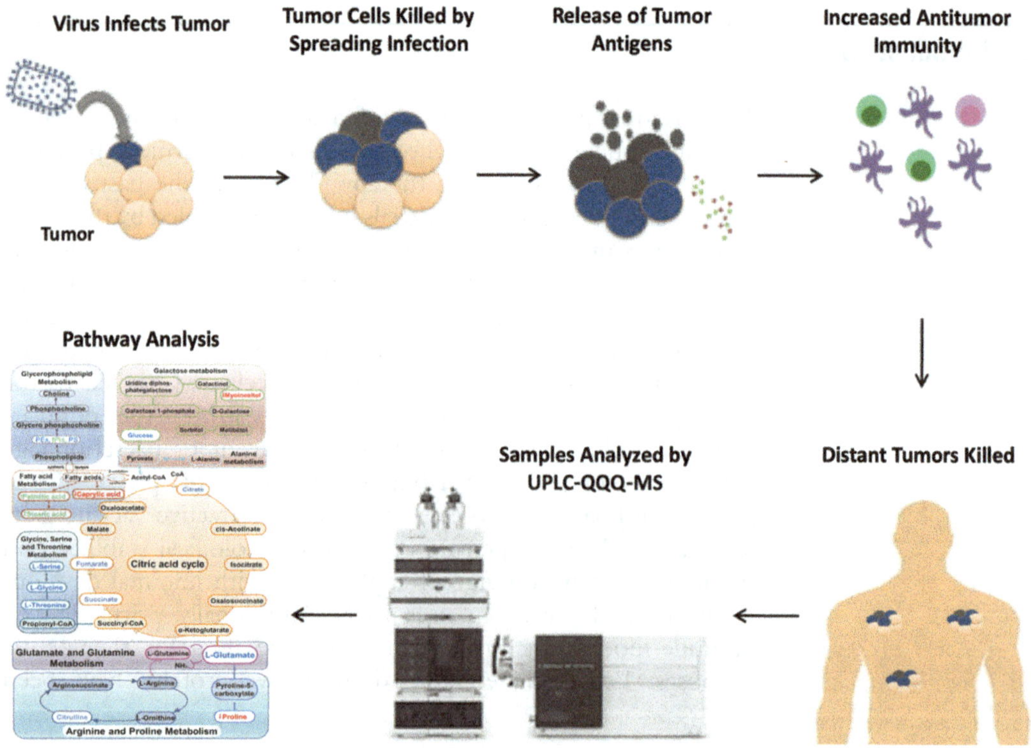

Fig. 1 Summary of cancer virotherapy mechanism and LC-MS/MS metabolomics analysis

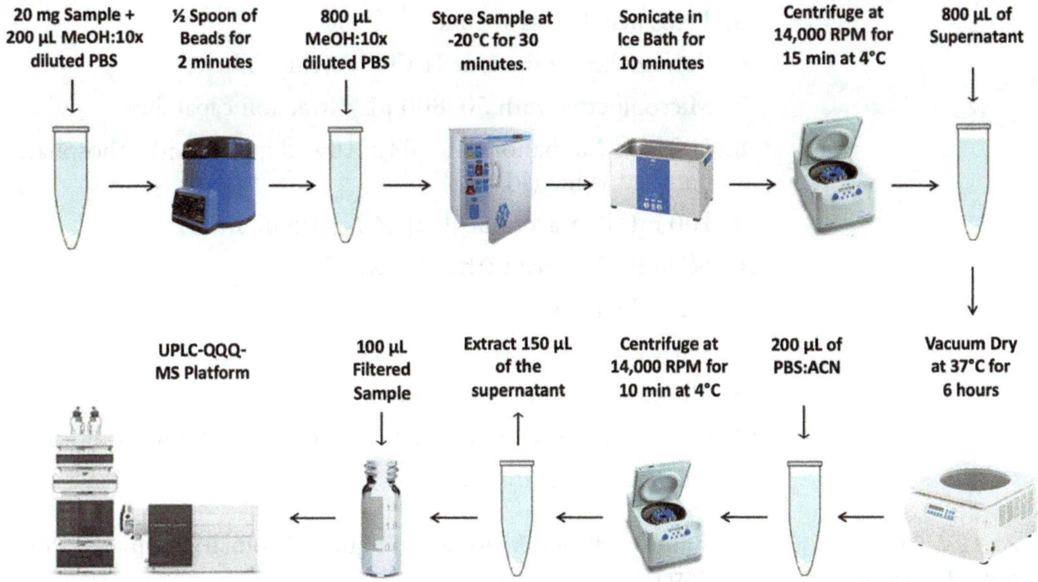

Fig. 2 Summary of LC-MS/MS sample preparation method

hepatocellular carcinoma [30], viruses of the gastrointestinal microbiome [31], inflammatory cytokines involved in H1N1 influenza [32], as well as fingerprinting of HIV-1 and -2 infection in macrophages [33].

Herein, we present a sample preparation protocol and associated liquid chromatography-tandem mass spectrometry (LC-MS/MS) conditions for the detection of aqueous metabolites in tissue (*see* Fig. 2). The targeted LC-MS/MS method used here was modeled and developed after those used in a growing number of studies [18, 34–39].

2 Materials

Prepare all solutions using ultrapure water (*see* **Note 1**). Ensure that all reagents used are at least analytical grade (≥99% purity). Prepare and store all reagents at room temperature unless indicated otherwise. Follow all pertinent waste disposal protocols as they apply. Thaw samples as directed. Check pipettes to confirm readiness for use. This targeted LC-MS/MS protocol does not provide absolute concentrations of detected metabolites.

2.1 Preparation of Tissue Samples

1. 1.5 mL Eppendorf tubes.
2. Bullet beads.
3. 2 mL glass vials.
4. 300 µL glass inserts.

5. 1 mL needle-less syringes.
6. Polyvinylidene fluoride (PVDF) syringe filters.
7. Micropipettes with 50–800 μL extraction capability.
8. 100 mL Methanol (MeOH): 10× dilution with phosphate-buffered saline (PBS).
9. 100 mL PBS:acetonitrile ([ACN] 4:6 dilution).
10. Scientific balance with mg readability.
11. Bullet blender.
12. Vortex mixer.
13. Ultrasonic liquid processor.
14. Centrifuge with speed, time, and temperature control.
15. Vacuum concentrator.

2.2 LC-MS/MS Targeted Analysis

1. Suitable ultra-performance liquid chromatography (UPLC) system.
2. Triple-quadrupole (QQQ)-MS instrument.
3. Electrospray ionization (ESI) source.
4. Hydrophilic interaction chromatography (HILIC) column.
5. Mobile-phase solvents:
   (a) 10 mM ammonium acetate, 10 mM ammonium hydroxide in 95% H_2O/5% ACN.
   (b) 10 mM ammonium acetate, 10 mM ammonium hydroxide in 95% ACN/5% H_2O.

3 Methods

In an effort to identify and overcome potential issues, please see Subheading 4.

3.1 Preparation of Tissue Samples

1. Weigh out 20 mg ± 1.5 mg of biological sample in a 1.5 mL Eppendorf tube.
2. Homogenize biological sample in 200 μL MeOH:10× diluted PBS in 1.5 mL Eppendorf tube (*see* **Note 2**).
3. Add ½ spoon of beads and disrupt the sample using bullet blender for 2 min.
4. Centrifuge for 1–2 min to remove liquid from the cap.
5. Add 800 μL MeOH:10× diluted PBS (*see* **Note 3**).
6. Vortex for 10 s.
7. Store sample at −20 °C for 30 min.
8. Sonicate the mixture in an ice bath for 10 min (*see* **Note 4**).

9. Centrifuge the mixture at 21,694 × *g* for 15 min at 4 °C (*see* **Note 5**).
10. Remove 800 μL supernatant and transfer to a new Eppendorf tube (*see* **Note 6**).
11. Completely dry the sample(s) using a vacuum concentrator at 37 °C for 6 ± 2 h (*see* **Note 7**).
12. Once sample(s) is/are completely dried, reconstitute by adding 200 μL of PBS:ACN solution.
13. Vortex the sample for 30 s.
14. Sonicate each sample for 5 min in a room-temperature water bath.
15. Centrifuge the samples at 21,694 × *g* for 10 min at 4 °C (*see* **Note 8**).
16. Extract 150 μL of the supernatant into a syringe and slowly plunge the sample through the filter and into a new 1.5 mL Eppendorf tube (*see* **Note 9**).
17. Extract the remaining supernatant from each sample and inject into a single 2 mL tube (larger tubes may be used if needed) for quality control (QC) analysis.
18. Transfer 100 μL of the filtered experimental sample into a glass insert inside a glass LC vial. The leftover may be used for backup.
19. Vortex the QC sample for 10 s and then filter it.
20. Transfer 1 mL of the QC sample into a glass LC vial (*see* **Note 10**).

3.2 LC-MS/MS Targeted Analysis

1. All experiments should be performed using an UPLC-QQQ-MS platform.
2. Dictate an appropriate worklist such that each sample is injected twice, 10 μL for analysis using negative ionization mode and 4 μL for analysis using positive ionization mode (*see* **Note 11**).
3. The flow rate is maintained at 0.3 mL/min, auto-sampler temperature is kept at 4 °C, while the column compartment is set to 40 °C.
4. *HILIC gradient parameters*: After an initial 1-min isocratic elution of 90% mobile phase solvent B, decrease the percentage of solvent B to 40% at $t = 11$ min. Maintain the composition of solvent B at 40% for 4 min ($t = 15$ min), after which the percentage of solvent B should gradually be returned to 90%, to prepare for the next injection (*see* **Note 12**). An example of a total ion chromatogram (TIC) can be found in Fig. 3.
5. *MS/MS parameters*: The mass spectrometer should be equipped with an electrospray ionization (ESI) source.

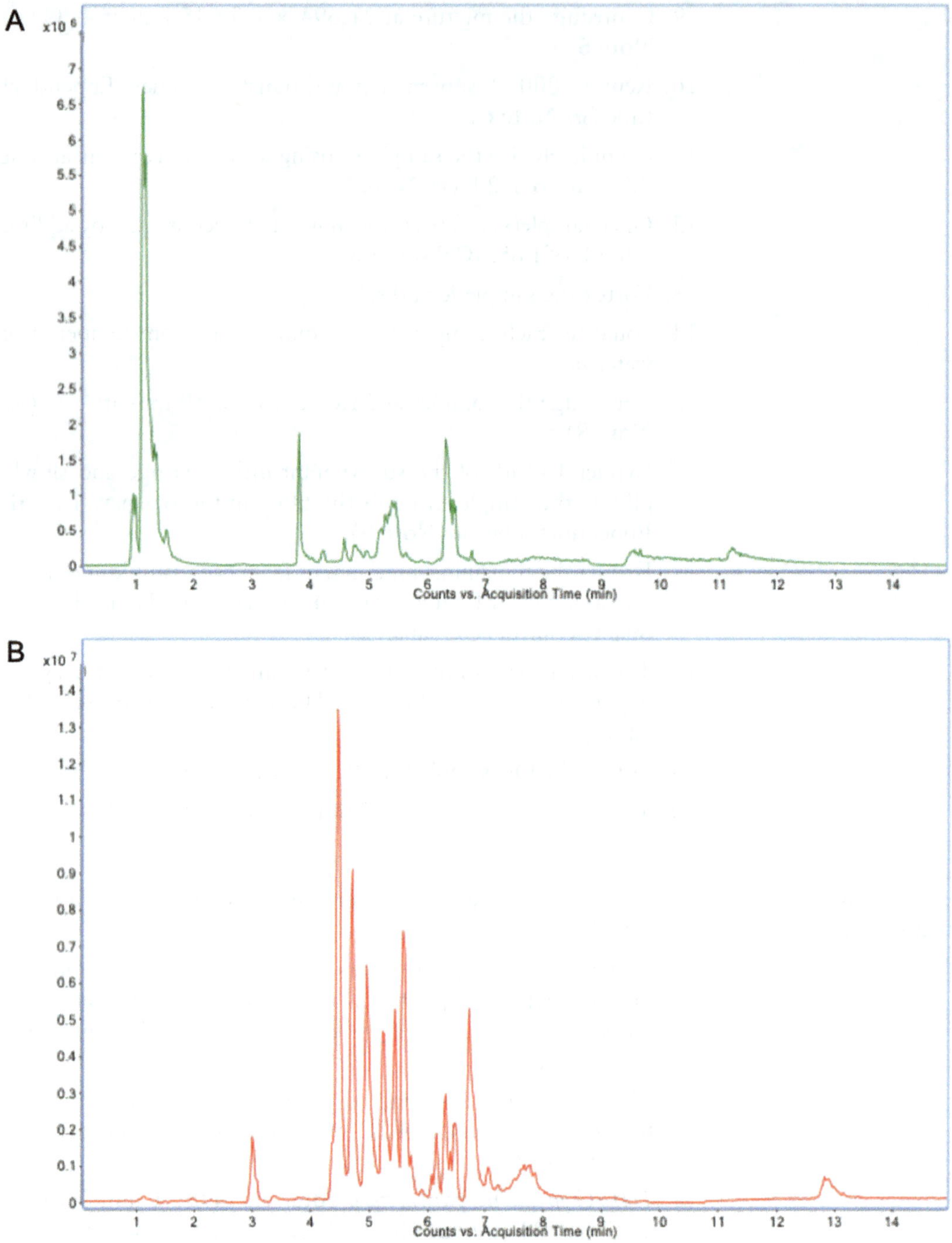

Fig. 3 LC-MS/MS total ion chromatogram: (**a**) negative and (**b**) positive modes

Targeted data acquisition should be performed in multiple reaction monitoring (MRM) mode.

6. Table 1 lists LC-MS parameters for the validated detection of 310 metabolites using reference standards. For additional parameter details, *see* **Note 13**.

Table 1
LC-MS parameters for the validated identification and detection of 310 metabolites

No.	HMDB/CAS	Compound name	Precursor ion	Product ion	Collision energy	Polarity	RT (min)
1	HMDB0031645	Acetamide	60.05	42.80	13	Positive	5.050
2	HMDB0001525	Imidazole	69.05	42.00	21	Positive	1.826
3	HMDB0000119	Glyoxylic acid	72.99	44.90	5	Negative	5.031
4	HMDB0001522	Methylguanidine	74.07	57.00	13	Positive	5.730
5	HMDB0000115	Glycolic acid	75.01	46.90	9	Negative	5.350
6	HMDB0014691	Acetohydroxamic acid	76.04	43.00	9	Positive	6.215
7	HMDB0000123	Glycine	76.04	30.10	13	Positive	6.598
8	HMDB0000925	TMAO	76.08	58.00	21	Positive	6.187
9	HMDB0002039	2-Pyrrolidinone	86.06	44.00	29	Positive	1.620
10	HMDB0000243	Pyruvate	87.00	43.00	5	Negative	5.45
11	HMDB0001873	Isobutyric acid	87.04	87.04	5	Negative	1.948
12	HMDB0000190	Lactate	89.02	43.00	9	Negative	4.1
13	HMDB0001414	Putrescine	89.11	72.00	9	Positive	6.118
14	HMDB0000161	Alanine	90.06	44.00	9	Positive	6.140
15	HMDB0000271	Sarcosine	90.06	44.10	9	Positive	6.184
16	HMDB0001882	Dihydroxyacetone	91.04	91.04	0	Positive	1.734
17	HMDB0000005	2-Ketobutyric acid	101.02	57.00	5	Negative	1.490
18	HMDB0000060	Acetoacetate	101.02	56.90	9	Negative	2.194
19	HMDB0002176	2-Methylbutyric acid	101.06	101.06	5	Negative	1.693
20	HMDB0000718	Isovaleric acid	101.06	101.06	5	Negative	6.064
21	HMDB0000691	Malonic acid	103.00	58.90	9	Negative	4.306
22	HMDB0000008	2-Hydroxybutyric acid	103.04	58.90	9	Negative	5.87
23	HMDB0000357	3-Hydroxybutyric acid	103.04	56.80	9	Negative	4.264
24	HMDB0002322	Cadaverine	103.13	85.90	5	Positive	6.506
25	HMDB0001906	2/3-Aminoisobutyric acid	104.07	57.90	9	Positive	5.295
26	HMDB0000452	2-Aminobutyric acid	104.07	57.90	13	Positive	5.706
27	HMDB0031654	3-Aminobutyric acid	104.07	85.90	5	Positive	5.934
28	HMDB0000112	4-Aminobutyric acid	104.07	87.00	9	Positive	6.596
29	HMDB0000092	Dimethylglycine	104.07	57.90	13	Positive	5.295
30	142-26-7	*N*-Acetylethanolamine	104.07	61.80	5	Positive	1.915

(continued)

Table 1
(continued)

No.	HMDB/CAS	Compound name	Precursor ion	Product ion	Collision energy	Polarity	RT (min)
31	HMDB0000097	Choline	104.11	58.00	33	Positive	7.000
32	HMDB0000139	Glyceric acid	105.02	75.00	9	Negative	4.843
33	HMDB0000187	Serine	106.05	60.00	9	Positive	6.824
34	HMDB0001169	4-Aminophenol	110.06	69.00	9	Positive	5.184
35	HMDB0000617	2-Furoic acid	111.01	66.90	5	Negative	1.760
36	HMDB0000630	Cytosine	112.05	94.90	21	Positive	3.445
37	HMDB0000870	Histamine	112.09	95.00	17	Positive	8.673
38	HMDB0000300	Uracil	113.04	71.60	25	Positive	8.707
39	HMDB0000562	Creatinine	114.07	44.10	21	Positive	2.988
40	HMDB0000134	Fumarate	115.00	71.00	5	Negative	2.130
41	HMDB0000176	Maleic acid	115.00	70.90	9	Negative	1.455
42	HMDB0000019	Alpha-ketoisovaleric acid	115.04	70.90	5	Negative	1.455
43	HMDB0000720	Levulinic acid	115.04	70.90	9	Negative	3.250
44	HMDB0000689	4-Methylvaleric acid	115.07	115.07	5	Negative	1.505
45	HMDB0000535	Hexanoic acid	115.07	115.07	5	Negative	1.471
46	HMDB0000162	Proline	116.07	70.00	13	Positive	5.408
47	HMDB0000202	Methylmalonic acid	117.02	73.00	5	Negative	4.622
48	HMDB0000254	Succinate	117.02	73.00	9	Negative	4.293
49	HMDB0000754	Beta-hydroxyisovaleric acid	117.05	71.00	9	Negative	2.132
50	HMDB0000532	Acetylglycine	118.05	75.90	5	Positive	4.335
51	HMDB0000128	Glycocyamine	118.06	42.90	45	Positive	6.344
52	HMDB0000738	Indole	118.07	85.80	13	Positive	5.385
53	HMDB0003355	Amino valerate	118.09	100.90	9	Positive	6.822
54	HMDB0000043	Betaine	118.09	58.00	29	Positive	4.791
55	HMDB0013716	Norvaline	118.09	72.00	5	Positive	5.019
56	HMDB0000883	Valine	118.09	72.00	5	Positive	5.224
57	HMDB0002649	Erythrose	119.03	42.90	17	Negative	4.519
58	HMDB0000719	Homoserine	120.07	73.90	9	Positive	6.343
59	HMDB0000167	Threonine	120.07	73.90	9	Positive	6.343
60	HMDB0011718	4-Hydroxy benzaldehyde	121.03	92.00	25	Negative	1.301
61	HMDB0001870	Benzoic acid	121.03	77.00	9	Negative	1.487

(continued)

Table 1 (continued)

No.	HMDB/CAS	Compound name	Precursor ion	Product ion	Collision energy	Polarity	RT (min)
62	HMDB0000574	Cysteine	122.03	80.90	13	Positive	13.454
63	HMDB0001406	Nicotinamide	123.06	79.80	25	Positive	1.746
64	HMDB0001488	Nicotinic acid	124.04	78.10	22	Positive	3.671
65	HMDB0002243	Picolinic acid	124.04	77.70	25	Positive	5.315
66	HMDB0000251	Taurine	126.02	44.10	17	Positive	5.223
67	HMDB0000898	Methylhistamine	126.11	109.10	13	Positive	4.879
68	HMDB0002024	4-Imidazoleacetic acid	127.05	80.90	13	Positive	5.497
69	HMDB0000634	Citraconic acid	129.02	85.10	5	Negative	1.487
70	HMDB0000620	Glutaconic acid	129.02	85.00	5	Negative	4.774
71	HMDB0000491	3-Methyl-2-oxovaleric acid	129.05	129.05	5	Negative	1.351
72	HMDB0000491	Ketoisoleucine	129.05	85.10	5	Negative	1.368
73	HMDB0000695	Ketoleucine	129.05	85.00	5	Negative	1.368
74	HMDB0000267	Pyroglutamic acid	130.05	83.80	9	Positive	4.675
75	HMDB0000070	Pipecolinic acid	130.09	84.00	13	Positive	5.405
76	HMDB0000223	Oxaloacetic acid	131.00	42.00	17	Negative	6.772
77	HMDB0000622	Ethylmalonic acid	131.03	87.00	5	Negative	3.282
78	HMDB0000661	Glutaric acid	131.03	86.90	9	Negative	6.179
79	HMDB0001844	Methyl succinate	131.03	98.90	5	Negative	2.011
80	HMDB0000665	Leucic acid	131.07	85.10	9	Negative	1.994
81	HMDB0001432	Agmatine	131.13	71.90	17	Positive	6.912
82	HMDB0001149	5-Aminolevulinic acid	132.07	85.90	13	Positive	6.521
83	HMDB0000725	Hydroxyproline	132.07	85.90	17	Positive	6.113
84	HMDB0000064	Creatine	132.08	44.00	21	Positive	6.204
85	HMDB0000172	Isoleucine	132.10	86.00	9	Positive	4.697
86	HMDB0000557	L-Alloisoleucine	132.10	86.00	9	Positive	4.468
87	HMDB0000687	Leucine	132.10	86.00	9	Positive	4.468
88	HMDB0001645	Norleucine	132.10	86.00	9	Positive	4.468
89	HMDB0000156	Malate	133.01	114.90	9	Negative	5.87
90	HMDB0000168	Asparagine	133.06	73.90	13	Positive	6.813
91	HMDB0003374	Ornithine	133.10	70.00	21	Positive	11.112
92	HMDB0000191	Aspartate	134.05	73.90	13	Positive	6.471
93	HMDB0000209	Phenylacetic acid	135.04	91.00	5	Negative	1.503

(continued)

Table 1
(continued)

No.	HMDB/CAS	Compound name	Precursor ion	Product ion	Collision energy	Polarity	RT (min)
94	HMDB0000034	Adenine	136.06	119.10	25	Positive	2.755
95	HMDB0001895	2-Hydroxybenzoic acid	137.02	93.10	21	Negative	1.215
96	HMDB0000710	4-Hydroxybenzoic acid	137.02	93.10	17	Negative	2.316
97	HMDB0014581	Allopurinol	137.05	54.20	29	Positive	2.070
98	HMDB0000157	Hypoxanthine	137.05	54.80	33	Positive	2.938
99	HMDB0001123	Anthranilic acid	138.06	120.00	9	Positive	1.568
100	20989-17-7	2-Phenylglycinol	138.09	103.00	21	Positive	2.549
101	HMDB0000301	Urocanic acid	139.05	92.90	21	Positive	4.444
102	HMDB0002658	6-Hydroxynicotinic acid	140.04	121.90	17	Positive	4.604
103	HMDB0002349	Muconic acid	141.02	97.00	5	Negative	5.684
104	HMDB0000393	3-Hexenedioic acid	143.03	99.00	5	Negative	6.042
105	HMDB0000482	Caprylic acid	143.11	143.11	5	Negative	1.350
106	HMDB0000208	Alpha-ketoglutaric acid	145.01	101.00	5	Negative	4.99
107	HMDB0000422	2-Methylglutaric acid	145.05	101.10	9	Negative	5.500
108	HMDB0000448	Adipic acid	145.05	101.00	9	Negative	6.025
109	HMDB0000895	Acetylcholine	146.12	86.80	9	Positive	3.439
110	HMDB0001257	Spermidine	146.17	72.00	17	Positive	6.499
111	HMDB0059655	2HG	147.03	129.00	5	Negative	5.49
112	HMDB0000641	Glutamine	147.08	84.00	17	Positive	6.673
113	HMDB0000182	Lysine	147.12	83.90	21	Positive	11.749
114	HMDB0003339	Glutamic acid	148.06	84.00	17	Positive	6.39
115	147-73-9	Meso-tartaric acid	149.01	86.90	13	Negative	11.000
116	HMDB0059916	Tartaric acid	149.01	87.00	9	Negative	11.000
117	HMDB0001587	Phenylglyoxylic acid	149.02	77.00	9	Negative	1.315
118	HMDB0000646	L-(+)-Arabinose	149.04	59.00	20	Negative	6.034
119	HMDB0000283	Ribose	149.04	89.00	5	Negative	4.4
120	HMDB0000098	Xylose	149.04	89.00	5	Negative	4.320
121	HMDB0002097	4-Ethylbenzoic acid	149.06	105.10	9	Negative	1.501
122	HMDB0032075	p-Tolylacetic acid	149.06	105.00	5	Negative	1.501
123	HMDB0001878	Thymol	149.09	87.00	13	Negative	11.100
124	HMDB0000696	Methionine	150.06	103.90	9	Positive	4.923

(continued)

Table 1 (continued)

No.	HMDB/CAS	Compound name	Precursor ion	Product ion	Collision energy	Polarity	RT (min)
125	HMDB0000669	2-Hydroxy phenylacetic acid	151.04	107.00	13	Negative	1.280
126	HMDB0000440	3-Hydroxyphenylacetic acid	151.04	107.00	5	Negative	1.280
127	HMDB0004815	4-Hydroxy-3-methylbenzoic acid	151.04	107.10	9	Negative	1.923
128	HMDB0000020	4-Hydroxyphenylacetic acid	151.04	107.00	9	Negative	1.907
129	HMDB0000703	Mandelic acid	151.04	107.00	5	Negative	1.907
130	HMDB0000508	Adonitol	151.06	70.80	16	Negative	4.149
131	HMDB0001851	L-(−)-Arabitol	151.06	89.10	8	Negative	4.335
132	HMDB0002917	Xylitol	151.06	59.00	20	Negative	4.390
133	HMDB0000132	Guanine	152.06	135.10	21	Positive	4.009
134	HMDB0000397	2,3-Dihydroxybenzoic acid	153.02	109.00	17	Negative	1.027
135	HMDB0000152	Gentisic acid	153.02	108.10	25	Negative	1.415
136	HMDB0001856	Protocatechuic acid	153.02	109.00	13	Negative	1.731
137	HMDB0000292	Xanthine	153.04	109.90	21	Positive	3.415
138	56-17-7	Cystamine	153.05	107.70	5	Positive	9.759
139	HMDB0001476	3-Hydroxyanthranilic acid	154.05	112.90	5	Positive	13.765
140	HMDB0000073	Dopamine	154.09	137.00	9	Positive	4.015
141	HMDB0000177	Histidine	156.08	110.10	13	Positive	7.182
142	HMDB0000555	3-Methyladipic acid	159.06	97.20	9	Negative	5.464
143	HMDB0000678	Isovalerylglycine	160.10	57.00	17	Positive	2.889
144	HMDB0000303	Tryptamine	161.11	144.00	9	Positive	3.433
145	HMDB0000510	2-Aminoadipic acid	162.08	98.00	13	Positive	6.839
146	HMDB0000062	Carnitine	162.12	43.00	25	Positive	6.550
147	HMDB0001713	m-Coumaric acid	163.04	119.00	17	Negative	1.499
148	HMDB0002035	p-Coumaric acid	163.04	119.00	17	Negative	1.849
149	HMDB0000205	Phenylpyruvic acid	163.04	91.00	9	Negative	1.279
150	HMDB0001890	Acetylcysteine	164.04	123.00	9	Positive	2.555
151	HMDB0002107	Phthalic acid	165.02	121.00	5	Negative	2.380

(continued)

Table 1
(continued)

No.	HMDB/CAS	Compound name	Precursor ion	Product ion	Collision energy	Polarity	RT (min)
152	HMDB0060256	D-Xylonic acid	165.04	75.00	16	Negative	5.218
153	HMDB0000779	3-Phenyllactic acid	165.05	147.10	9	Negative	1.499
154	HMDB0002072	4-Methoxyphenylacetic acid	165.05	106.00	9	Negative	1.499
155	HMDB0000159	Phenylalanine	166.09	120.20	17	Positive	4.373
156	HMDB0000263	Phosphoenolpyruvic acid	166.97	78.80	33	Negative	6.060
157	HMDB0000130	Homogentisic acid	167.03	108.00	13	Negative	5.765
158	HMDB0000484	Vanillic acid	167.03	152.00	13	Negative	2.000
159	HMDB0000232	Quinolinic acid	168.03	150.00	5	Positive	5.263
160	HMDB0001112	D-Glyceraldehyde 3-phosphate	168.99	79.00	33	Negative	5.800
161	HMDB0000289	Urate	169.04	141.00	13	Positive	4.768
162	HMDB0000239	Pyridoxine	170.08	134.10	17	Positive	2.295
163	HMDB0000001	1-Methylhistidine	170.10	95.90	21	Positive	6.930
164	HMDB0000511	Capric acid	171.14	171.14	5	Negative	1.261
165	HMDB0000072	Aconitic acid	173.01	85.00	9	Negative	6.191
166	HMDB0003070	Shikimic acid	173.04	93.10	9	Negative	5.801
167	HMDB0000893	Suberic acid	173.08	111.10	13	Negative	5.004
168	HMDB0000721	Glycylproline	173.09	116.10	13	Positive	6.701
169	HMDB0000044	L-Ascorbic acid	175.02	114.90	9	Negative	3.027
170	HMDB0003357	Acetylornithine	175.11	70.00	25	Positive	6.769
171	HMDB0000517	Arginine	175.12	70.00	25	Positive	11.665
172	HMDB0000197	Indole-3-acetic acid	176.07	158.10	13	Positive	6.193
173	HMDB0000904	Citrulline	176.11	69.90	25	Positive	7.020
174	HMDB0000259	Serotonin	177.10	160.00	9	Positive	4.326
175	HMDB0000707	4-Hydroxy phenylpyruvic acid	179.03	135.00	9	Negative	1.273
176	HMDB0000660	Fructose	179.05	89.00	5	Negative	4.716
177	HMDB0033704	Galactose	179.05	89.10	5	Negative	4.732
178	HMDB0000122	Glucose	179.05	89.00	5	Negative	5.410
179	HMDB0000169	Mannose	179.05	58.90	17	Negative	5.350

(continued)

Table 1 (continued)

No.	HMDB/CAS	Compound name	Precursor ion	Product ion	Collision energy	Polarity	RT (min)
180	HMDB0000211	Myoinositol	179.05	135.00	1	Negative	1.497
181	HMDB0000714	Hippuric acid	180.07	104.90	13	Positive	2.499
182	HMDB0006479	Glucosamine	180.09	162.00	5	Positive	6.791
183	HMDB0000118	Homovanillic acid	181.05	136.80	5	Negative	2.268
184	HMDB0003269	Nicotinuric acid	181.06	134.80	17	Positive	4.234
185	HMDB0000765	D-Mannitol	181.07	101.10	20	Negative	5.213
186	HMDB0000107	Dulcitol	181.07	58.90	24	Negative	5.294
187	HMDB0000247	Sorbitol	181.07	58.70	17	Negative	5.223
188	HMDB0000158	Tyrosine	182.08	90.90	33	Positive	5.284
189	HMDB0000017	4-Pyridoxic acid	184.06	166.10	9	Positive	1.288
190	HMDB0000068	Epinephrine	184.10	166.10	5	Positive	5.366
191	HMDB0000819	Normetanephrine	184.10	166.10	5	Positive	4.462
192	HMDB0000807	3-Phosphoglyceric acid	184.98	79.00	50	Negative	7.013
193	HMDB0000784	Azelaic acid	187.09	125.10	13	Negative	4.020
194	HMDB0006029	Acetyl-L-glutamine	189.09	130.00	13	Positive	4.651
195	HMDB0000715	Kynurenic acid	190.05	144.00	25	Positive	2.338
196	HMDB0002302	3-Indolepropionic acid	190.09	130.10	25	Positive	1.562
197	HMDB0000094	Citrate	191.02	111.00	9	Negative	7
198	HMDB0000193	Isocitrate	191.02	111.00	13	Negative	6.41
199	HMDB0000763	HIAA	192.07	146.10	13	Positive	4.507
200	HMDB0002545	D-Galacturonic acid	193.03	58.70	17	Negative	6.662
201	HMDB0000127	Glucuronic acid	193.03	113.00	9	Negative	6.493
202	HMDB0000954	Ferulic acid	193.05	134.00	13	Negative	1.512
203	HMDB0029965	Methyl alpha-D-glucopyranoside	193.07	102.80	13	Negative	4.320
204	HMDB0029965	Methyl-D-mannopyranoside	193.07	58.80	21	Negative	3.044
205	HMDB0000565	Galactonic acid	195.05	75.00	21	Negative	6.831
206	HMDB0000625	Gluconic acid	195.05	129.00	13	Negative	6.900
207	HMDB0001847	Caffeine	195.09	138.00	17	Positive	1.447
208	HMDB0000609	DOPA	198.08	152.10	5	Positive	5.838
209	HMDB0004063	Metanephrine	198.12	180.10	5	Positive	3.959

(continued)

Table 1 (continued)

No.	HMDB/CAS	Compound name	Precursor ion	Product ion	Collision energy	Polarity	RT (min)
210	HMDB0000624	Lauric acid	199.17	199.17	5	Negative	1.224
211	HMDB0000792	Sebacic acid	201.11	139.10	17	Negative	3.612
212	HMDB0003334	Dimethylarginine	203.15	70.00	21	Positive	10.306
213	HMDB0060484	Indole-3-pyruvic acid	204.07	163.20	5	Positive	12.162
214	HMDB0000201	Acetylcarnitine	204.13	85.00	17	Positive	5.661
215	HMDB0000929	Tryptophan	205.10	187.90	5	Positive	4.437
216	HMDB0000881	Xanthurenic acid	206.05	131.90	29	Positive	2.245
217	HMDB0000671	Indole-3-lactic acid	206.08	118.10	25	Positive	4.453
218	HMDB0000639	Mucic acid	209.03	84.90	13	Negative	7.078
219	HMDB0000684	Kynurenine	209.09	192.00	5	Positive	4.483
220	HMDB0001511	Phosphocreatine	212.05	45.00	17	Positive	1.674
221	2280-85-5	6-Methyl-DL-tryptophan	219.12	202.10	5	Positive	4.071
222	HMDB0000210	Pantothenic acid	220.12	90.00	9	Positive	4.071
223	HMDB0000472	5-Hydroxytryptophan	221.09	204.00	5	Positive	5.236
224	HMDB0000215	Acetylglucosamine	222.10	138.10	17	Positive	4.505
225	HMDB0000853	*N*-Acetyl-D-galactosamine	222.10	204.10	4	Positive	4.387
226	HMDB0000215	*N*-Acetyl-D-glucosamine	222.10	204.30	0	Positive	4.487
227	HMDB0000742	Homocysteine	223.08	88.00	29	Positive	8.604
228	HMDB0000732	3-Hydroxykynurenine	225.09	207.00	5	Positive	1.102
229	2387-23-7	*N,N*-Dicyclohexylurea	225.20	100.10	13	Positive	1.239
230	HMDB0001904	3-Nitrotyrosine	227.07	181.00	5	Positive	4.801
231	HMDB0000033	Carnosine	227.12	110.00	21	Positive	7.655
232	HMDB0000806	Myristic acid	227.20	227.20	5	Negative	1.189
233	HMDB0000014	2-Deoxycytidine	228.10	112.00	9	Positive	3.591
234	4300-28-1	D-Ribose 5-phosphate	229.01	78.80	45	Negative	6.850
235	66768-39-6	D-Xylose 5-phosphate	229.01	78.90	45	Negative	5.200
236	HMDB0000012	2-Deoxyuridine	229.08	113.00	9	Positive	2.289
237	HMDB0001923	Naproxen	229.08	169.00	33	Negative	1.358
238	HMDB0014732	Amiloride	230.06	171.00	13	Positive	4.192

(continued)

Table 1
(continued)

No.	HMDB/CAS	Compound name	Precursor ion	Product ion	Collision energy	Polarity	RT (min)
239	HMDB0001389	Melatonin	233.13	174.00	13	Positive	1.307
240	HMDB0000192	Cystine	241.03	74.00	21	Positive	7.980
241	HMDB0000826	Pentadecanoic acid	241.21	241.21	5	Negative	1.189
242	HMDB0000089	Cytidine	244.10	112.00	9	Positive	4.229
243	HMDB0000296	Uridine	245.08	113.00	5	Positive	2.950
244	HMDB0000030	Biotin	245.10	227.00	9	Positive	3.955
245	HMDB0000101	2-Deoxyadenosine	252.11	136.10	13	Positive	2.585
246	HMDB0000845	Neopterin	254.09	206.00	21	Positive	5.621
247	HMDB0000220	Palmitic acid	255.23	255.23	5	Negative	1.189
248	HMDB0000982	5-Methylcytidine	258.11	126.10	9	Positive	3.932
249	HMDB0001078	D-Mannose 6-phosphate	259.02	79.10	52	Negative	7.170
250	HMDB0000124	Fructose 6-phosphate	259.02	78.80	41	Negative	6.87
251	HMDB0001586	Glucose 1-phosphate	259.02	79.10	29	Negative	6.889
252	HMDB0001401	Glucose 6-phosphate	259.02	78.80	41	Negative	7.300
253	HMDB0001254	Glucosamine 6-phosphate	260.06	126.10	9	Positive	7.177
254	HMDB0001849	Propranolol	260.17	116.00	17	Positive	2.516
255	94-24-6	Tetracaine	265.19	176.00	17	Positive	1.534
256	HMDB0000085	2-Deoxyguanosine	268.11	151.90	17	Positive	4.000
257	HMDB0000050	Adenosine	268.11	136.00	13	Positive	3.018
258	HMDB0000195	Inosine	269.09	137.00	21	Positive	3.794
259	HMDB0002259	Heptadecanoic acid	269.25	269.25	5	Negative	1.154
260	HMDB0001316	6-Phosphogluconic acid	275.01	78.90	49	Negative	13.882
261	506-24-1	9-Octadecynoic acid	279.23	279.23	5	Negative	1.188
262	HMDB0003331	1-Methyladenosine	282.12	150.00	29	Positive	6.397
263	HMDB0000827	Stearic acid	283.26	283.26	5	Negative	1.137
264	HMDB0000133	Guanosine	284.10	152.10	13	Positive	4.616
265	HMDB0000299	Xanthosine	285.09	153.00	9	Positive	4.844
266	HMDB0060493	*N*-Acetylmuramic acid	292.10	89.00	8	Negative	3.895
267	HMDB0000772	Nonadecanoic acid	297.28	297.28	5	Negative	1.120

(continued)

Table 1
(continued)

No.	HMDB/CAS	Compound name	Precursor ion	Product ion	Collision energy	Polarity	RT (min)
268	HMDB0001409	dUMP	307.03	195.10	17	Negative	7.235
269	HMDB0000125	Glutathione reduced	308.09	84.10	25	Positive	6.070
270	HMDB0014950	Phenylbutazone	309.16	76.90	45	Positive	1.030
271	HMDB0000230	*N*-Acetylneuraminic acid	310.12	274.00	5	Positive	6.350
272	HMDB0000651	Decanoylcarnitine	316.25	84.90	25	Positive	2.332
273	HMDB0001227	dTMP	323.07	80.80	21	Positive	6.033
274	HMDB0001058	Fructose 1,6 biphosphate (F16BP)	338.99	78.90	49	Negative	7.9
275	HMDB0003514	Glucose 1,6 biphosphate (G16BP)	338.99	241.10	17	Negative	8.000
276	HMDB0000055	D-(+)-Cellobiose	341.11	161.00	4	Negative	6.869
277	HMDB0005826	Galactinol dihydrate	341.11	179.20	16	Negative	8.318
278	HMDB0000186	Lactose	341.11	161.00	5	Negative	7.065
279	HMDB0000258	Sucrose	341.11	59.00	50	Negative	6.474
280	HMDB0000975	Trehalose	341.11	178.90	9	Negative	7.014
281	HMDB0003559	Gibberellic acid	345.13	143.00	37	Negative	1.559
282	HMDB0001314	cGMP	346.06	152.00	21	Positive	4.63
283	HMDB0000045	AMP	348.10	136.20	17	Positive	6.544
284	HMDB0001220	Prostaglandin E2	351.21	315.20	9	Negative	1.474
285	HMDB0000939	Adenosyl-L-homocysteine	385.13	136.00	25	Positive	6.875
286	HMDB0001245	DCDP	388.03	112.00	17	Positive	6.922
287	HMDB0000774	Pregnenolone sulfate	395.19	96.80	45	Negative	1.017
288	HMDB0000295	UDP	402.99	78.90	50	Negative	6.79
289	HMDB0001341	ADP	428.04	136.10	21	Positive	6.900
290	HMDB0000121	Folic acid	442.15	295.00	25	Positive	6.806
291	HMDB0001201	GDP	444.03	152.10	33	Positive	7.255
292	HMDB0001056	Dihydrofolic acid	444.17	297.10	17	Positive	6.783
293	HMDB0000797	SAICAR	455.08	110.00	44	Positive	6.947
294	HMDB0000536	Adenylosuccinate	464.08	252.00	25	Positive	8.036

(continued)

Table 1 (continued)

No.	HMDB/CAS	Compound name	Precursor ion	Product ion	Collision energy	Polarity	RT (min)
295	HMDB0000998	dCTP	468.00	111.90	17	Positive	7.5
296	HMDB0001191	dUTP	468.98	80.90	13	Positive	7
297	HMDB0001562	Folinic acid	474.18	327.10	17	Positive	1.464
298	HMDB0003213	Raffinose	503.16	178.90	25	Negative	7.912
299	HMDB0000538	ATP	508.01	136.00	45	Positive	7.350
300	HMDB0001178	ADP ribose	560.10	136.20	21	Positive	6.020
301	HMDB0000290	UDP-GlcNAc	608.09	204.30	9	Positive	7.010
302	HMDB0003337	Glutathione oxidized	613.16	484.00	17	Positive	7.267
303	HMDB0000902	NAD	664.12	136.00	41	Positive	6.767
304	HMDB0003553	Stachyose hydrate	665.21	383.20	40	Negative	8.960
305	HMDB0001487	NADH	666.10	348.10	17	Positive	5.700
306	HMDB0000217	NADP	745.10	604.10	17	Positive	7.200
307	HMDB0000221	NADPH	746.10	746.10	17	Positive	6.500
308	HMDB0001206	Acetyl-CoA	810.14	303.10	37	Positive	6
309	HMDB0001243	Isobutyryl-CoA	838.17	331.20	37	Positive	7.000
310	HMDB0001166	Hydroxybutyryl coenzyme A	854.16	347.10	37	Positive	6.954

4 Notes

1. Render ultrapure water by filtering deionized water.
2. 4:6 dilution of PBS:ACN contains 4:1 (v:v) solution of ^{13}C-lactate and ^{13}C-glutamic acid.
3. Added solutions must be cooled at 4 °C.
4. Prepare ice for use prior to sonication.
5. Ensure that the centrifuge is balanced.
6. Resulting protein pellets may be extracted and saved for protein analysis as needs arise.
7. Drying time is contingent on the number of samples and type of tissue used. It is permitted to dry samples overnight. If the extracted supernatant cannot be dried immediately, samples may be kept at −20 °C for short-term storage or at −80 °C for long-term storage.

8. During this time, you may prepare the vial inserts and syringe filters.
9. It may be necessary to plunge multiple times in order to filter all of the sample. At least 50 μL of filtered sample is required for LC-MS/MS analysis.
10. A glass insert is not needed for analysis of QC sample given sufficient volume of liquid.
11. Both chromatographic separations should be performed in HILIC mode.
12. Targeted data acquisition should be performed in multiple-reaction-monitoring (MRM) mode.
13. MS parameters were optimized using standard references. All LC separation was performed using basic HILIC. The dwell time should be configured to 5 s for each ion signal. Unit resolution is required for both MS1 and MS2.

References

1. Finer M, Glorioso J (2017) A brief account of viral vectors and their promise for gene therapy. Gene Ther 24:1–2
2. Wirth T, Parker N, Ylä-Herttuala S (2013) History of gene therapy. Gene 525:162–169
3. Clark JR, March JB (2006) Bacteriophages and biotechnology: vaccines, gene therapy and antibacterials. Trends Biotechnol 24:212–218
4. Celec P, Gardlik R (2017) Gene therapy using bacterial vectors. Front Biosci 22:81–95
5. Naso MF, Tomkowicz B, Perry WL et al (2017) Adeno-associated virus (AAV) as a vector for gene therapy. BioDrugs 31:317–334
6. Zinn E, Vandenberghe LH (2014) Adeno-associated virus: fit to serve. Curr Opin Virol 8:90–97
7. Zhou Q, Kunder N, De la Paz JA et al (2018) Global pairwise RNA interaction landscapes reveal core features of protein recognition. Nat Commun 9:2511
8. Mattiacio JL, Brewer M, Dewhurst S (2017) Methods in molecular biology (Clifton, N.J.) 1581:245–253
9. Huey R, Hawthorne S, McCarron P (2017) The potential use of rabies virus glycoprotein-derived peptides to facilitate drug delivery into the central nervous system: a mini review. J Drug Target 25:379–385
10. Zheng D, Chen H, Bartee MY et al (2012) Virus-derived anti-inflammatory proteins: potential therapeutics for cancer. Trends Mol Med 18:304–310
11. Patti GJ, Yanes O, Siuzdak G (2012) Metabolomics: the apogee of the omics trilogy. Nat Rev Mol Cell Biol 13:263–269
12. Everett JR, Holmes E, Veselkov KA et al (2019) A unified conceptual framework for metabolic phenotyping in diagnosis and prognosis. Trends Pharmacol Sci 40:763–773
13. Chong J, Wishart DS, Xia J (2018) MetaboAnalyst 4.0: towards more transparent and integrative metabolomics analysis. Nucleic Acids Res 46:W486–W494
14. Ismail IT, Showalter MR, Fiehn O (2019) Inborn errors of metabolism in the era of untargeted metabolomics and lipidomics. Metabolites 9:242
15. Rinschen MM, Ivanisevic J, Giera M et al (2019) Identification of bioactive metabolites using activity metabolomics. Nat Rev Mol Cell Biol 20:353–367
16. Gowda GN, Raftery D (2013) Biomarker discovery and translation in metabolomics. Curr Metabolomics 1:227–240
17. Gu H, Gowda GN, Raftery D (2012) Metabolic profiling: are we en route to better diagnostic tests for cancer? Future Oncol 8:1207–1210
18. Jasbi P, Wang D, Cheng SL et al (2019) Breast cancer detection using targeted plasma metabolomics. J Chromatogr B 1105:26–37
19. Yin P, Xu G (2017) Methods in molecular biology (Clifton, N.J.) 1619:467–475
20. Kaysen GA, Johansen KL, Chertow GM et al (2015) Associations of trimethylamine

N-oxide with nutritional and inflammatory biomarkers and cardiovascular outcomes in patients new to dialysis. J Ren Nutr 25:351–356
21. Bowers J, Hughes E, Skill N et al (2014) Detection of hepatocellular carcinoma in hepatitis C patients: biomarker discovery by LC–MS. J Chromatogr B 966:154–162
22. Ahn J, Kim J, Hwang J et al (2017) Urinary metabolomic profiling to identify potential biomarkers for the diagnosis of Behcet's disease by gas chromatography/time-of-flight–mass spectrometry. Int J Mol Sci 18:2309
23. Madsen R, Lundstedt T, Trygg J (2010) Chemometrics in metabolomics—A review in human disease diagnosis. Anal Chim Acta 659:23–33
24. Ene IV, Brunke S, Brown AJP et al (2014) Metabolism in fungal pathogenesis. Cold Spring Harb Perspect Med 4: a019695–a019695
25. Poddighe S, Murgia F, Lorefice L et al (2017) Metabolomic analysis identifies altered metabolic pathways in multiple sclerosis. Int J Biochem Cell Biol 93:148–155
26. Monnerat G, Seara FAC, Evaristo JAM et al (2018) Aging-related compensated hypogonadism: role of metabolomic analysis in physiopathological and therapeutic evaluation. J Steroid Biochem Mol Biol 183:39–50
27. Zhu J, Djukovic D, Deng L et al (2015) Targeted serum metabolite profiling and sequential metabolite ratio analysis for colorectal cancer progression monitoring. Anal Bioanal Chem 407:7857–7863
28. Manchester M, Anand A (2017) Metabolomics: strategies to define the role of metabolism in virus infection and pathogenesis. Adv Virus Res 98:57–81
29. Gray DW, Welsh MD, Mansoor F et al (2018) DIVA metabolomics: differentiating vaccination status following viral challenge using metabolomic profiles. PLoS One 13:e0194488
30. Zhang L, Huang Y, Lian M et al (2017) Metabolic profiling of hepatitis B virus-related hepatocellular carcinoma with diverse differentiation grades. Oncol Lett 13:1204–1210
31. Barko PC, McMichael MA, Swanson KS et al (2018) The gastrointestinal microbiome: a review. J Vet Intern Med 32:9–25
32. Chandler JD, Hu X, Ko EJ et al (2016) Metabolic pathways of lung inflammation revealed by high-resolution metabolomics (HRM) of H1N1 influenza virus infection in mice. Am J Physiol Integr Comp Physiol 311:R906–R916
33. Hollenbaugh JA, Montero C, Schinazi RF et al (2016) Metabolic profiling during HIV-1 and HIV-2 infection of primary human monocyte-derived macrophages. Virology 491:106–114
34. Zhu J, Djukovic D, Deng L et al (2014) Colorectal cancer detection using targeted serum metabolic profiling. J Proteome Res 13:4120–4130
35. Carroll PA, Diolaiti D, McFerrin L et al (2015) Deregulated Myc requires MondoA/Mlx for metabolic reprogramming and tumorigenesis. Cancer Cell 27:271–285
36. Gu H, Zhang P, Zhu J et al (2015) Globally optimized targeted mass spectrometry: reliable metabolomics analysis with broad coverage. Anal Chem 87:12355–12362
37. Gu H, Carroll PA, Du J et al (2016) Quantitative method to investigate the balance between metabolism and proteome biomass: starting from glycine. Angew Chemie Int Ed 55:15646–15650
38. Li R, Grimm SA, Mav D et al (2018) Transcriptome and DNA methylome analysis in a mouse model of diet-induced obesity predicts increased risk of colorectal cancer. Cell Rep 22:624–637
39. Buas MF, Gu H, Djukovic D et al (2017) Candidate serum metabolite biomarkers for differentiating gastroesophageal reflux disease, Barrett's esophagus, and high-grade dysplasia/esophageal adenocarcinoma. Metabolomics 13:23

Chapter 11

Detecting Necroptosis in Virus-Infected Cells

Samantha M. Cotsmire, Mateusz Szczerba, and Bertram L. Jacobs

Abstract

Necroptosis has been implicated as a critical cell death pathway in cancers, Alzheimer's and other neurodegenerative diseases, and virus-infected cells. Necroptosis occurs when mixed-lineage kinase domain-like protein (MLKL) punctures the cytoplasmic membrane allowing a rapid influx of water leading to a loss of cellular integrity. As its role in human disease becomes apparent, methods identifying necroptosis will need to be further developed and optimized. Here we describe identification of necroptosis through quantifying cell death with pathway inhibitors and using western blots to identify end points of MLKL activation and protein-protein interactions leading to it.

Key words Necroptosis, Cell death, Vaccinia, Dye exclusion, Immunoblots, DSP-cross-linking

1 Introduction

Necroptosis is a relatively newly identified form of programmed cell death that, unlike apoptosis and pyroptosis, is independent of caspases [1, 2]. Necroptotic cell death has been implicated in the response to infection, death receptor activation, and activation of toll-like receptors 3 and 4 (TLR 3 and TLR 4). Viruses have evolved necroptotic inhibitor proteins to retain pathogenesis during infection [3, 4]. Additionally, necroptosis has been detected in brains of patients with Alzheimer's disease and Parkinson's disease and has been implicated in age-related macular degeneration and in skin inflammation [5–7]. Additionally, a key regulator of necroptosis, receptor-interacting protein kinase 3 (RIPK3), has been implicated in tumor invasion during colorectal cancers and in response to chemotherapeutic agents [8, 9]. While necroptosis inhibits viral pathogenesis in animal models, its involvement in human cancers and neurological diseases has not been elucidated. By studying the regulation of necroptosis during viral infection and other human diseases, we hope to identify targets for oncolytic viruses and viral protein therapies.

Alexandra R. Lucas (ed.), *Viruses as Therapeutics: Methods and Protocols*, Methods in Molecular Biology, vol. 2225, https://doi.org/10.1007/978-1-0716-1012-1_11,

While independent of caspase activation, necroptosis is dependent on the serine/threonine protein kinase, RIPK3. In fact, active caspase 8, one key initiator caspase for apoptosis, can inhibit necroptosis by cleaving RIPK3 [10]. Once activated, RIPK3 can phosphorylate mixed-lineage kinase domain like pseudokinase (MLKL), the executioner of necroptosis [11]. Phosphorylation of MLKL on serine 358 in humans and S345 in mice is necessary but not sufficient to induce necroptotic cell death [12, 13]. A second step, trimerization, is dependent on highly phosphorylated inositol phosphates and on tyrosine phosphorylation (T357) of MLKL by the tyro3-Axl-Myr (TAM) kinases [14–16].

There are three sensors, receptor-interacting serine/threonine protein kinase 1 (RIPK1), DNA-dependent activator of interferon-regulatory factors ((DAI, also known as Z-DNA binding protein 1 (ZBP1) and DLM-1), (Note: *DLM*-1 does not have a definition, it is a protein expressed from the Dlm1 gene), and TIR-domain-containing adapter-inducing interferon-β (TRIF) that act upstream of RIPK3 that can lead to RIPK3 activation. RIPK1 senses activation of death receptors [17]. DAI senses Z-form nucleic acid, associated with viral infection, while TRIF senses signaling by TRL3 and TLR4 [18, 19]. RIPK1, DAI, and TRIF can bind to RIPK3 through interaction of mutual RIP, receptor interacting protein, is defined above. Homotypic interaction motifs, (RHIM) RHIMs, leading to RIPK3 phosphorylation and activation [20]. Once activated, RIPK3 can phosphorylate MLKL, initiating the pathway leading to necroptotic cell death. In this manuscript, we describe methods to detect cell death and determine if the cell death is RIPK3-kinase dependent and associated with MLKL-serine and -tyrosine phosphorylation and MLKL trimerization (*see* Fig. 1). While the requirement for RIPK1 kinase activity can be evaluated using RIPK1-specific kinase inhibitors and RIPK1 phospho-specific antiserum, no similar reagents are available to evaluate activation of DAI or TRIF. However, co-immune precipitation can be used to ask if DAI and TRIF are bound to RIPK3, which is presumed to be necessary for DAI- and TRIF-dependent activation of RIPK3. We have also recently shown that DAI-dependent necroptosis is associated with formation of large cross-linkable complexes that contain DAI and RIPK3 [21]. Cross-linking can be done in living cells and in tissue from experimental animals and necropsy tissue from humans. Thus, cross-linking can be used to determine if DAI activation is associated with disease.

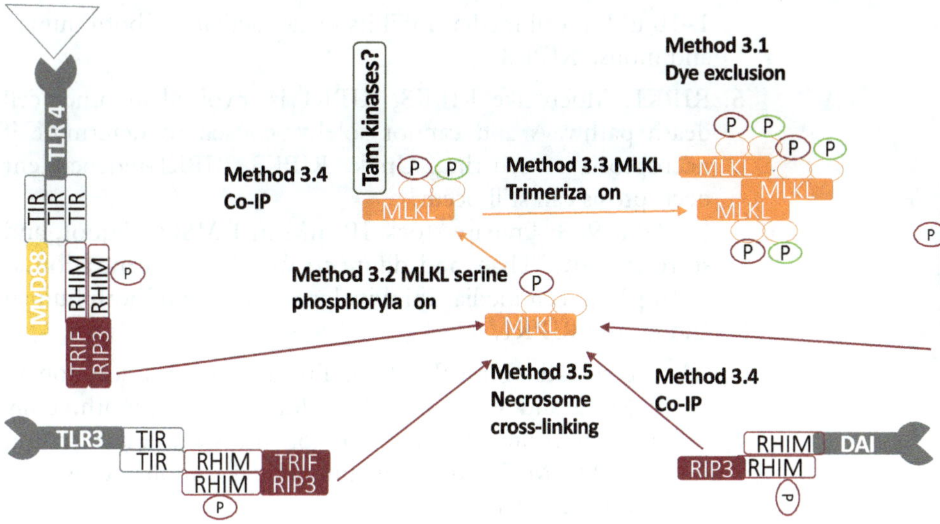

Fig. 1 Overview showing how necroptosis can be induced and detected. This chapter describes how to detect the different stages of necroptotic death. We can detect initiation of RIPK3 activation with CoIP (*see* Subheading 3.4). Then we can detect RIPK3 induction of the necrosome with cross-linking (*see* Subheading 3.5). Next, we can detect MLKL serine phosphorylation (*see* Subheading 3.2) and tyrosine phosphorylation (*see* Subheading 3.4). We also describe detection of MLKL trimerization (*see* Subheading 3.2) and cell death (*see* Subheading 3.1). The inhibitors we have described in Subheading 3.1 are critical for determining how the cells are dying

2 Materials

2.1 Cell Viability by Dye Exclusion with Inhibitors

Prepare all inhibitor stocks in sterile DMSO or water, as indicated. Dilute into the working concentration in warm medium. There are many alternative live/dead nucleic acid stains, but these are our preference due to cost and availability.

1. Hoechst: Allows for nuclear staining in blue fluorescent channels. Prepare stock solution at 1 mM in dimethyl sulphoxide (DMSO), storing at −20 C until use at 1 mM in dimethyl sulphoxide.
2. Propidium iodide (PI): Stains nucleic acids, but is excluded from living cells. Detectable in the red channel. Prepare stock at 1 mg/mL in water and use at 1 μg/mL.
3. Fluorescent microscope or cell counter with red fluorescent protein (RFP) and 4′,6-diamidino-2-phenylindole (DAPI) channels.
4. RIPK3: While lack of RIPK3 involvement can rule out necroptosis, RIPK3 has been shown to be involved in other death pathways and cannot be used to conclusively confirm it.

   GSK 863: Prepare 10 mM stock in DMSO, aliquot, and store at −80C. Thaw and dilute to working concentration of

1–10 μM in cell media. Inhibits kinase activity of both human and mouse RIPK3.

5. RIPK1: Much like RIPK3, RIPK1 is involved in other cell death pathways and cannot solely be used to determine if necroptosis is occurring. Unlike RIPK3, RIPK1-independent necroptosis can still occur.

   GSK 963: Prepare stock 10 mM in DMSO, aliquot, and store at -80C. Thaw and dilute to working concentration at 1–10 μM in cell media. Inhibits kinase activity of both human and mouse RIPK1.

   (a) Necrostatin-1 s (Nec-1 s): Prepare 10 mM solution in DMSO and store at −20 C for up to 1 month. Fully bring to room temperature before using at a concentration of 100 μM. Validated in human cells, mouse cells, and murine models.

6. MLKL: At the cell membrane MLKL executes necroptosis by forming pores allowing for unregulated ion transfer. It requires serine phosphorylation, homologous trimerization, cell membrane relocation, and a tyrosine phosphorylation before executing necroptosis. Inhibition of MLKL can be used to positively determine that necroptosis occurred.

   (a) Necrosulfonamide (NSA) inhibits only human MLKL after serine phosphorylation but before pore formation. Prepare a stock solution of 10 mM in DMSO and store at −20 C. Thaw and use at 1–10 μM.

2.2 MLKL Serine Phosphorylation

1. Approximately 1×10^6 cells per sample to be tested.
2. Halt™ Protease and Phosphatase Inhibitor Cocktail (100×) (we use 78446, Thermo Fisher or equivalent).
3. Dithiothreitol (DTT) prepared at 500 mM in DMSO and stored at −20 C.
4. Triton-X lysis buffer: Prepare a Triton-X lysis buffer to the following final concentrations in sterile water: 5 mM Ethylenediaminetetraacetic acid (EDTA), 10 mM Tris (hydroxymethyl) aminomethane-Cl (Tris–Cl) pH 8.0, 50 mM NaCl, 1% (v/v) Triton-X 100 (*see* **Note 1**).
5. 4× Sodium dodecyl sulfate-polyacrylamide gel electrophoresis (SDS-PAGE) loading buffer: Prepare the loading buffer to the following concentrations: 0.24% SDS, 0.2% bromophenol blue, 20% glycerol in sterile water (*see* **Note 2**).
6. 10% Polyacrylamide gels can be purchased or prepared.
7. Nitrocellulose membranes.

8. CAPS transfer buffer: Resuspend 10 mM CAPS (3-(cyclohexylamino)-1-propane sulfonic acid) in sterile water and adjust pH to 11. Just before use add 20% v/v methanol.
9. Tris-buffered saline with Tween (TBST): Dissolve NaCl (137 mM), KCl (2.7 mM), and Tris base (19 mM) in sterile water. Titrate pH to 7.4 and add Tween-20 to 0.05% (v/v).
10. Membrane-blocking buffer (MBB): Nonfat milk (NFM) diluted to 5% (w/v) in TBST.
11. BSA dilution buffer (BDB): 3% (v/v) Bovine serum albumin (BSA) in TBST.
12. Antibody dilution buffer (ADB): Nonfat milk (NFM) diluted to 3% (w/v) in TBST.
13. Reagents for detecting secondary antibodies.
14. Stripping buffer: 7.25 mM Tris–Cl, 0.1% sodium dodecyl sulfate, with 0.1% (v/v) β-mercaptoethanol added just before use.
15. Primary and secondary antibodies (*see* **Note 3**).

2.3 MLKL Trimerization

1. Approximately 1×10^6 cells per sample to be tested.
2. Halt™ Protease and Phosphatase Inhibitor Cocktail (100×) (we use 78446, Thermo Fisher or equivalent).
3. DTT prepared at 500 mM in DMSO and stored at −20 C.
4. 4× SDS-PAGE loading buffer (*see* **Note 2**).
5. 7.5% Polyacrylamide gels can be purchased or prepared.
6. Nitrocellulose membranes.
7. CAPS transfer buffer.
8. TBST.
9. Primary and secondary antibodies and appropriate buffers (*see* **Note 3**).
10. Reagents for detecting secondary antibodies.
11. Stripping buffer.

2.4 Co-Immunoprecipitation

1. Approximately 1×10^6 cells per sample to be tested.
2. Halt™ Protease and Phosphatase Inhibitor Cocktail (100×) (we use 78446, Thermo Fisher or equivalent).
3. DTT prepared at 500 mM in DMSO and stored at −20 C.
4. Non-denaturing lysis buffer (NDB): Dilute Tris–HCl, pH = 7.5 to 10 mM, $MgCl_2$ to 2 mM, KCl to 50 mM, and Triton X-100 to 1% (v/v) in sterile water (*see* **Note 1**).
5. Buffer I: Prepare Tris–HCl, pH = 7.5 at 20 mM, NaCl at 0.4 mM, EDTA at 1 mM Triton X-100 at 1% (v/v), glycerol at 20%(v/v), and Na_3VO_4 at 2 mM in sterile water.

6. Protein A/G PLUS-Agarose (Santa Cruz, sc-2003), 40 μL of 50% slurry/sample (20 μL for lysate pre-clearing and 20 μL for immunoprecipitation).
7. Primary antibody to pull down protein complexes (*see* **Note 3**).
8. 10% Polyacrylamide gels can be purchased or prepared.
9. Nitrocellulose membranes.
10. CAPS transfer buffer.
11. TBST.
12. Primary and secondary antibodies and appropriate buffers (*see* **Note 3**).
13. Reagents for detecting secondary antibodies.
14. Stripping buffer.

2.5 Necrosome Cross-Linking

1. Approximately 1×10^6 cells or 5–25 mg of animal tissue per sample to be tested.
2. Dithiobis[succinimidyl propionate] (DSP) crosslinker (we use 22585, Thermo Scientific): Prepare 100 μL of 25 mM DSP, by resuspending 1 mg of DSP in 100 μL DMSO (*see* **Note 4**).
3. Halt™ Protease and Phosphatase Inhibitor Cocktail (100×) (we use 78446, Thermo Fisher or equivalent).
4. Radioimmunoprecipitation assay (RIPA) lysis buffer: 150 mM NaCl, 25 mM Tris, pH 7.4, NP40, 0.1% SDS, 0.5% sodium deoxycholate.
5. 4× SDS-PAGE loading buffer. DO NOT ADD DTT.
6. 10% Polyacrylamide gels can be purchased or prepared.
7. Nitrocellulose membranes.
8. CAPS transfer buffer.
9. TBST.
10. Primary and secondary antibodies and appropriate buffers (*see* **Note 3**).
11. Reagents for detecting secondary antibodies.
12. Stripping buffer.

3 Methods

3.1 Cell Viability by Dye Exclusion

Cell viability and dye exclusion can only be used to identify necroptosis in conjunction with necroptosis inhibitors. It is critical to set up inhibitors for this experiment according to their use as described in the Subheading 2.1 and demonstrated in Fig. 2. Furthermore, establishing the time of membrane disruption as indicated by dye

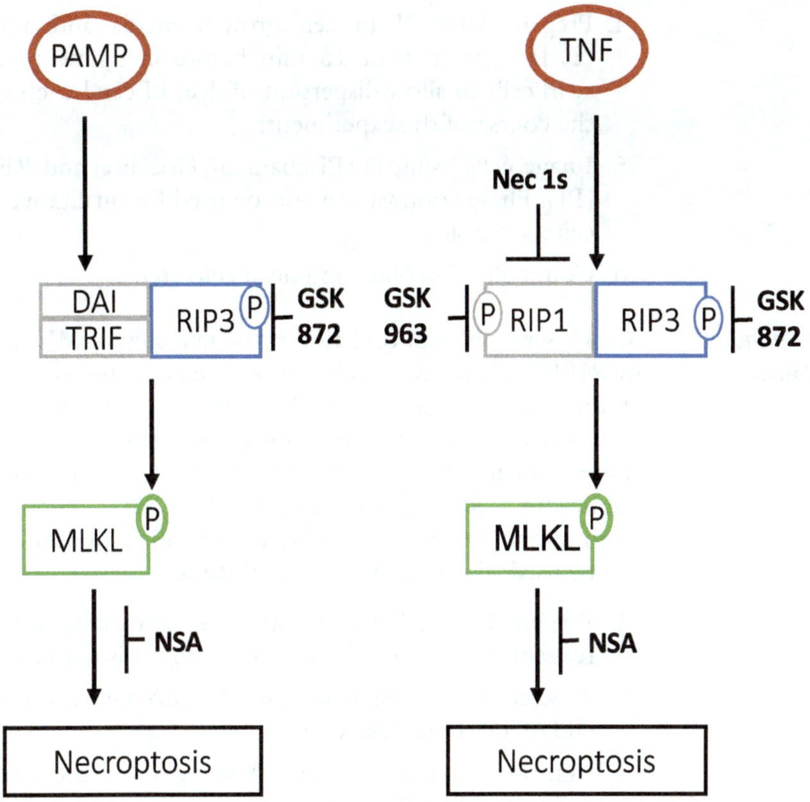

Fig. 2 Inhibitors used to determine mechanism of cell death. Membrane permeabilization is the final step before cell death. Detecting membrane permeabilization in the presence of kinase inhibitors of different proteins involved in necroptotic death is a critical first step for determining whether or not necroptosis might be involved and how long it takes for cells to die. Nec-1s and GSK 963 both inhibit RIPK1 kinase activity. GSK 872 inhibits RIPK3 kinase activity. Necrosulfonamide (NSA) inhibits MLKL post-serine phosphorylation, and ZVAD inhibits pan-caspase activity

exclusion makes determining the timelines for Subheadings 3.2, 3.3, 3.4, and 3.5 much simpler (*see* **Note 5** for more information).

1. One day prior to testing, plate cells in a tissue culture dish suitable for fluorescent imaging.
2. Prepare working concentrations of Hoechst dye and inhibitors of the necroptosis pathway by diluting into cell growth media. Pretreat cells by aspirating media and replacing with dye and inhibitors and incubating for 1 h under normal cell growth conditions (i.e., 37 °C, 5% CO_2). All samples should have Hoechst dye; inhibitors should be used as appropriate.
3. Begin the experiment by aspirating dye media and adding a chemical death inducer or infecting with virus. If using inhibitors, include them in death inducer and viral preparations (*see* **Note 6**).

4. Prepare 10× PI in cell growth media and add to cells (1/10 v/v) at least 15 min before first time point. Gently swirl cells to allow dispersion of dye. PI can be left on cells for the course of the experiment.
5. Image cells using DAPI channel (Hoechst) and RFP channel (PI). Phase contrast can also be used for qualitative analysis of cell morphology.
6. Count all cells (blue) and dead cells (red).

3.2 MLKL Serine Phosphorylation

MLKL serine phosphorylation is the first step of MLKL activation by RIPK3. MLKL is also phosphorylated on a tyrosine residue, but at the time of this writing there are no commercially available antibodies specific to this phosphorylation site. To look at subsequent activation, Subheading 3.1 Cell death by dye exclusion, Subheading 3.3 MLKL Trimerization, and Subheading 3.4 Co-immunoprecipitation (with anti-tyrosine phospho antibody) can be used. For time points, *see* **Note 5**.

1. Prepare 1 × 10^6 cells by infecting or treating with chemical reagents suspected of inducing necroptosis (*see* **Note 6**).
2. At various time points scrape cells into media using the blunt end of a P1000 pipette tip.
3. Centrifuge @ 1000 × *g* for 10′ and gently remove media from pellet (*see* **Note 7**).
4. Resuspend pellet thoroughly in 60 μL of Triton-X lysis buffer with protease and phosphatase inhibitors.
5. Incubate cells on ice for 5′.
6. Centrifuge cells for 15′ at max speed at 4 °C.
7. Keep the supernatant and add to 4× MLKL loading buffer + DTT (*see* **Note 8**).
8. Heat all samples for at least 10′ at 95 °C.
9. Load onto a 10% polyacrylamide gel (*see* **Note 9**).
10. Run slowly (70 V) until the ladder is in the separating gel. After this, the speed can be increased to 100 V until the 50 kDa marker is 2/3 of the way down the gel.
11. Transfer to nitrocellulose membrane in 1× CAPS transfer buffer with added methanol for 1 h at 100 V.
12. Block for 1 h with MBB. Rinse with TBST until no longer cloudy.
13. Add primary antibody (serine phospho-MLKL) overnight at 4 °C, and rinse 1× with TBST (*see* **Note 3**).
14. Wash with TBST 3× at RT for 5 min.
15. Add secondary antibody and incubate for 2 h at RT in ADB.

16. Wash with TBST 3× at RT for 15 min.
17. Expose and image blot.
18. Strip the blot by incubating in stripping buffer (*see* **Note 10**) for 15 min at 56 °C.
19. Block with MBB for 1 h at RT. Rinse with TBST until no longer cloudy.
20. Add primary MLKL antibody overnight at 4 °C (*see* **Note 3**).
21. Rinse 3× with TBST at RT for 5 min.
22. Add secondary antibody. Incubate for 2 h at RT in ADB.
23. Wash with TBST 3× at RT for 5 min.
24. Expose and image blot.

3.3 MLKL Trimerization

MLKL trimerization is a straightforward way to measure one of the final steps of MLKL activation before execution of cell death. After MLKL is trimerized, it moves to the cell surface creating pores which facilitate rapid cell death. This method takes advantage of MLKL trimers being resistant to denaturization by SDS and boiling while fully denaturing with DTT. For time points, *see* **Note 5**.

1. Prepare 1 × 10^6 cells by infecting or treating with chemical reagents suspected of inducing necroptosis (*see* **Note 6**).
2. At various time points scrape cells into media using the blunt end of a p1000 pipette tip.
3. Centrifuge @ 1000 × *g* for 10′ and gently remove media from pellet (*see* **Note 6**).
4. Resuspend pellet thoroughly in 100 μL of 1× MLKL loading buffer *without DTT*. Separate half of sample and add DTT (*see* **Note 11**).
5. Add all of the sample to the Qiashredder and spin at maximum speed for 3′.
6. Heat all samples for at least 10′ at 95 °C.
7. Load onto a 7.5% polyacrylamide gel (*see* **Note 9**).
8. Run slowly (70 V) until the 50 kDa marker is 2/3 of the way down the gel.
9. Transfer to nitrocellulose membrane for 18 h at 15 V in 4 °C in CAPS transfer buffer with magnetic spinner turned to slow.
10. Block for 1 h with MBB.
11. Add primary antibody (MLKL *see* **Note 3**) overnight at 4 °C (rinse with TBST).
12. Wash with TBST 3× at RT for 5 min.

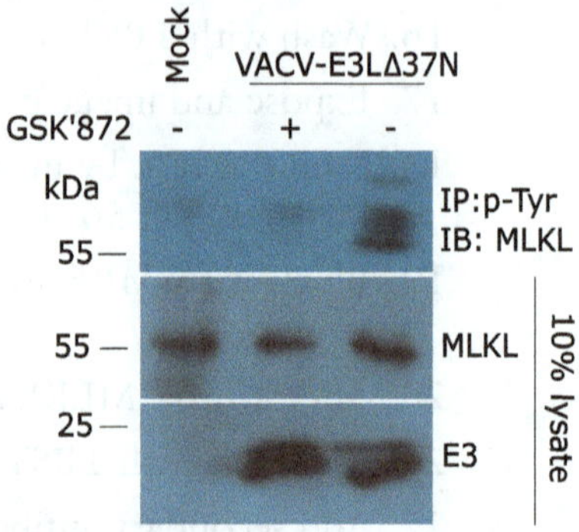

Fig. 3 Co-IP to measure MLKL tyrosine phosphorylation. L929 cells pretreated with type I interferon (IFN) were infected with VACV-E3LΔ37N and harvested at 4 h postinfection. Lysates then were immunoprecipitated as described in Subheading 3.3 using phosphotyrosine-specific antibody. Immunoprecipitated proteins and input lysate were run on a Western blot and probed for MLKL

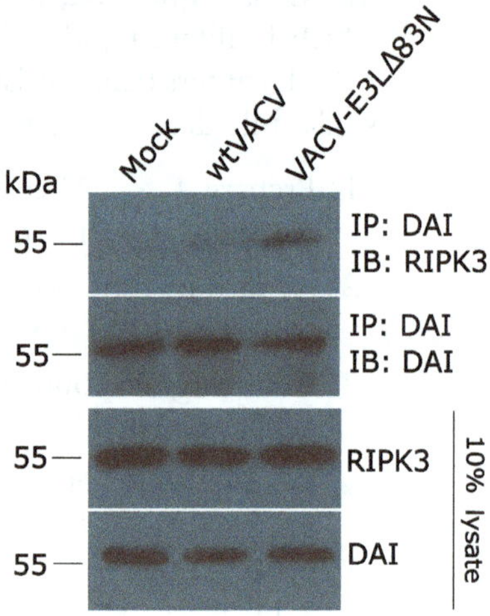

Fig. 4 Co-IP to determine activator of necroptosis. L929 cells pretreated with type I IFN were infected with VACV-E3LΔ83N. At 2 h postinfection, lysates were harvested and subjected to immunoprecipitation by antibodies specific for DAI as described in Subheading 3.3. Immunoprecipitated protein and input lysate were run on Western blots and probed with antibody for RIPK3

13. Add secondary antibody and incubate for 2 h at RT in ADB.
14. Wash with TBST 3× at RT for 15 min.
15. Expose and develop blot.

3.4 Co-Immunoprecipitation

We have validated this co-immunoprecipitation protocol to identify interactions between DAI and RIPK3 (*see* Fig. 4) and to detect MLKL tyrosine phosphorylation (*see* Fig. 3). Each of these pull-downs will require a different timeline (*see* **Note 5**).

1. Prepare 2×10^6 cells by infecting or treating with chemical reagents suspected of inducing necroptosis (*see* **Note 6**).
2. At 2 h posttreatment or postinfection, aspirate the media and wash cells with 1 mL cold PBS.
3. Scrape cells into cold PBS using the blunt end of a p1000 and transfer the suspension to a microcentrifuge tube.
4. Centrifuge at 1000 × *g* for 10 min at 4 °C.
5. Remove PBS (supernatant) carefully, not disturbing the pellet (*see* **Note 6**).
6. Lyse cells by resuspending the pellet in 100 μL ice-cold NDB and incubating on ice for 5 min.
7. Add 120 μL ice-cold Buffer I (*see* **Note 12**).
8. Measure 80 μL of 25% (v/v) raw bead slurry per sample (protein A/G agarose) and spin the slurry at 500 × g for 5 min (*see* **Note 13**).
9. Wash the pellet of beads with 1 mL Buffer I and spin again.
10. Perform a total of three washes (*see* **Note 12**).
11. Pre-clear samples, by adding 20 μL of washed 50% protein A/G PLUS-Agarose slurry to the collected cell lysates.
12. Incubate with rotation at 4 °C for 30–60 min.
13. Pellet beads (500 × g for 5 min) (*see* **Note 14**).
14. Transfer 200 μL of the supernatant into a fresh tube for immunoprecipitation (discard beads used for pre-clearing; the other 40 μL of the lysate will be the total lysate control) (*see* **Note 14**).
15. Add the bait antibody (*see* **Note 3**) to the 200 μL of cell lysate.
16. Incubate with rotation overnight at 4 °C.
17. Add 20 μL of the previously prepared 50% protein A/G PLUS-Agarose slurry to each sample.
18. Incubate with rotation for 60–90 min at 4 °C.
19. Pellet the bead-antibody-protein complexes by centrifugation at 500 × *g* for 5 min and set aside the supernatant to be kept as a control (*see* **Note 14**).
20. Wash the beads to remove proteins with nonspecific binding by adding 1 mL of Buffer I.
21. Centrifuge at 500 × *g* for 5 min. Preserve a fraction of the supernatant as a control (*see* **Note 14**).
22. Repeat the wash for a total of three times.
23. When removing supernatant after last wash, leave about 15 μL of Buffer I on top of the bead slurry in the tube. Add at least

15 μL of 2× denaturing loading buffer containing 20–50 μM DTT.

24. To release the antibody-protein complexes from the beads, heat the samples at 95 °C for 10–15 min.
25. Pellet the beads by centrifugation at 500 × *g* for 10 min. Transfer the supernatant to a fresh tube. The pellet can be kept as a control for successful release of antibody-protein complexes (*see* **Note 14**).
26. Run two 10% PAGE-SDS gels at 100 V. One gel should be run for the immunoprecipitated fraction, and another one for the total lysate control from **step 14**.
27. Transfer to nitrocellulose membrane in 1× CAPS transfer buffer with added methanol for 1 h at 100 V.
28. Block for 1 h with membrane blocking buffer (MBB). Rinse with TBST until no longer cloudy.
29. Add primary antibody overnight at 4 °C in antibody dilution buffer (ADB) (*see* **Notes 3** and **15**) (rinse with TBST).
30. Rinse 3× with TBST at RT for 5 min.
31. Add secondary antibody. Incubate for 2 h at RT in ADB.
32. Wash with TBST 3× at RT for 5 min.
33. Expose and image blot.
34. Strip the blot by incubating in stripping buffer (*see* **Note 9**) for 15 min at 56 °C.
35. Block for 1 h at RT. Rinse with TBST until no longer cloudy.
36. Add primary antibody overnight at 4 °C (rinse with TBST) (*see* **Notes 3** and **15**).
37. Rinse 3× with TBST at RT for 5 min.
38. Add secondary antibody. Incubate for 2 h at RT in ADB.
39. Wash with TBST 3× at RT for 5 min.
40. Expose and image blot.

3.5 Necrosome Cross-Linking

Necrosome cross-linking uses an amine-reactive crosslinker, DSP, to covalently cross-link all proteins in the necrosome shifting their weight on a PAGE gel to >250 kDa (*see* Fig. 5). It is helpful for looking at interactions between proteins that cannot easily be detected by Co-IP (*see* Subheading 3.4) and can readily be used to analyze animal and human tissue samples frozen in liquid nitrogen. It is not as specific as Co-IP but can be used with other methods to demonstrate protein inclusion in the necrosome. For timelines, *see* **Note 5**.

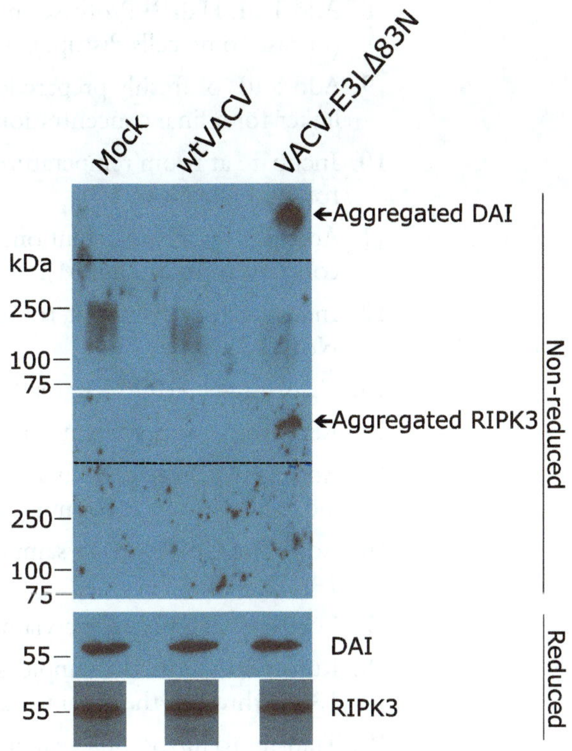

Fig. 5 Cross-linking to determine the activator of necroptosis. L929 cells pretreated with type I IFN were infected with VACV-E3LΔ83N. At 2 h postinfection, cells were harvested and cross-linked according to the protocol described in Subheading 3.4. Lysates were then run on Western blots and probed with antibodies against DAI and RIPK3

1. For tissue culture experiments prepare 2×10^6 cells by infecting or treating with chemical reagents suspected of inducing necroptosis (*see* **Note 6**).
2. At 2 h posttreatment or postinfection, aspirate the media and wash cells with 1 mL room-temperature reaction buffer (1× PBS) (*see* **Note 16**).
3. Scrape cells into the PBS using the blunt end of a p1000 and transfer the suspension to a microcentrifuge tube.
4. Centrifuge at $1000 \times g$ for 10 min.
5. Remove PBS (supernatant) carefully, not disturbing the pellet (*see* **Note 7**).
6. For treatments done in vivo (mice) or tissues collected from a diseased individual (human), homogenize about 5 mg tissue per sample by grinding in liquid nitrogen. The protocol can be scaled up for 25 mg (*see* **Note 17**).
7. Resuspend pelleted cells in 100 μL or ground tissue in PBS at a concentration of 5 mg/100 μL.

8. Add 1 μL Halt™ Protease and Phosphatase Inhibitor Cocktail (in case some cells disrupt).
9. Add 8 μL of freshly prepared (*see* **Note 4**) 25 mM DSP cross-linker for a final concentration of 2 mM (*see* **Note 18**).
10. Incubate at room temperature for 30 min or 4 °C for 2 h with rotation (*see* **Note 17**).
11. Add 2 μL of stop solution: 1 M Tris pH 7.5 (for a final concentration of 20 mM).
12. Incubate for 15 min at room temperature with rotation (*see* **Note 17**).
13. Centrifuge at 1000 × *g* for 10 min at 4 °C.
14. Remove the supernatant carefully (*see* **Note 7**).
15. Add 120 μL of ice-cold RIPA lysis buffer (*see* **Note 1**) and incubate on ice for 10 min (*see* **Note 17**).
16. Add 60 μL of 4× SDS sample buffer *without DTT* (*see* **Note 14**).
17. Load onto a 10% polyacrylamide gel (*see* **Note 9**).
18. Run at 80 V until the sample is past the stacking gel and then at 130 V through the separating gel.
19. Transfer to nitrocellulose membrane for 18 h at 20 V at 4 °C in CAPS transfer buffer with a low-speed magnetic stirrer (*see* **Note 19**).
20. Block for 1 h at RT.
21. Add primary antibody overnight at 4 °C (rinse with TBST) (*see* **Note 3**).
22. Rinse 3× with TBST at RT for 5 min.
23. Add secondary antibody. Incubate for 2 h at RT in ADB.
24. Wash with TBST 3× at RT for 5 min.
25. Expose and develop blot (*see* **Note 20**).
26. For total lysate controls, reduce 50 μL of the cross-linked samples by adding DTT to a final concentration of 50 mM (5.5 μL for 50 μL samples) and incubating the sample first at 37 °C for 1 h and then at 95 °C for 5–10 min.
27. Load on a 10% polyacrylamide gel.
28. Run at 80 V until the sample is past the stacking gel and then at 130 V through the separating gel.
29. Transfer to nitrocellulose membrane in 1× CAPS transfer buffer with added methanol for 1 h at 100 V.
30. Block for 1 h with MBB. Rinse with TBST until no longer cloudy.

31. Add primary antibody overnight at 4 °C (rinse with TBST); *see* **Note 3**.
32. Rinse 3× with TBST at RT for 5 min.
33. Add secondary antibody. Incubate for 2 h at RT in ADB.
34. Wash with TBST 3× at RT for 5 min.
35. Expose and image blot (*see* **Note 20**).

4 Notes

1. Just before lysing cells, add protease and phosphatase inhibitors.
2. Just before denaturing protein, add DTT to 4× loading buffer to a final concentration of 4 mM.
3. Table 1 contains information on the antibodies we used for these protocols.
4. DSP is very moisture sensitive and must be stored desiccated at 4C and be equilibrated to room temperature before opening. Because the NHS-ester moiety readily hydrolyzes and becomes nonreactive, prepare each use fresh and discard any unused solution.
5. Once necroptosis is initiated, cells die very quickly. As such, the time points used to harvest protein to study these phenomena are critical. Each time a new cell line or treatment will be studied, the first step is to establish a timeline. The fastest way to do this is to use Subheading 3.1 dye exclusion to take images of the cells at 30-min intervals to determine the time of membrane rupture. Table 2 contains data from our observations on the best times to perform these assays in L929 cells.
6. If infecting cells with virus, follow virus-specific infection protocols to infect >99% of cells and grow under standard culture conditions until desired time point.
7. Dead volumes of 10–50 μL should be left on pellets to avoid disturbing fragile cells.
8. The samples for this experiment do not freeze well. Upon thawing, phosphorylated MLKL will barely be distinguishable between positive and negative controls. Have the gel ready to run the same day as the experiment is conducted.
9. If you are pouring the gel, the interface between the stacking and separating gels must be perfect. Isopropanol can be substituted for saturated isobutanol for a cleaner transition.
10. Add 0.1% β-mercaptoethanol to stripping buffer immediately before use.

Table 1
Antibodies used in these protocols

Protein	Antibody clone	Species detected	Assays	Blocking buffer	Concentration	Dilution buffer
DAI	Zippy-1 AG-20B-0010-C100	Mouse, human	WB Co-IP	MBB	500 ng/mL (1:2000) 5 ng/μL	ADB Buffer I
RIPK3	CST D4G2A	Mouse	WB	MBB	85 ng/mL (1:2000)	ADB
Phospho-tyrosine	Sigma 05–321×	Any	Co-IP		2.5 ng/μL	Buffer I
MLKL (serine phospho)	CST D6E3G	Mouse	WB	MBB	28 ng/mL (1:1000)	BDB
MLKL (total)	CST D6W1K	Mouse	WB	BDB	313 ng/mL (1:1000)	BDB

Table 2
Optimal time points (in hours posttreatment) for these assays in L929 cells

Subheading	Assay	TNF + ZVD	VACV E3Δ83N
3.1	Dye exclusion	4–6	6–8
3.2	MLKL serine phosphorylation	2–4	4–5
3.3	MLKL trimerization	4	5–6
3.4	Tyrosine phospho-MLKL co-IP	4	4–5
3.4	DAI-RIP3 co-IP	N/A	1–3
3.5	Necrosome aggregation	1–3	1–3

11. Very fresh β-mercaptoethanol can be used in place of DTT. However, we have seen that this application is highly sensitive to oxidation and recommend using DTT for more consistent results.

12. If necessary, lysates can be stored at -80 °C until a more suitable time for processing.

13. The commercial beads we use come as 25% (v/v) slurry. They need to be resuspended to a 50% (v/v) final concentration of beads. For this experiment, each sample requires 20 μL of the 50% slurry for both pre-clearing and immunoprecipitation. For ten samples, that will be 20 μL × 2 × 10 = 400 μL of 50% final slurry concentration. Therefore, you will need 400 μL x 2 = 800 μL of the 25% slurry. After washing the 800 μL of 25% slurry, you have roughly 200 μL of solid beads and add

200 μL buffer I to dilute to the final concentration of 400 μL 50% slurry.

14. These can be kept at −80 °C as a control.
15. When probing the co-immunoprecipitation blots, you will look for both the protein targeted by the bait antibody in the IP and associated proteins. The first primary antibody probe should be against associated proteins. Stripping and re-probing for other associated proteins can be done up to two more times. The final probe should be against the protein pulled down during the IP.
16. Alternatively, 10–25 mM HEPES, bicarbonate/carbonate, and borate buffers pH 7–9 can be used as the reaction buffer.
17. For 25 mg of tissue several of the times need to be extended. **Step 10**: Cross-link by rotating for 1 h at RT or 4 h at 4 °C. **Step 12**: Incubate with the stop solution for 30 min at RT with rotation. **Step 15**: Lyse with RIPA for 30 min at 4 °C (with rotation).
18. A final concentration of 1–2 μM is recommended. 2 μM works for L929 cells and brain tissue samples, but the concentration may need adjusting for other tissues and cell types.
19. Do not discard the stacking gel; many macromolecular complexes will be present in this part of gel. The entire gel (stacking + separating) should be transferred.
20. Blots can be stripped and probed up to three times. The cross-linked and total lysate control blots should be probed with the same antibodies.

Acknowledgments

We would like to thank Nobuko Fukushima and James Bonner for technical assistance and Dr. Heather Koehler and Dr. Edward Mocarski for sharing expertise in necroptosis through consultation and protocol optimization.

References

1. Holler N, Zaru R, Micheau O et al (2000) Fas triggers an alternative, caspase-8–independent cell death pathway using the kinase RIP as effector molecule. Nat Immunol 1:489–495. https://doi.org/10.1038/82732
2. Vercammen D, Beyaert R, Denecker G et al (1998) Inhibition of caspases increases the sensitivity of L929 cells to necrosis mediated by tumor necrosis factor. J Exp Med 187:1477–1485. https://doi.org/10.1084/jem.187.9.1477
3. Huang Z, Wu S-Q, Liang Y et al (2015) RIP1/RIP3 binding to HSV-1 ICP6 initiates necroptosis to restrict virus propagation in mice. Cell Host Microbe 17:229–242. https://doi.org/10.1016/j.chom.2015.01.002
4. Petrie EJ, Sandow JJ, Lehmann WIL et al (2019) Viral MLKL homologs subvert Necroptotic cell death by sequestering cellular

RIPK3. Cell Rep 28:3309–3319.e5. https://doi.org/10.1016/j.celrep.2019.08.055

5. Zhang S, Tang M, Luo H et al (2017) Necroptosis in neurodegenerative diseases: a potential therapeutic target. Cell Death Dis 8: e2905–e2905. https://doi.org/10.1038/cddis.2017.286
6. Hanus J, Anderson C, Wang S (2015) RPE necroptosis in response to oxidative stress and in AMD. Ageing Res Rev 24:286–298. https://doi.org/10.1016/j.arr.2015.09.002
7. Bonnet MC, Preukschat D, Welz P-S et al (2011) The adaptor protein FADD protects epidermal keratinocytes from necroptosis in vivo and prevents skin inflammation. Immunity 35:572–582. https://doi.org/10.1016/j.immuni.2011.08.014
8. Moriwaki K, Bertin J, Gough PJ et al (2015) Differential roles of RIPK1 and RIPK3 in TNF-induced necroptosis and chemotherapeutic agent-induced cell death. Cell Death Dis 6: e1636–e1636. https://doi.org/10.1038/cddis.2015.16
9. Bozec D, Iuga AC, Roda G et al (2016) Critical function of the necroptosis adaptor RIPK3 in protecting from intestinal tumorigenesis. Oncotarget 7:46384–46400. https://doi.org/10.18632/oncotarget.10135
10. Oberst A, Dillon CP, Weinlich R et al (2011) Catalytic activity of the caspase-8-FLIPL complex inhibits RIPK3-dependent necrosis. Nature 471:363–367. https://doi.org/10.1038/nature09852
11. Wang H, Sun L, Su L et al (2014) Mixed lineage kinase domain-like protein MLKL causes necrotic membrane disruption upon phosphorylation by RIP3. Mol Cell 54:133–146. https://doi.org/10.1016/j.molcel.2014.03.003
12. Tanzer MC, Tripaydonis A, Webb AI et al (2015) Necroptosis signalling is tuned by phosphorylation of MLKL residues outside the pseudokinase domain activation loop. Biochem J 471:255–265. https://doi.org/10.1042/BJ20150678
13. Rodriguez DA, Weinlich R, Brown S et al (2016) Characterization of RIPK3-mediated phosphorylation of the activation loop of MLKL during necroptosis. Cell Death & Differ 23:76–88. https://doi.org/10.1038/cdd.2015.70
14. Najafov A, Mookhtiar AK, Luu HS et al (2019) TAM kinases promote necroptosis by regulating oligomerization of MLKL. Mol Cell 75:457–468.e4. https://doi.org/10.1016/j.molcel.2019.05.022
15. Dondelinger Y, Declercq W, Montessuit S et al (2014) MLKL compromises plasma membrane integrity by binding to phosphatidylinositol phosphates. Cell Rep 7:971–981. https://doi.org/10.1016/j.celrep.2014.04.026
16. Dovey CM, Diep J, Clarke BP et al (2018) MLKL requires the inositol phosphate code to execute necroptosis. Mol Cell 70:936–948.e7. https://doi.org/10.1016/j.molcel.2018.05.010
17. de Almagro MC, Goncharov T, Izrael-Tomasevic A et al (2017) Coordinated ubiquitination and phosphorylation of RIP1 regulates necroptotic cell death. Cell Death Different 24:26–37. https://doi.org/10.1038/cdd.2016.78
18. Kaiser WJ, Sridharan H, Huang C et al (2013) Toll-like receptor 3-mediated necrosis via TRIF, RIP3, and MLKL. J Biol Chem 288:31268–31279. https://doi.org/10.1074/jbc.M113.462341
19. Upton JW, Kaiser WJ, Mocarski ES (2012) DAI/ZBP1/DLM-1 complexes with RIP3 to mediate virus-induced programmed necrosis that is targeted by murine cytomegalovirus vIRA. Cell Host Microbe 11:290–297. https://doi.org/10.1016/j.chom.2012.01.016
20. Baker MODG, Shanmugam N, Pham CLL et al (2020) RHIM-based protein:protein interactions in microbial defence against programmed cell death by necroptosis. Semin Cell Dev Biol 99:86–95. https://doi.org/10.1016/j.semcdb.2018.05.004
21. Jang T, Zheng C, Li J et al (2014) Structural study of the RIPoptosome core reveals a helical assembly for kinase recruitment. Biochemistry 53:5424–5431. https://doi.org/10.1021/bi500585u

Chapter 12

Topical Application of Virus-Derived Immunomodulating Proteins and Peptides to Promote Wound Healing in Mouse Models

Liqiang Zhang, Jordan R. Yaron, Qiuyun Guo, Jacquelyn Kilbourne, Enkidia A. Awo, Michelle Burgin, Lauren N. Schutz, Sarah E. Wallace, Kenneth M. Lowe, and Alexandra R. Lucas

Abstract

Immune modulators play critical roles in the progression of wounds to normal or conversely delayed healing, through the regulation of normal tissue regrowth, scarring, inflammation, and growth factor expression. Many immune modulator recombinants are under active preclinical study or in clinical trial to promote improved acute or chronic wound healing and to reduce scarring. Viruses have evolved highly efficient immune modulators for the evasion of host-defensive immune responses that target and kill invasive viruses. Recent studies have proven that some of these virus-derived immune modulators can be used to promote wound healing with significantly improved speed and reduced scarring in rodent models. Mouse full-thickness excisional wound model is one of the most commonly used animal models used to study wound healing for its similarity to humans in the healing phases and associated cellular and molecular mechanisms. This chapter introduces this mouse dermal wound healing model in detail for application in studying viral immune modulators as new treatments to promote wound healing. Details of hydrogel, protein construction, and topical application methods for these therapeutic proteins are provided in this chapter.

Key words Viral immune modulator, Wound healing, Full-thickness excision, Hydrogel

1 Introduction

Large surface wounds remain a major health issue, after trauma and burns, often increasing mortality and prolonging morbidity. With poor healing, large wounds impair skin function and increase risk of hemorrhage, infection, sepsis, and death. Diabetic wounds are at particularly high risk for chronic infections and poor healing, requiring surgical debridement or even amputations. The wound healing process generally includes three steps: [1] hemostasis and inflammation, [2] new tissue generation, and [3] remodeling.

Alexandra R. Lucas (ed.), *Viruses as Therapeutics: Methods and Protocols*, Methods in Molecular Biology, vol. 2225, https://doi.org/10.1007/978-1-0716-1012-1_12,

Wound healing is affected by many factors, such as infection and advanced age, as well as diabetes and other related diseases with suppressed immunity. Multiple inflammatory, immune, and fibrotic pathways are sequentially activated in wound repair beginning from the time of injury to new tissue regeneration. Among these are the coagulation (thrombotic) and hemorrhagic (thrombolytic) serine protease pathways, chemokines, inflammasomes, and apoptotic pathway activation. On the other hand, although inflammation is necessary for early wound healing, sustained and excess inflammation and immune cell activation after severe wound can also prolong healing and cause excess scarring.

The Lucas lab and other research teams have developed a series of highly active biological therapeutics, immune-modulating proteins, and peptides derived from virus where these proteins have evolved over millions of years to a high level of efficacy for modulating key immune and inflammatory pathways [1]. Several of these new biologics have proven beneficial in animal models of wound healing, including Serp-1 [2], viral IL-10 (vIL-10), and viral VEGF (VEGF-E) [3]. Based on our work and prior experience in studying new therapeutic approaches, specifically virus-derived immune-modulating protein biologics, we will use myxoma virus-derived Serp-1 as a typical protein to demonstrate the method of preclinical testing on a full-thickness wound healing mouse model. The development of a hydrogel for delivery in these immunomodulating proteins as topical treatments for wound healing is also discussed.

The mouse full-thickness excisional wound model has apparent advantages over other larger animal models, including lower cost, easier maintenance, and ability to standardize the experimental conditions and analyses, and additionally transgenic strains available for analysis of mechanism studies [4, 5]. Considering the distinct and overlapping phases of highly complex cellular and molecular events, wound healing is considered to be similar in humans and mice. One major difference or problem in the mouse model with the use of a full-thickness excisional model is the marked contraction of the skin caused by the mouse panniculus carnosus which differs from human wounds and does not completely reproduce human wound healing. Researchers have developed a splint model using a silicon ring attached around the wound that prevents excess skin contraction, allowing the wound tissue edge regrowth to be monitored and measured [5, 6]. With transparent Tegaderm films covering the top of the silicon ring, this becomes a very convenient method to observe and analyze all phases of wound healing. In this chapter, we describe the use of this splinted full-thickness skin excisional mouse model for the topical application of viral immune-modulating protein biologics to promote wound healing (*see* Fig. 1).

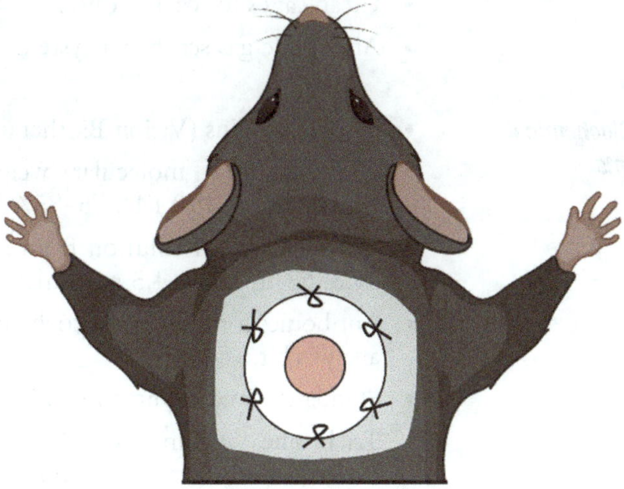

Fig. 1 Splinted full-thickness excisional mouse wound model

2 Materials

2.1 Materials for Mouse Skin Wound Surgery

All materials are sterile prior to use. All animal models must be reviewed and approved by the local IACUC committee prior to proceeding with any surgical studies. Animals are monitored daily and during surgery by the local veterinary staff.

- Silicone sheet (0.5 mm thick/clear, Grace Bio-Labs).
- Fiskars flower lever punch (small, 1.5 cm).
- Krazy Glue.
- Electronic hair clippers for small animal.
- Puralube vet ointment.
- Tegaderm transparent roll (3 M).
- Biopsy punch (3–6 mm).
- Ethilon suture (we use Ethicon sutures, size 4–0, needle FS-2).
- Small-animal surgical tools (forceps, surgical tweezers, scissors, needle holder).
- Heat pad.
- Insulin syringes.
- Anesthetic (ketamine (100 mg/mL, 120 mg/kg) and xylazine (20 mg/mL, 6 mg/kg) mixture).
- Saline.
- 70% Ethanol.
- Surgical gauze.
- Isoflurane.

- Clean cages for each mouse.
- Anesthetic gas scavenger system.

2.2 Biochemical Reagents

- Serp-1 proteins (Virion Biotherapeutics, 1.8 mg/mL).
- Chitosan, low molecular weight (we use Sigma-Aldrich, #448869, 50–190 KD, 75–85% deacetylated).
- Collagen, type I solution from rat tail (we use Sigma-Aldrich, #C3867, about 4–4.5 mg/mL).
- Antibodies for immunohistochemical (IHC) analysis and ELISA analysis kit.
- Standard hematoxylin and eosin (H&E) staining solutions.
- Trichrome staining solutions: Phosphomolybdic/phosphotungstic acid, Weigert's iron hematoxylin A and B, Bouin's fixative, Biebrich scarlet solution (1%), aniline blue solution, acetic acid (v/v, 1%). All are available from VWR.

2.3 Animals

All animal procedures should be approved by the Institutional Animal Care and Use Committee (IACUC). Male and female wild-type C57BL6/J mice aged 8–12 weeks are randomly selected and equal numbers of male and female mice used where possible. Mice are kept on a standard 12/12 light-dark cycle in specific pathogen-free housing conditions and given food and water ad libitum. Mice are single housed after the wounding procedure to prevent biting and trauma from other mice and potential interference or further damage and delayed healing during the wound healing.

3 Methods

3.1 Splint Preparation

1. Cut the silicone sheet into disks with Fiskars punch (diameter = 1.5 cm).
2. Make a silicone ring (donut shape) by cutting the middle of silicone disks with 3.5 mm biopsy punch (*see* **Note 1**).
3. Sterilize the silicone rings by rinsing in 70% ethanol and allow them to dry on clean gauze.
4. Attach Tegaderm film on one side of ring and trim the film along the outside edge of silicone ring with sterile scissors.

3.2 Chitosan Hydrogel Preparation

1. Swell low-molecular-weight chitosan by adding 10 mg chitosan to 10 mL of deionized water and rotating overnight at 4 °C.
2. Centrifuge the mixture at 1000 × *g* for 15 min and remove the excess water.

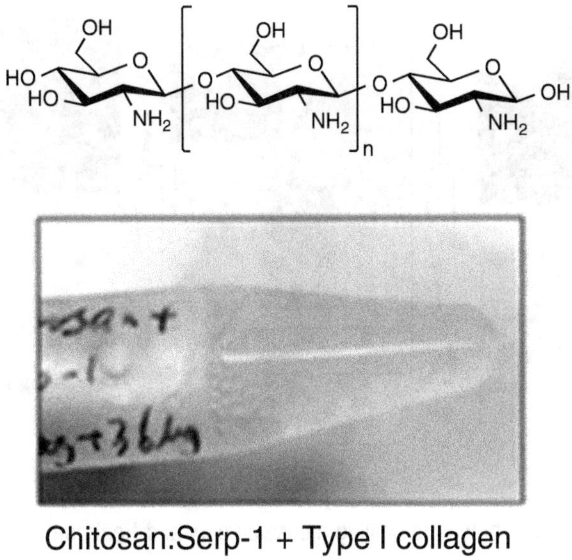

Fig. 2 Chitosan-collagen hydrogel containing Serp-1 protein (Zhang et al. J Clin Med 2019, 8, 1626.)

3. Freeze the wet chitosan at −20 °C for 8 h followed by incubation overnight at 4 °C.
4. Add specific amounts of purified protein solutions to the resuspended mixture and rotate the tube at 4 °C for 8 h and then lyophilize the mixture overnight. Stock the dried powder at −20 °C.
5. Before application on wound, add type I collagen solution (Sigma Life Science, C3867) to a total volume of 300 μL to form a chitosan-collagen/proteinhydrogel at a certain concentration of protein (*see* Fig. 2) (*see* **Note 2**).

3.3 Wound Procedure and Treatments

As discussed above, this protocol uses the splinted full-thickness excisional wound healing mouse model, where the contraction exhibited in mouse skin is prevented by holding the wound open with a donut-shaped splint (*see* Fig. 3).

1. **Anesthetize** mice with intraperitoneal injection of a cocktail of 120 mg/kg ketamine and 6 mg/kg xylazine, 0.1 mL/25 g body weight.
2. **Shave** an area of approximately 1 × 1 in. spanning from between the ears to the peak of the spine, centered between each shoulder, and sterilize the area with two washes with 2% chlorhexidine gluconate solution (Dyna-Hex 2®, Xttrium Laboratories) and 70% ethanol using cotton swabs, respectively.

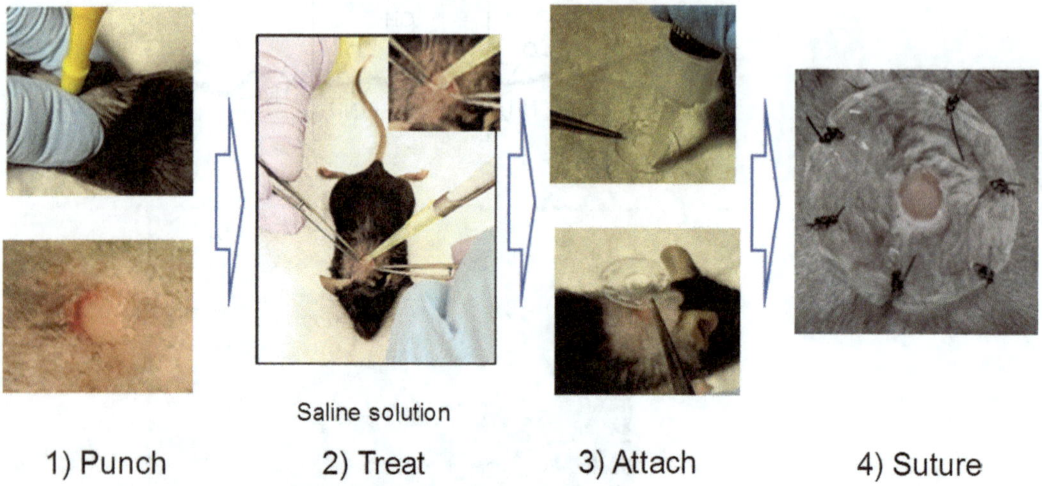

Fig. 3 Procedure of wound healing mouse surgery and treatment

3. **Punch**: Cut one full-thickness wound with a sterilized biopsy punch in the middle of shaved area, and keep the panniculus carnosus beneath the skin.
4. **Treatment**: Drop the treatment in 20 μL PBS, saline, or prepared gel using a micropipette into the punch hole while the edge of wound is pulled up slightly by two forceps (*see* **Note 3**).
5. **Attach**: Wait for 3–5 min to let the application spread and become absorbed in the local wound tissue. Spread Krazy glue carefully on the outside edge of silicone ring while keeping the other side of the splint with the Tegaderm film clean. Attach the splint gently on the top of the wound to locate the inside hole of splint around the wound and cover the wound with Tegaderm film on the top side of the attached splint.
6. **Suture**: Place 6 or 8 interrupted sutures (4–0 black Ethilon monofilament with a FS-2 reverse cutting needle) around the outer edge of the splint.
7. Recover the mice on a heated pad and return them to single-house new cages for the remainder of experiment.
8. Remove splint at day 7 when mice were anesthetized by 1–3% isoflurane (*see* **Note 4**).
9. If there is necessity to apply the new treatment solution before day 7, mice will be anesthetized by 1–3% isoflurane. Solution is injected through the silicone splint with an insulin needle and loaded on the top of the wound without disturbance of new tissue growth in the wound area. It is not suggested to inject solution through the Tegaderm film-covered area that does not have the splint support.

3.4 Observation (Measurement and Intermediate Tissue Collection)

1. Digitally image the wound bed daily beginning from the day of the procedure (day 0) for a total of 16 days; a length reference should be included in the image. Planimetrically measure wound healing progress using ImageJ/FIJI [7].
2. Euthanize each mouse at day 15 by asphyxiation in a CO_2 chamber at a fill rate of 2.0 L/min, followed by cervical dislocation.
3. Collect wound-healing skin at the site of the puncture and treatment and fix the tissue with 3 mL of 10% buffered formalin in a 15 mL centrifuge tube for later analysis.
4. For the intermediate assay, cut off wound-healing skin tissue covered by splint with scissors after asphyxiation and dislocation.
5. Slice the tissue through the middle of wound, put a half tissue into 3 mL of 10% buffered formalin in a 15 mL centrifuge tube for histology analysis later, and freeze other half on dry ice immediately for biochemistry analysis later.

3.5 Pre-Histology Tissue Processing

Follow the tissue processing protocols as described in Chapter 16.

3.6 Hematoxylin and Eosin (H&E) Staining

Follow the tissue processing protocol in Chapter 16.

3.7 Immunohistochemistry (IHC) Staining

IHC is commonly used to characterize inflammation and angiogenesis during wound healing using markers including Ly6G (neutrophils), F4/80 and iNOS (M1 macrophage), arginase-1 (M2 macrophage), and CD31 (endothelial cell). Most operations are standard IHC methods as described in Chapter 15, except for the use of heat-induced epitope retrieval (HIER) steps.

HIER overnight: Sandwich tissue slide with a clean slide in citrate buffer/epitope buffer and clip with metal clips. Keep clips away from buffer to prevent rust stains. Place clipped slides in tray with citrate buffer (covered with foil to prevent dehydration). Place in incubator at 60 °C overnight. Save water bath for the next day. Place in water bath next day and proceed to **step 5**.

3.8 Masson's Trichrome Staining

Masson's trichrome staining is one of most common methods for qualitative histological observation used in wound healing research to characterize collagen fibers which stained blue color. Using this staining method, the thickness and alignment of collagen fibers can be measured and analyzed by using image analysis software [2, 8].

1. Do the same HIER steps overnight as the method for H&E and IHC (*see* Subheading 3.7). Keep slides in tap water until ready for Bouin's trichrome steps.

2. Place slides in Bouin's solution at room temperature overnight in the hood. Save water bath for the next day.
3. Mix equal parts of Weigert's iron hematoxylin A and B (good for 10 days). Place slides in mixture for 10 min in Coplin jars (50 mL t.v.).
4. Wash in gently running distilled water for 5 min.
5. Stain with Biebrich scarlet-acid for 5 min.
6. Rinse in distilled water.
7. Place slides in phosphomolybdic/phosphotungstic acid (PAPA) for 10 min. Discard used acid every time.
8. Place slides in aniline blue for 5 min.
9. Rinse in distilled water.
10. Place in acetic acid for 3 min. Discard used acid every time.
11. Rinse in distilled water.
12. 95% Ethanol two times, 3 min each time.
13. 100% Ethanol two times, 3 min each time.
14. Xylene two times, 3 min each time.
15. Mount and dry for imaging analysis.

3.9 Herovici's Polychrome Staining

Herovici's staining method can specifically stain type I collagen (mature) which gives a red color and type III collagen (young) is stained a blue color [9, 10]. This method is a highly specific and reproducible staining method for collagen assessment in wound healing. It is particularly useful in wound experiments to identify the maturation status of collagen.

1. Do the same HIER steps overnight as the method for H&E and IHC (*see* Subheading 3.7). Keep slides in tap water until ready for the next step.
2. Place slides in plastic Coplin jar containing Weigert's hematoxylin working solution (equal parts of Weigert's hematoxylin A and B) for 5 min.
3. Wash with tap water by placing in Coplin jar with water. Replace water until it runs clear and let slides sit for 1 min in clean tap water.
4. Place slides in Coplin jar with Herovici's working solution for 2 min (50 mL t.v. of equal parts Herovici's stain solutions A and B).
5. Tap slides on paper towel and place in Coplin jar with 1% acetic acid for 2 min.
6. Place slides in Coplin jar containing absolute alcohol, three changes at 1 min each.
7. Place slides in Coplin jar containing xylene for 1 min.

8. Transfer slides to metal rack and place in xylene stations 16 and 17 in chemical fume hood for 1 min each.
9. Mount and dry for imaging analysis.

4 Notes

1. We suggest to cut the hole of splint with a punch. The size of splint should be larger (0.5 mm) than wound size. In this way, the wound area will not have contact with the glue added on the inner side of splint. A too large splint hole will result in inefficient function for holding the mouse skin and preventing contraction. A unified size of splint hole can be used as size reference when the wound size is analyzed later by imaging.
2. The final protein concentration can be adjusted in **step 4** of this procedure. The Serp-1 concentration in the hydrogel we tested for wound healing was 3 μg/20 μL gel [2]. When the collagen solution is mixed well with chitosan/Serp-1 and degassed, the material is a semitransparent and homogeneous gel (*see* Fig. 2).
3. The volume of application (solution or gel) can be adjusted according to the wound size prepared in the specific experiment. Twenty microliters of application is good for wound with 3.0 mm diameter.
4. According to our observation on wild-type C57BL6/J mouse, the mice try to scratch off the splint on its back between days 6 and 7, causing serious secondary damage around the wound. This will affect the healing effect and experimental results. We removed the splints from each mouse wound area carefully after cutting sutures with scissors when mice were anesthetized by 1–3% isoflurane. If the splint is removed from the wound area before day 7, a strong contracting effect is observed, which will also affect the experiment result.

References

1. Lucas A, McFadden G (2014) Secreted immunomodulatory viral proteins as novel biotherapeutics. J Immunol 173:4765–4774
2. Zhang L, Yaron JR, Tafoya AM et al (2019) A virus-derived immune modulating serpin accelerates wound closure with improved collagen remodeling. J Clin Med 8:1626
3. Wise LM, Bodaan CJ, Stuart GS et al (2018) Treatment of limb wounds of horses with orf virus IL-10 and VEGF-E accelerates resolution of exuberant granulation tissue, but does not prevent its development. PLoS One 13: e0197223
4. Chen JS, Longaker MT, Gurtner GC (2013) Murine models of human wound healing. Methods Mol Biol 1037:265–274
5. Zomer HD, Trentin AG (2018) Skin wound healing in humans and mice: challenges in translational research. J Dermatol Sci 90:3–12
6. Wong VW, Sorkin M, Glotzbach JP et al (2011) Surgical approaches to create murine models of human wound healing. J Biomed Biotechnol 2011:1–8
7. Schindelin J, Arganda-Carreras I, Frise E et al (2012) Fiji: an open-source platform for biological-image analysis. Nat Methods 9:676–682

8. Suvik A, Effendy AWM (2012) The use of modified Masson's trichrome staining in collagen evaluation in wound healing study. Malaysian J Vet Res 3:39–47
9. Levame M, Meyer F (1987) Herovici's picropolychromium. Application to the identification of type I and III collagens. Pathol Biol (Paris) 35:1183–1188
10. Turner NJ, Pezzone MA, Brown BN et al (2013) Quantitative multispectral imaging of Herovici's polychrome for the assessment of collagen content and tissue remodelling. J Tissue Eng Regen Med 7:139–148

Chapter 13

Neurologic and Histologic Tests Used to Measure Neuroprotective Effectiveness of Virus-Derived Immune-Modulating Proteins

Jacek M. Kwiecien, Jordan R. Yaron, Kathleen H. Delaney, and Alexandra R. Lucas

Abstract

Severe inflammatory disease initiated by neurotrauma and stroke is of primary concern in these intractable pathologies as noted in recent studies and understanding of the pathogenesis of spinal cord injury (SCI) in the rat model. Successful anti-inflammatory treatments should result in neuroprotection and limit the loss of neurological function to injury caused by the initial damage. Continuous subdural infusion offers direct access to the cavity of injury (COI) that forms after balloon crush SCI deep in the spinal cord. Some anti-inflammatory compounds are not likely capable of crossing the blood-spinal cord barrier. Subdural infusion of myxoma virus-derived Serp-1, an anti-thrombotic/anti-thrombolytic, and also of M-T7, a chemokine inhibitor, improved the locomotor scores and pain sensation scores as well as reduced the numbers of macrophages in the COI by 50 and 80%, respectively, while intraperitoneal infusion of either protein had little effect. Injection of a chitosan hydrogel loaded with Serp-1 into the dorsal spinal column crush also resulted in improved neurological deficits and in reduction of the size of the crush lesion 4 weeks after injury. While neurological scores in a simplified hind-end (HE) locomotor test together with a toe-pinch withdrawal test demonstrated improvement in all balloon crush injury and dorsal spinal crush injury rats, a severe inflammation is induced by the injury indicating additional damage to the spinal cord. Thus neurological function testing can be contradictory, rather than corresponding, to the pathogenesis of SCI. The count of macrophages in the COI offers a precise, reliable method of measuring the effectiveness of a neuroprotective treatment of SCI in preclinical studies.

Key words Serp-1, M-T7, Spinal cord injury, Neuroinflammation, Cavity of injury, Subdural infusion, Spinal cord implantation, Anti-inflammatory treatment, Neuroprotection, Macrophage quantification

1 Introduction

Spinal cord injury (SCI) causes local hemorrhages, massive necrosis, and edema leading to severe, macrophage-rich inflammation starting at day 3 that lasts for over 16 weeks and considerably adds to destruction of the spinal cord [1]. Immunomodulatory proteins

Alexandra R. Lucas (ed.), *Viruses as Therapeutics: Methods and Protocols*, Methods in Molecular Biology, vol. 2225, https://doi.org/10.1007/978-1-0716-1012-1_13,

derived from myxoma virus, Serp-1 and M-T7 [2, 3], have been shown to have a potent anti-inflammatory therapeutic effect on animal models of vasculitis [4], organ transplant [5], and spinal cord injury (SCI) ([6, 7], *see* **Note 1**). Continuous subdural infusion of 0.2 mg/rat of Serp-1 or M-T7 from an Alzet 2ML1 osmotic pump for 7 days resulted in improvement in ***hind-end (HE) locomotor test*** and ***toe-pinch withdrawal test*** and in a 50–80% reduction of macrophage infiltration in the cavity of injury (COI) considered as a neuroprotective effect ([6], *see* **Note 1**). Intraperitoneal infusion of either protein from the 2ML1 osmotic pump resulted in no (for Serp-1) or little (for M-T7) effect [6], indicating that a systemic administration of both proteins is not effective in the neuroprotective treatment of SCI and that a continuous intralesional delivery is required.

In a more recent experiment, a chitosan hydrogel loaded with Serp-1 was implanted into the dorsal spinal column crush [8] to determine whether the intralesional delivery would result in neuroprotection. 100 μg of Serp-1 in 50 μL of hydrogel injected into the crush lesion was associated with better initial neurological recovery on both HE locomotor and toe-pinch withdrawal tests and also with reduction of the size of the crush lesion within 28 days of the study indicating an early neuroprotective effect (*see* **Note 2**). The intralesional implant of a hydrogel loaded with an anti-inflammatory compound may potentially lead to second-generation treatments of the SCI where neuroprotective effect of the drug released from the hydrogel is combined with the function of the hydrogel to act as a bridge for axonal regeneration across the SCI [8]. In a previous study, the implantation of choroid plexus cells into the acute spinal crush lesion in the dysmyelinated Long-Evans shaker (LES) rat resulted in exogenous ependymal cells guiding regenerating axons across the COI, their support, and integration into the peri-lesional white matter [9] indicating that such a therapeutic strategy is feasible.

The studies using subdural infusion or intra-lesional implantation of viral immunomodulatory proteins as anti-inflammatory/neuroprotective agents in SCI [6, 7] offered an opportunity to develop simplified and improved neurological tests to measure the success of a neuroprotective treatment and also to understand their limitations (*see* **Note 3**). While a BBB locomotor test, established by Basso, Beatie, Bresnahan et al. [10], has been widely used as a fundamental and often the sole test to measure the effects of experimental manipulations and treatments in rodent models of SCI, the BBB test has some limitations, specifically due to previous lack of sufficient understanding of pathogenesis of this very complex disease [1]. The BBB test uses a 21-point rising scale of neurological improvements in the function of the hip, knee, and hock joints; positioning of a hind leg; and its support of the body [10]. This test is performed in the open field, requires preoperative training, and requires at least 4 min of standard videotaping and also two independent observers to complete the scoring. These requirements for the BBB test are time consuming and thus the test is generally performed every 3–4 days or once a week. Since the BBB test is considerably time consuming and cumbersome, a simplified locomotor test of hind-end (HE) locomotor function was developed by my lab that is easy to perform in the cage; it is fast requiring 20–30 s, and uses a rising 6-point scale; therefore it can be used to collect neurological score from each study animal every day ([6, 7], *see* **Note 3**) that can be analyzed statistically. The 6-score-point HE locomotor test and the 21-score-point BBB test are presented in Table 1 side by side for comparison. Detailed analysis of a degree of movement of 1–3 joints, hip, knee, and hock, in the BBB test [10] requires considerable experience multiplied by the need for two independent observers to analyze the video recordings of individual rats. This imposes a demand for considerable resources to perform the test and the time required makes the BBB test impractical to perform it every day on all study rats. The HE locomotor test however can be performed in many rats every day [6, 7].

In addition, the HE test was supplemented by a *toe-pinch withdrawal test* that allows for measurement of the pain sensation and the strength of the hind limbs which adds information on neurological deficits in SCI, information not provided in the BBB test (*see* Table 2).

It needs to be emphasized however that the scores in all three neurological tests, BBB, HE locomotor, and toe-pinch withdrawal, while improving during the initial 3 weeks post-SCI, invariably stabilize during the fourth week ([7, 10], *see* **Note 3**) which poses two important problems. Firstly, MRI, histologic, and molecular data [1] indicate that initial damage to the spinal cord caused by trauma is remarkably augmented by the severity and destructive nature of subsequent inflammation and edema. Therefore, improvements in neurological scores in SCI rats fail to reflect the main mechanisms in the pathogenesis of this disease and thus are difficult to interpret. Secondly, the improvements in neurologic scores occur in all untreated and treated rats but they end during the fourth week after SCI and do not change thereafter ([11], *see* **Note 3**) despite the progression of inflammatory pathology for >3 additional months [1]. Therefore, all three neurologic tests are not suitable for determination of the therapeutic effect during the required course of an anti-inflammatory/ neuroprotective treatment required for elimination of inflammation post-SCI. In the case of Serp-1 therapy when infused subdurally, the required course of continuous treatment is at least 8 weeks [11] which exceeds the usefulness of current available neurological tests discussed above.

Table 1
Comparison of hind-end Locomotor test and BBB locomotor test

Score	Hind-end locomotor test	Score	BBB locomotor test
0	Both hind legs have no motion	0	No observable hind limb movement
1	One hind leg has flexing motion caudal to the level of the hip joint, with the plantar surface of the foot up, no weight support	1, 2	Slight movement of one or two joints, usually the hip and/or kneeextensive movement of one joint or extensive movement of one joint and slight movement of one other joint
2	Both legs have flexing motion caudal to the hip, with the plantar surface of the foot up, no weight support, or one leg has flexing motion beyond the hip, no body support, the other leg no motion	2, 3	Extensive movement of one joint or extensive movement of one joint and slight movement of one other joint Extensive movement of two joints
3	One leg has flexing motion beyond the hip, with the dorsal surface of the foot up, no weight support, the other leg has flexing motion caudal to the hip, with the plantar surface of the foot up, or one leg has flexing motion beyond the hip, with dorsal surface of the foot up, with body weight support but the other leg has no motion	2, 3, 4, 5, 6, 7, 8, 9, 10, 11, 12, 13, 14, 15, 16, 17, 18, 19, 20, 21	Extensive movement of one joint or extensive movement of one joint and slight movement of one other joint Extensive movement of two joints **Weight support**
4	Both legs have flexing motion beyond the hip, with the dorsal surface of the foot up, but no body weight support, or one leg with the flexing motion beyond the hip with body support and the other leg with flexing motion caudal to the hip, with the plantar surface of the foot up, but no body weight support	6, 7, 8, 9, 10, 11, 12, 13, 14, 15, 16, 17, 18, 19, 20, 21	Extensive movement of two joints and slight movement of the third **Weight support**
5	One leg has flexing motion beyond the hip with body weight support, the other leg flexing motion beyond the hip, with the dorsal surface of the foot up, but no body support	6, 7, 8, 9, 10, 11, 12, 13, 14, 15, 16, 17, 18, 19, 20, 21	Extensive movement of two joints and slight movement of the third **Weight support**
6	Normal gait, no apparent weakness or proprioceptive deficits	21	Consistent plantar stepping and consistent coordinated gait, consistent toe clearance predominant paw position is parallel throughout stance, and consistent trunk stability; tail consistently up

Table 2
Scoring of the toe-pinch withdrawal test in SCI rats

Score	Description
0	No toe retraction
1	Weak retraction, no jerking
2	Weak retraction with jerking
3	Strong/normal retraction with jerking

Since areas of necrosis deep in the spinal cord are converted into the COI where inflammatory exudate is sequestered along with astroglial reaction [1, 6, 12] it is possible to count macrophages in the COI in histologic sections of the spinal cord in a standardized fashion ([1, 6], *see* **Note 4**). Subdural infusion of dexamethasone for 1–2 weeks [10, 12] and of Serp-1 and M-T7 for 1 week [6] resulted in 50–80% reduction of macrophages in the COI which indicates an anti-inflammatory and neuroprotective effect. Although 1–2-week-long infusion is not sufficient to eliminate destructive inflammation, the *macrophage count test* (macrophage quantification) can serve in preliminary preclinical studies to determine (a) an optimal route of administration, (b) an optimal concentration of the drug administered, and (c) the minimum duration required to completely eliminate macrophages or destructive inflammation after SCI in preclinical studies ([11], *see* **Note 4**). For clinical studies however, the neurological tests and macrophage count test used in the rat model of SCI are not applicable and other in vivo tests need to be developed to determine the inflammatory damage to the spinal cord and also the severity of inflammation (*see* **Note 4**). Once available, such analyses will assist with testing of anti-inflammatory drugs such as Serp-1 [11, 13] and M-T7 for their neuroprotective efficacy in human patients of SCI, traumatic brain injury, and stroke.

2 Materials

2.1 Model Animals

2.1.1 Lab Animal Care

Spinal cord injury induces remarkable negative effects on experimental rats requiring well-justified ethical approval. All animal studies must be performed by national and international guidelines and after approval by the local IACUC (institutional animal care committee). Given the invasive nature of the spinal cord injury the following end point, lethargy, marked dehydration, hypothermia, or ruptured urinary bladder was followed by humane euthanasia and rat was excluded from the study. An end-point monitoring sheet serves also to record the data from the hind-end locomotor test and the hind-limb pinch withdrawal test (*see* Fig. 1).

Endpoint Monitoring Sheet

	Subdural infusion of Serp-1 or of M-T7 in balloon crush SCI rat.				-	-
Principal Investigator	**Rat #**	**Sx. Date:**	**Tx.**	*AUP #*		

Date	A	B	C	D	E	F	G	H	I	J	K	L	M	Comments	Initial

Legend

A – **Hind End**; 6 = normal; 0 = complete lack of motion. See table I.	D – **% b.w. loss**	G – S/Q saline in mL, once daily .	J – **left hind leg**, as I
B – **dehydration**; 0 = none, 1 = mild, 2 = moderate, 3 = marked.	E – **Urinary bladder**; 0 = undetectable, 1 = moderately distended, easy to express, 2 = distended, some blood, 3 = blood in urine, can express; 4 =, cannot express or bladder ruptured upon expression.	H – Baytril 50μL I/M → Y	K – micturition → Y
C – **body weight**, recorded every 3rd day in grams.	F – Anafen, 5 mg/kg b.w. once daily → Y.	I – **right hind leg**, pinch withdrawal reflex; 0 – none, 1 - week, 2 – moderate, 3 – normal, flexing of knee and hip joints with considerable force	L – other

Fig. 1 Endpoint monitoring sheet used for SCI rats

Long-Evans rats: Long-Evans (LE), male, 16-week-old rats that weigh approximately 370–420 g and are of similar anatomical qualities are used for the SCI models [14]. A recently completed systematic study on the pathogenesis of SCI was conducted on this animal model [1].

2.2 Preoperative Procedures

1. *Anesthetic*: 5% and 4% isoflurane admixtures in oxygen flowing at a rate of 1.0 L/min.
2. Three working stations: (i) induction chamber, (ii) pre-surgical prep station with the nose cone, (iii) surgery station with the heating pad and the nose cone.
3. Peanut hair clippers.
4. Lacri-Lube to cover the eyes.
5. 70% Ethyl alcohol, iodine solution.

2.3 Surgical Supplies, Sterile

1. Surgical gloves.
2. Scalpel.
3. Fine Rongeur forceps.
4. Rat tooth forceps.
5. Curved jeweller's forceps for the dorsal column crush [8, 9].
6. Gauze.

7. Cotton Q-tips.
8. 25 ga needle.
9. Needle driver.
10. Absorbable sutures (Vicryl, 4.0).
11. Stainless steel skin staples.
12. 3 Fogarty catheter (Balton, Warsaw, Poland) to create the epidural balloon crush injury [1, 5, 10–12].
13. Rat intrathecal catheter, short (Alzet) for the subdural infusion.
14. Osmotic pump of 2 mL volume with the delivery volume of 2.5–10 mL per hour (2ML4, 2ML2, 2ML1 Alzet).
15. Chitosan-collagen hydrogel [7, 15].
16. 1 mL Syringe with blunted 20 ga needle for hydrogel intralesional injection.
17. Analgesic: Ketoprofen (10 mg/mL, we use Anafen®, Merial Canada, Inc., Baie d'Urfe).
18. Sterile saline: For postoperative rehydration.

2.4 Pathology Supplies

1. Heparin sodium (1000 I.U. per mL).
2. Lactated Ringer's solution.
3. Neutral buffered formalin.
4. Ethylenediaminetetraacetic acid (EDTA).
5. Sodium hydroxide.
6. 50 mL Plastic Falcon tubes for spine decalcification.
7. Orbital motion shaker for decalcification.
8. Tissue histology cassettes.
9. Paraffin wax.
10. Tissue processing and embedding and tissue sectioning (microtome) equipment.

3 Methods

3.1 The Clinical Endpoint

The clinical endpoint at which the animal with inadvertent excessive damage to the spinal cord should be removed from the study and promptly, humanely euthanized is as follows: urinary bladder is impossible to void manually or is ruptured; there is marked dehydration, lethargy, and hypothermia. The clinical observations and scoring of neurologic testing are performed each day starting at 1 day post-surgery and recorded on an endpoint monitoring (EPM, Fig. 1) sheet for each experimental rat.

3.2 Spinal Cord Injury

1. *Anesthesia*: Induce and maintain the surgical level of anesthesia with (a) 5% isoflurane in 95% oxygen in the induction chamber and (b) 4% isoflurane, respectively, flowing at a rate of 1.0 L/ min supplied via the nose cone.
2. *Surgical skin preparation*: Shave the hair over the dorsum and flanks closely with peanut clippers, and then vacuum hair off, followed by swabbing with 70% EtOH and then applying of iodine solution on the shaved area.
3. *Laminectomy*: Perform the laminectomy on the thoracic 10 (T10) vertebral arch. Cut the skin over several vertebral arches of the caudal thoracic and lumbar spinal cord and dissect it to unveil the spine and spinal muscles. Isolate the dorsal spines of T9 and T10 vertebrae by dissecting the spinal muscles with the scalpel blade and firmly grasp the spinal process of the T9 vertebrae with the rat tooth forceps. With the fine Rongeur forceps, gently break off in small fragments and remove the caudal wings of the arches of T10 vertebra to create a window of laminectomy over the dorsal caudal thoracic spinal cord. Blot off hemorrhagic exudate, if any, with surgical gauze and cotton Q-tips.
4. *Balloon catheter injury and subdural infusion*: Insert the 3 Fogarty catheter under the dorsal arches of the vertebrae over the dura mater of the dorsal midline of the thoracic spinal cord, rostral to the site of laminectomy with the caudal margin of the balloon 0.5 cm rostral to the rostral margin of laminectomy. Inflate the balloon with 20 μL saline for 5 s, deflate, and remove the catheter. Make a small cut in the dura over the dorsal spinal cord in the middle of laminectomy with a 25 ga needle. Insert a rat short intrathecal catheter cut at 1.5 cm through the cut opening under the dura rostrally to approximate the end of the catheter with the site of the balloon crush injury. Fix the thickened portion of the catheter to a spinal muscle at the site of laminectomy with a simple interrupted suture and close the muscles with a continuous pattern using an absorbable suture. Remove the guiding steel wire from the catheter, insert the other end over the metal regulator of the osmotic pump, and place it under the skin on the caudal flank. Close the skin with stainless steel staples.
5. *Dorsal column crush and hydrogel injection*: In the laminectomy at T10 cut the dura over the dorsal spinal cord and insert the curved jeweller's forceps 0.5 mm apart, 0.5 mm deep into the dorsal spinal cord, and close for 10 s (*see* Fig. 2). Remove the forceps and stabilize the resulting hemorrhage. Gently inject 50 μL of liquid hydrogel via 20 ga blunted needle into the spinal cord lesion. Close the spinal muscles with a continuous

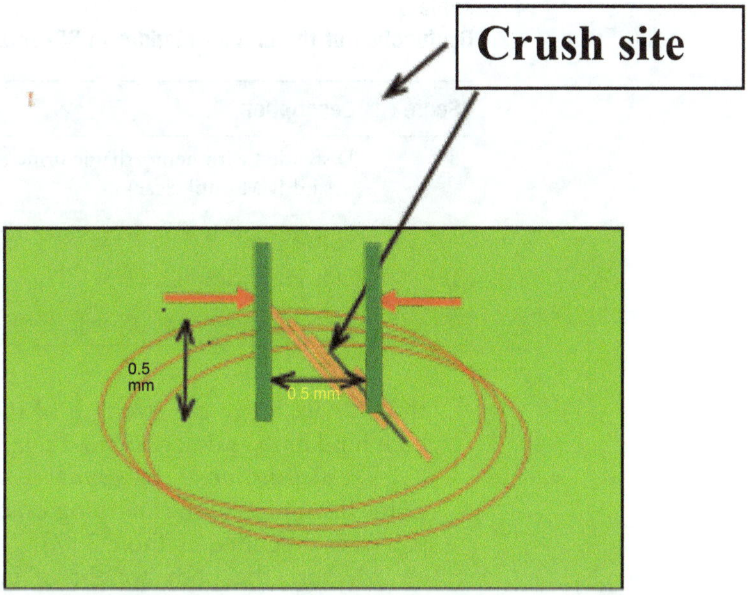

Fig. 2 Dorsal column crush spinal cord injury. A crush with jeweller's forceps was performed in the rats at thoracic 10 vertebra (T10) after dura was cut open by insertion at 0.5 mm in the spinal cord and 0.5 mm apart. The length of the crush lesion was 1 mm. This crush lesion obliterated most of the dorsal column and adjacent area of both dorsal horns

pattern of absorbable suture and the skin with stainless steel staples.

6. *Postoperative treatment*: Administer the subcutaneous injections of 3 mL saline and of 0.4 mL ketoprofen (10 mg/mL) for analgesia prior to the recovery from anesthesia. Administer ketoprofen once daily for 2 more days and saline 3–10 mL to treat dehydration as needed. Void rats with dilated urinary bladder by gentle manual pressure once daily until the neurological control of the bladder is restored. Void rats with hemorrhagic urine in a distended bladder and administer 50 microL of enrofloxacin (50 mg/mL) intramuscularly once daily until the hemorrhagic exudate clears from urine.
7. *Neurological analyses*: House rats individually in oversized, breeding cages for the ease of observation of locomotor deficits. The revised neurological tests used in our recent studies (*see* Table 1) can be performed after little training and fast allowing for testing of a large number of SCI rats every day.

   Perform *hind-end (HE) locomotor test* in rats moving freely in the cage and score locomotor deficits as described in Table 1 [7].

   Perform *toe-pinch withdrawal test* on rats immobilized by holding the tail and moderately pinching the toe with the

Table 3
Dysfunction of the urinary bladder in SCI rats

Score	Description
3	Distended with hemorrhagic urine (treated with 50 μL Baytril s.i.d I/M until clear)
2	Distended with some blood in urine at the beginning of voiding
1	Distended, clear urine
0	Normal function, not distended

thumb nail. Score the strength of the withdrawal motion for each hind limb as described in Table 2 [7].

The dysfunction of the urinary bladder is considered related to the disruption of the neurological control caused by SCI and scored as described in Table 3 [7].

Measure the body weight of individual rats before the surgery and then every third day to monitor its postsurgical decline and then recovery starting at 3–4 weeks post-SCI.

8. *Pathology*:
   (a) *Perfusion and decalcification*: Overdose the rats by intraperitoneal injection of sodium pentobarbital, 80 mg/kg body weight to effect deep anesthesia. Open the chest, inject 100 IU of heparin sodium into the left ventricle, then insert a cannula into the left ventricle with flowing lactated Ringer's solution, and cut open the right auricle to effect the outflow of blood. After the blood is washed out from the carcass, switch the flow to phosphate-buffered formalin for 8 min and transfer the fixed carcass to the refrigerator for 2–3 h. Remove the spine with the spinal cord, trim off spinal muscles, and postfix the spine in formalin overnight. Place the spine in a 50 mL plastic tube with formalin supplemented with 8% EDTA, pH 7.2, a decalcifying solution, and place it on the rotating shaker. Replace the decalcifying solution with fresh one, every second day, and the spines become soft within 3–4 weeks.
   (b) *Histopathology*: For the histologic analysis and macrophage count test, cut the spines perpendicularly every 3 mm and place the rostral cut face down in the tissue cassette to position them as they are to appear on the slide. The tissues are processed, embedded in paraffin wax, cut 5 μm thick, mounted on glass slides for staining with luxol fast blue, and counterstained with hematoxylin and eosin (LFB + H&E). The selected paraffin blocks are used to prepare sections for immunohistochemistry using a

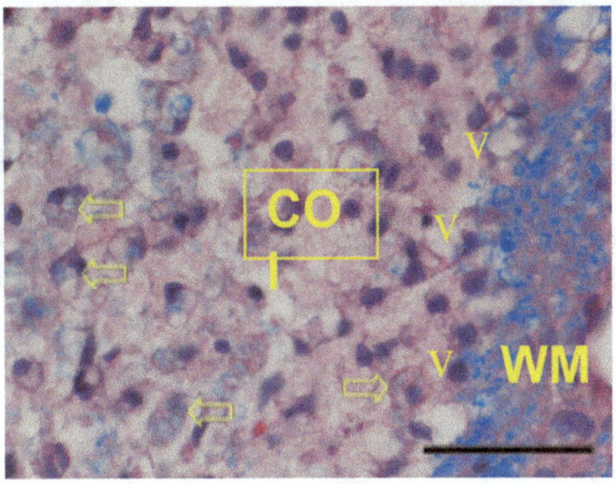

Fig. 3 Phagocytizing macrophages in the cavity of injury following the spinal cord injury. The micrograph of the balloon crush spinal cord injury in the mid-thoracic spinal cord 7 days post-injury. A cavity of injury (COI) is surrounded by white matter (WM) with blue myelin sheaths delineated by arrowheads. The COI contains numerous large cells with a round or oval nucleus and a large amount of vacuolated cytoplasm that often contains blue granules of phagocytized myelin (open arrows). Luxol fast blue counterstained with hematoxylin and eosin. Size bar, 50 μm

primary anti-CD68 antibody for macrophages and anti-glial fibrillary acidic protein (GFAP) for astrocytes. For the analysis of the area of the dorsal column crush cut a sagittally 2.5 cm long segment of the spine involving laminectomy and process histologically as above.

- *Microscopic histopathological analysis.* Perform light microscopic analysis of sections of eight segments of the spinal cord involving the balloon crush and photograph representative changes at 2–60× magnification.
- *Macrophage count in the cavity of injury (COI), balloon crush injury model.* From each rat, take light microscopic images of 3–6 separate cross sections of the spinal cord stained with LFB + H&E with the 40× objective. The images measure 225 × 300 μm and include the periphery of the COI with adjacent spinal cord tissue taking 20% of the area of the image (*see* Fig. 3). Count macrophages, large cells with a large oval nucleus and abundant vacuolated cytoplasm with blue granules of myelin and/or with red blood cells, and average them and determine the standard deviation for each rat [1]. Reduction in numbers of macrophages is indicative of anti-inflammatory effect of a treatment and is associated with greater amount of un-phagocytized, free myelin and membranous debris and red blood cells in the COI.

- *Area of the crush, the dorsal column crush model.* In sagittal sections of the dorsal column crush injury delineate the edge of surrounding spinal cord and measure the area using ImageJ/FIJI [7]. The delineated area contains granulomatous infiltration by macrophages, hyperplastic blood vessels, fibroblasts, and fragments of implanted material. Reduction in the size of the area of the crush is related to inhibition of inflammation.

4 Notes

1. The short-term, 7-day treatment [6] results indicate that the subdural infusion of viral serpin, Serp-1 and of M-T7, has potent inhibitory effect on the infiltration of the cavity of injury (COI) by macrophages at the dose of 0.2 mg/week/rat. There are no clinical signs of toxicity in treated rats and longer duration of subdural administration to eliminate inflammation from the COI can be considered [11].
2. Implantation of chitosan hydrogel loaded with 100 μg Serp-1/ 50 μL hydrogel resulted in reduction of the size of the crush after 4 weeks of injury indicating an anti-inflammatory/ neuroprotective effect [7]. This is a practicable method of intra-lesional delivery for treatment of SCI, traumatic brain injury, and stroke.
3. The BBB test and new, rapidly performed neurological tests, the hind-end (HE) locomotor test and the toe-pinch withdrawal test, are difficult to interpret in view of elucidated pathogenesis of SCI [1] and their respective scores stabilize after the first 3–4 weeks of SCI rendering them not useful for more chronic neuroprotective treatments.
4. Newly developed, macrophage count in the COI test is performed in histologic sections of the spinal cord and allows for detection of the effect of an anti-inflammatory treatment beyond 4 weeks post-SCI until complete elimination of macrophages. While useful in preclinical studies [11], this test is not adaptable in clinical trials.

References

1. Kwiecien JM, Dabrowski W, Dąbrowska-Bouta B et al (2020) Protracted inflammation extends damage after spinal cord injury. PLoS One 15 (3):e0226584. https://doi.org/10.1371/journal.pone.0226584
2. Bartee MY, Dai E, Liu L, Munuswamy-Ramanujam G, Macaulay C, McIvor D, McFadden G, Lucas AR (2009) 10 M-T7: measuring chemokine-modulating activity. Methods Enzymol 460:209–228. https://doi.org/10.1016/S0076-6879(09)05210-0
3. Rahman MM, Lucas AR, McFadden G (2009) Viral TNF inhibitors as potential therapeutics. Adv Exp Med Biol 666:64–77

4. Viswanathan K, Bot I, Liu L et al (2012) Viral cross-class serpin inhibits vascular inflammation and T lymphocyte fratricide; a study in rodent models in vivo and human cell lines in vitro. PLoS One 7(9):e44694. https://doi.org/10.1371/journal.pone.0044694
5. Dai E, Liu LY, Wang H et al (2010) Inhibition of chemokine-glycosaminoglycan interactions in donor tissue reduces mouse allograft vasculopathy and transplant rejection. PLoS One 5 (5):e10510. https://doi.org/10.1371/journal.pone.0010510
6. Kwiecien JM, Dabrowski W, Marzec-Kotarska B et al (2019) Myxoma virus derived immune modulating proteins, M-T7 and Serp-1, reduce early inflammation after spinal cord injury in the rat model. Folia Neuropathol 57:41–50. https://doi.org/10.5114/fn.2019.83830
7. Kwiecien JM, Zhang L, Yaron JR et al (2020) Local chitosan-serpin injection after spinal cord injury reduces inflammatory damage and improves neurologic function. J Clin Med 9 (4). pii: E1221. https://doi.org/10.3390/jcm9041221
8. Kwiecien JM (2016) Tissue reaction to acellular implants in the acute spinal cord injury in the dysmyelinated rat. In: Berhardt LV (ed) Advances in medicine and biology, vol 99. Nova Science Publishers, Inc., Hauppauge, NY, pp 111–123
9. Kwiecien JM (2013) Cellular mechanisms of white matter regeneration in adult dysmyelinated rat model. Folia Neuropathol 51:189–202
10. Basso DM, Beattie MS, Bresnahan JC (1995) A sensitive and reliable locomotor rating scale for open field testing in rats. J Neurotrauma 12:1–21
11. Kwiecien JM, Dabrowski W, Kwiecien-Delaney BJ, Kwiecien-Delaney CJ, Siwicka-Gieroba D, Yaron JR, Zhang L, Delaney KH, Lucas AR. Neuroprotective effect of subdural infusion of Serp-1 in spinal cord trauma. Biomedicines (in revisions)
12. Kwiecien JM, Jarosz B, Machova-Urdzikova L et al (2015) Subdural infusion of dexamethasone inhibits leukomyelitis after acute spinal cord injury in a rat model. Folia Neuropathol 53:41–51
13. Kwiecien JM, Jarosz B, Oakden W et al (2016) An in vivo model of anti-inflammatory activity of subdural dexamethasone following the spinal cord injury. Pol J Neurol Neurosurg 50:7–15. https://doi.org/10.1016/j.pjnns.2015.10.006
14. Tardif J C, L'Allier PL, Grégoire J et al (2010) A randomized controlled, phase 2 trial of the viral serpin Serp-1 in patients with acute coronary syndromes undergoing percutaneous coronary intervention. Circ Cardiovasc Interv 3:543–548. https://doi.org/10.1161/CIRCINTERVENTIONS.110.953885
15. Kwiecien JM (2018) Methods for assessing serpins as neuroprotective therapeutics. In: Lucas A (ed) Serpins. Methods in molecular biology, vol 1826. Humana Press, New York, NY

Chapter 14

Preclinical Testing of Viral Therapeutic Efficacy in Pristane-Induced Lupus Nephritis and Diffuse Alveolar Hemorrhage Mouse Models

Qiuyun Guo, Liqiang Zhang, Jordan R. Yaron, Michelle Burgin, Lauren N. Schutz, Enkidia A. Awo, and Alexandra R. Lucas

Abstract

Systemic lupus erythematosus (SLE) is a multifactorial and heterogeneous autoimmune disease involving multiple organ systems and tissues. Lupus nephritis occurs in approximately 60% of patients with SLE and is the leading cause of morbidity. Diffuse alveolar hemorrhage (DAH) is a rare but very serious complication of SLE with a greater than 50% associated mortality. The etiology of SLE is unclear but has proposed genetic, hormonal, and environmental aspects. Pristane is a saturated terpenoid alkane and has become the most popular laboratory model for inducing lupus in mice. The pristane model of SLE has the capacity to reproduce many components of the human presentation of the disease. Previous studies have demonstrated that virus-derived immune-modulating proteins have the potential to control inflammatory and autoimmune disorders. Serp-1, a 55 kDa secreted and highly glycosylated immune modulator derived from myxoma virus (MYXV), has potent immunomodulatory activity in models of vasculitis, viral sepsis, collagen-induced arthritis, and transplant rejection. This chapter describes the mouse preclinical pristane lupus model as a method to examine virus-derived protein efficacy for treating autoimmune diseases and specifically lupus nephritis and DAH.

Key words Virus-derived immune-modulating protein, Serp-1, Systemic lupus erythematosus (SLE), Lupus nephritis (LN), Diffuse alveolar hemorrhage (DAH), Pristane

1 Introduction

Systemic lupus erythematosus (SLE) is an autoimmune disease characterized by T- and B-cell hyperactivity, and production of antibodies against self-antigens such as double-stranded (ds) DNA, histones, and ribonucleoprotein (RNP). The etiology of SLE is still incompletely defined. SLE is considered a multifactorial disease associated with abnormal genetic, hormonal, and environmental responses. Patients with SLE have a wide range of clinical symptoms including nephritis, nonerosive arthritis, serositis, and hematological and respiratory disorders. Lupus nephritis is

Alexandra R. Lucas (ed.), *Viruses as Therapeutics: Methods and Protocols*, Methods in Molecular Biology, vol. 2225, https://doi.org/10.1007/978-1-0716-1012-1_14,

detected in approximately 60% of patients with SLE and is the main source of morbidity [1]. Pulmonary abnormalities are common in SLE [2]. Among SLE pulmonary disorders, diffuse alveolar hemorrhage (DAH) occurs in about 3% of patients with lung manifestations. DAH has an associated very high mortality that is greater than 50% in most analyses [3–7]. The current treatment options for SLE include steroids, immunosuppressive agents, respiratory support, etc. The efficacy of these treatments is limited and there are many significant side effects. Therefore, there is an urgent, unmet need for new drugs to improve the treatment for SLE and in particular to improve outcomes and reduce mortality for DAH.

Pristane (2,6,10,14-tetramethylpentadecane, TMPD) is an isoprenoid alkane found at high concentration in mineral oil and low concentration in vegetables [8]. It is also found in the liver of some sharks [9]. Intraperitoneal injections of pristane can induce in mice a wide range of autoantibodies specific to or associated with SLE [10, 11], making pristane a popular method to establish mouse models of SLE. The symptoms induced by pristane and the severity of disease are dependent on the mouse strain treated with pristane. Pristane administration in BALB/c mice induces nephritis and arthritis, and is characterized by production of autoantibodies and high levels of serum inflammatory cytokines similar to those detected in SLE [10–13]. Intraperitoneal (IP) injection of pristane in C57BL/6J (B6) mice causes severe alveolar hemorrhage within 2 weeks, manifested by alveolar and perivascular inflammation (capillaritis, small-vessel vasculitis), hemorrhage, endothelial injury, and about 50% mortality [3, 7, 14, 15].

The pathogenesis of SLE is complicated. Genetic variants or environment triggers result in either increased production or decreased clearance of nuclear particles. Antigen-presenting cells recognize the nuclear particles through the Toll-like receptors on the cell surface. The immune system loses tolerance to autoantigens, leading to IFN-α production, lymphoproliferation, lymphocyte differentiation, proinflammatory cytokine and chemokine release, polyclonal autoantibody production, immune complex deposits, and ultimately organ damage (*see* Fig. 1). This procedure partially resembles immune reactions to viral infection. Throughout millions of years of evolution, viruses have developed highly effective immune-modulating proteins designed to evade the immune responses of the host. These immune-modulating proteins can bind and inhibit proteases, cytokines, and chemokines, while yet other proteins are homologues or decoys for cytokines and cytokine receptors. These virus-derived proteins provide a natural source of tools that may lead to cures and/or treatments for autoimmune disorders. Previous studies have found that treatment with M-T1, a chemokine-binding protein, deriving from myxoma virus, had some benefit in atherosclerosis and arthritis models [16, 17]. Intravenous injections of vMIP-II, a macrophage

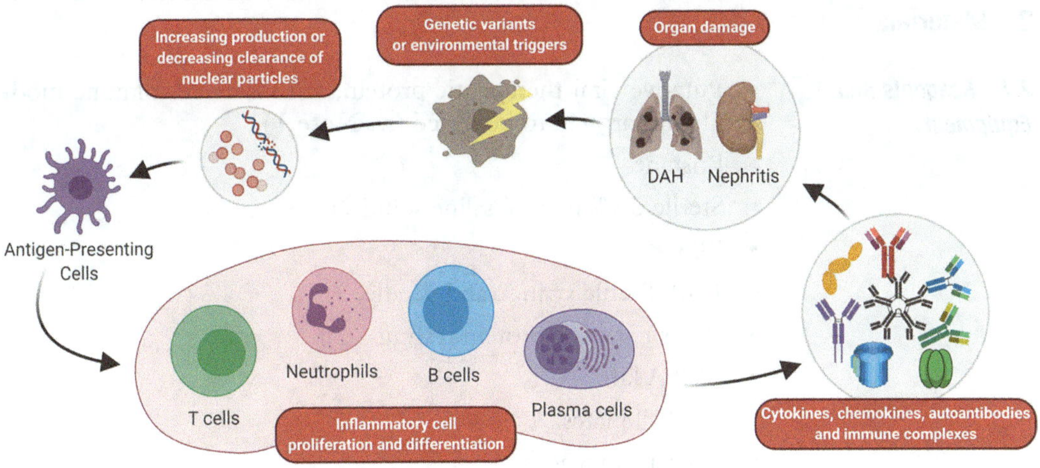

Fig. 1 The pathogenesis of SLE. Genetic variants or environment trigger results in either increased production or decreased clearance of nuclear particles. Antigen-presenting cells recognize the nuclear particles. The immune system loses tolerance, leading to proliferation and differentiation of inflammatory and autoimmune cells, releasing of cytokines and chemokines, production of polyclonal autoantibody, deposit of immune complex, and ultimately organ damage

inflammatory protein-1α mimic, derived from human herpesviruses, protected rats in a model of anti-glomerular basement membrane-induced glomerulonephritis [18]. Our research has shown that Serp-1, a 55 kDa secreted viral glycoprotein derived from MYXV, belonging to the *serpin* (*ser*ine *p*rotease *in*hibitor) superfamily can reduce mononuclear cell infiltration into transplanted hearts, kidneys, and aorta in rodent models, with improved histopathological evidence of chronic rejection [19–21]. In a vasculitis mouse model induced by mouse herpesvirus-68 (MHV-68) infection and also in aortic transplant models, Serp-1 significantly reduced inflammation and arterial plaque growth. Additionally, Serp-1 treatment reduced lung hemorrhage and consolidation and also improved survival in the MHV8 infection mouse model. Serp-1 has been examined in phase I and IIa clinical trials in patients with coronary stent implantation for unstable angina pectoris. Serp-1 treatment significantly reduced markers for myocardial damage after coronary stent implant. There was no significant increase in adverse reactions and no neutralizing antibody was detected [22].

These prior studies indicate that treatment with virus-derived biologic protein may prove to be very effective and safe for treating lupus. This chapter provides some methods to test the therapeutic efficacy of these proteins in pristane-induced lupus nephritis and diffuse alveolar hemorrhage mouse models.

2 Materials

2.1 Reagents and Equipment

- Putative viral therapeutic protein, virus-derived immune-modulating protein to be tested (*see* **Note 1**).
- Pristane.
- Sterile 0.9% normal saline solution.
- PBS.
- 1 mL Sterile syringe and needle.
- Strips for urine protein testing.
- ELISA kit.
- 96-Well plates.
- Microplate reader.
- Antibody for the viral therapeutic protein for further analysis (WB, ELISA, etc.).
- Antibodies, choose the right primary antibody or secondary antibody.
- Hep-2 ANA slide.
- Bovine serum albumin (BSA).
- Methylated BSA.
- Fluoromount.
- Immulon 2HB plate.
- dsDNA (50 μg/mL in PBS).
- Tween 20.
- 0.1% Gelatin containing 3% BSA, 3 mM EDTA.
- 0.1% Gelatin containing 2% BSA, 3 mM EDTA, and 0.05% Tween 20.
- 0.025 M Bicarbonate buffer.
- 1 M $MgCl_2$.
- Para-nitrophenyl phosphate (*pNPP*) tablet (5 mg pNPP per tablet).
- Xylene.
- 100% Ethanol, 95% ethanol,70% ethanol.
- Tap water.
- Hematoxylin solution.
- Acid alcohol (0.5% hydrochloric acid in 70% ethanol).
- Ammonium water (2% ammonium hydroxide in deionized water).
- Acidic, aqueous Eosin Y solution.
- *Deionized* water (DI water).

- 3% H_2O_2 in PBS.
- TBST (NaCl 8 g/L, KCl 0.2 g/L, Tris base 3 g/L, Tween 20 1 mL/L in ddH_2O, adjust pH to 7.4).
- ImmPACT DAB (1 mL dilution buffer + 1 drop DAB).
- Triton X-100.
- Prolong Gold Antifade Reagent or Prolong Gold Antifade Reagent with DAPI.
- Avertin solution (2,2,2-tribromoethanol).
- Fluorescence microscope, image analysis software.
- Tubes to prepare standard dilutions and to aliquot samples.
- Precision pipettes P10, P100, P200, and P1000, and corresponding sterilized tips.
- Disposable reagent reservoirs.
- Paper towel.
- Laboratory timer.
- 60 °C oven or air convection incubator.
- Chemical fume hood.

2.2 Animal Models

All mice strains are susceptible to pristane-induced lupus, but the spectrum of autoantibodies and clinical manifestations differ from strain to strain [23]. In BALB/c mice, pristane induces glomerulonephritis, arthritis, and production of several characteristic autoantibodies found in SLE, including anti-dsDNA and anti-Sm. The BALB/c mouse model is the most commonly used lupus nephritis model [10–13]. IP injection of pristane in C57BL/6 (B6) mice causes severe alveolar hemorrhage within 2 weeks [3, 7, 14, 15]. Severe pulmonary hemorrhage was not observed in other types of mice, making B6 the unique animal model of lupus-associated DAH [3, 7, 14, 15]. Like human SLE patients, pristane-induced lupus is more severe in females than in males [24], so female mice are more commonly used in experiments.

Prior to use, all mouse experiments must be approved by the appropriate Institutional Animal Care and Use Committees and should conform to local, national, and international guidelines for animal care.

- Mice model of nephritis: Six- to 8-week-old female BALB/c mice are randomly assigned into pristane-induced or control groups (6 mice each group, $n = 6$). Mice of pristane-induced group have one IP injection of 0.5 mL pristane at day 0. Control mice will be treated with one IP injection of 0.5 mL saline each mouse at day 0.
- Mice model of DAH: Six- to 8-week-old female C57BL/6 (B6) mice are randomly assigned into pristane-induced or

control groups (6 mice each group, $n = 6$). Mice of pristane-induced group have one IP injection of 0.5 mL pristane at day 0. Control mice will be treated with one IP injection of 0.5 mL saline each mouse at day 0.

3 Methods

3.1 Methods of Nephritis

Pristane can induce lupus-associated glomerulonephritis in one-third of BALB/c mice [12, 14] beginning 4–6 months after pristane injection, manifested as proteinuria, abnormal production of autoantibody, and proinflammatory cytokines. The schedule of the experiments can be seen in Fig. 2.

3.1.1 Administration of Viral Therapeutic Protein

- IP inject 100 μL, or appropriate volume, of putative therapeutic protein in saline solution into protein-treated group of BALB/c mice right after pristane treatment described in Subheading 2.2. IP inject 100 μL of saline in saline-treated group (*see* **Note 2**).
- Repeat the treatments daily till the end of the experiment.

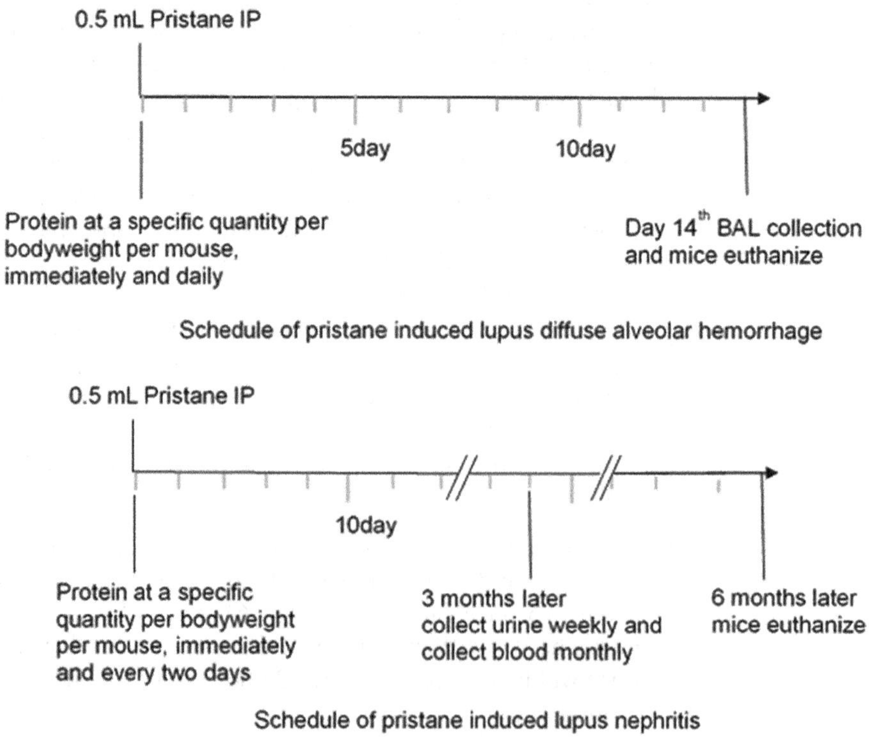

Fig. 2 Schedule of pristane-induced lupus diffuse alveolar hemorrhage and nephritis

- The primary endpoint of the experiment is 6 months of treatment. If an animal develops severe disease, or shows two successive >300 mg/dL proteinuria scores, and/or a body weight reduction >15% within a week, the animal will be sacrificed (*see* **Note 3**).

3.1.2 Monitoring Proteinuria

Proteinuria normally begins 4–6 months after pristane injection [12]. Mouse proteinuria levels can be monitored weekly after 3-month induction by a semiquantitative method with strips for urine protein testing according to the standard protocol provided by the manufacturer. Scales can be set for easy comparison, such as using a 0–4 scale (0, negative; 1, 30 mg/dL; 2, 100 mg/dL; 3, 300 mg/dL; and 4, over 2000 mg/dL of urinary protein).

Method of urine collection:

1. Gently grasp the base of the mouse's tail with dominant hand.
2. Place thumb and index finger on either side of the neck and draw up the loose skin with nondominant hand.
3. Hold the mouse over the collection device (e.g., tube, strips).
4. Lightly stroke the belly of the animal.
5. Collect the urine into the tube or onto a test strip.
6. If the above method does not work, then do the following steps. Place the mouse in a clean, dry, empty cage or on a tabletop cover with a nonabsorbable, sanitizable material (i.e., plastic wrap).
7. Aspirate urine using a pipette after the mouse urinates on it, and drip the urine onto the strip.

3.1.3 Blood Collection

Detection of antibodies and cytokines in mouse serum is often used to assess the severity of lupus and explore the pathogenesis of the disease. Venous blood draws from the lateral tail vein or the submandibular vein are usually taken during the experiment while the mouse is restrained. Tail vein can only provide about 0.1 mL blood, while submandibular vein can provide 0.2–0.5 mL blood. The so-called terminal or non-survival blood collection can be taken via cardiac puncture and can collect about 1 mL blood which is usually done at the end of the experiment. Excellent serum collection method: Use a blood separator tube, allow sample to clot for 30 min, and then centrifuge for 10 min at 1000 × *g*. Run assay immediately; otherwise aliquot and store sample below −20 °C. Avoid repeat thaw-freeze cycle.

3.1.4 Detection of Cytokines and Chemokines

In SLE, endogenous nucleic acids or immune complexes activate mononuclear macrophages by complement-receptor binding or dendritic cells by acting on Toll-like receptors or Fcγ receptors. These cells present antigens to T cells, and activated T cells promote

B cells to differentiate into plasma cells. During the process, large amounts of cytokines and chemokines are released, such as IFN-I, TNF-α, IL-6, IL-12, CCL3, CCL2, and CXCL1 [25–31]. These cytokines can further promote the production of autoantibodies and immune complexes [32–35]. Production of cytokines and chemokines is already detectable as early as 2 weeks after induction [36].

Below is provided a direct ELISA method to detect cytokines or chemokines; for the specific experiment you should refer to the manufacturer's instructions (*see* **Note 4**).

1. Standard preparation: prepare standard curves by stepwise dilution (*see* **Note 5**).
2. Pipette 100 μL of assay diluent to each well of the plate.
3. Pipette 100 μL of prepared standards and samples to different wells.
4. Incubate the plate at 4 °C for 1 h.
5. Aspirate the solution from the wells.
6. Wash the plate with prepared wash solution four times.
7. Invert the plate and pound it vigorously on clean paper towels to remove excess liquid.
8. Pipette 100 μL detection reagent to each well.
9. Incubate the plate at 4 °C for 1 h.
10. Aspirate the solution from the wells.
11. Wash the plate with prepared wash solution four times.
12. Invert the plate and pound it vigorously on clean paper towels to remove excess liquid.
13. Pipette 100 μL of substrate from the reagent reservoir to each well (*see* **Note 6**).
14. Incubate in the dark at room temperature (RT) for 15–20 min.
15. Pipette 50 μL of stop solution to each well.
16. Read the plate at 450 nm with a microplate reader to obtain absorbance values.
17. Calculate the mean absorbance for each set of duplicate standards and samples.
18. Subtract the zero mean absorption from each mean absorbance to receive adjusted absorbance value.
19. Plot a standard curve and calculate the cytokine amount in sample.

3.1.5 Detection of Autoantibodies

Intraperitoneal injections of pristane in BALB/c, SJL/J, and C57BL/6 mice can develop autoantibodies, including anti-ANAs, anti-dsDNA, anti-ssDNA, anti-Sm, anti-RNP, and anti-ribosomal

P around the third or fourth month after induction [10, 12, 36, 37].

Antinuclear antibodies (ANA) are a hallmark of lupus but are not very specific which are usually used as a screening marker. Anti-dsDNA in high titer is nearly specific to lupus. The following are examples of common detection methods for these two markers.

For detection of antinuclear antibodies (ANAs) in mouse serum samples using indirect immunofluorescence detection, all steps should be performed at RT.

1. Dilute mouse serum to 1:40 with PBS.
2. Carefully add 20 μL of the above sample or control to each well on a pre-coated Hep-2 ANA slide, cover with a lid, and incubate for 30 min.
3. Remove the coating solution and wash carefully three times with PBS (*see* **Note 7**).
4. Dilute fluorescence-conjugated goat anti-mouse IgG with 0.1% bovine serum albumin in PBS buffer according to the instructions, add 20 μL to each well, and incubate in the dark for 30 min (or as required).
5. Carefully aspirate and wash three times with PBS.
6. Place a drop of Fluoromount in the center of each well, cover the slide with a 60 mm coverslip, and then press gently to avoid air bubbles.
7. Capture images with a fluorescence microscope and analyze the fluorescence intensity with the ImageJ program.

Detection of anti-dsDNA IgG by ELISA

1. Add 100 μL of 1 mg/mL methylated BSA in PBS to the Immulon 2HB plate, and incubate for 2 h at 37 °C.
2. Wash with PBS three times.
3. Coat the Immulon 2HB plate with 50 μg/mL of dsDNA in PBS, and incubate overnight at 4 °C.
4. Wash with PBS three times (*see* **Note 8**).
5. Apply 0.1% gelatin containing 3% BSA and 3 mM EDTA, and incubate at 37 °C for 1.5 h (or at 4 °C overnight).
6. Wash with PBS three times.
7. Add diluted sample serum (e.g., 1:2400 dilutions in 0.1% gelatin containing 2% BSA, 3 mM EDTA, and 0.05% Tween 20), and incubate at 37 °C for 1.5 h (*see* **Note 9**).
8. Wash with PBS three times.
9. Add HRP-conjugated goat anti-mouse IgG (15,000 dilution in PBS containing 1% BSA and 0.05% Tween 20), and incubate at 37 °C for 1.5 h.

10. Wash with PBS three times.
11. Mix 0.025 M bicarbonate buffer, 10 μL 1 M MgCl2, and 1 pNPP tablet, and add to the plate.
12. Read O.D. at 405 nm.

3.1.6 Renal Pathological Examinations

Hematoxylin and eosin (H&E) and periodic acid-Schiff (PAS) staining can detect changes of renal pathological structure, such as mesangial and endocapillary proliferation and increased thickness of mesangial membranes in many glomeruli with mononuclear cell infiltration. PAS is commonly used for detecting the deposition of immune complexes in the renal mesangium and small vessel walls. Immunohistochemistry (IHC) is used to detect infiltration of inflammatory cells and expression of specific proteins to further speculate the molecular mechanisms involved. Immunofluorescence can show large amounts of IgG, IgM, C3 deposited in the glomerular mesangium, subendothelium, and capillary walls [13, 38].

For H&E staining, PAS staining, and IHC, the kidneys are fixed in 10% buffered formalin, embedded in paraffin, and cut into 4 μm thick sections. For detecting IgG and IgM by immunofluorescence, the kidney is embedded in ornithine carbamoyl transferase and cut into 5 μm thick sections.

Steps for H&E staining:

1. Deparaffinize sections, two changes of xylene, 5 min each.
2. Rehydrate in two changes of 100% ethanol, 2 min each.
3. 95% Ethanol for 2 min and 70% ethanol for 2 min.
4. Water wash for 2 min.
5. Stain in hematoxylin solution for 5 min.
6. Quick dip in water.
7. Wash in clean water for 1 min.
8. Change to clean water and wash for 5 min.
9. Differentiate in acid alcohol for 12 dips.
10. Wash in clean water for 2 min.
11. Stain in 0.2% ammonia water for 30 s.
12. Wash in clean water for 2 min. Counterstain in acidic and aqueous eosin Y solution for 5 min.
13. Dehydrate through two changes of 95% ethanol, 2 min each.
14. Dehydrate through two changes of 100% ethanol, 2 min each.
15. Clear in two changes of xylene, 5 min each.
16. Let slides air-dry.
17. Mount with coverslips.

Steps for IHC staining:

1. The formalin-fixed, paraffin-embedded tissue sections are kept in 60 °C incubator oven for 30 min.
2. **Steps 1–4** of H&E staining.
3. Heat-induced epitope retrieval: Sandwich slides with a clean slide, then boil for 30 min, and cool for 20 min. Unsandwich and place in DI water.
4. Use PAP pen to circle tissue. Tap off water, place tray filled with di-water, and cover with TBST.
5. Block with 3% H_2O_2 in PBS for 15 min at room temperature, and rinse with TBST.
6. Block with 5% BSA in TBST solution for 1 h at RT.
7. Add primary antibody solution according to the manufacturer's instruction. Cover slides with parafilm. Incubate overnight at 4 °C. Rinse with TBST five times.
8. Add secondary antibody 1:500, and incubate at room temperature for 1 h.
9. Rinse thoroughly with TBST five times.
10. ImmPACT DAB (1 mL dilution buffer + 1 drop DAB) develops for ~10 min or less; rinse DAB off into biohazard box directly.
11. Place slides in tap water.
12. **Steps 5–16** of H&E staining.

Steps for indirect immunofluorescence:

1. Prepare blocking buffer (1× PBS/5% normal serum/0.3% Triton X-100): e.g., to prepare 10 mL, add 0.5 mL normal serum from the same species as the secondary antibody to 9.5 mL PBS. While stirring, add 30 μL Triton X-100.
2. Prepare antibody dilution buffer: (1× PBS/1% BSA/0.3% Triton X-100): e.g., to prepare 10 mL, add 30 μL Triton X-100 to 10 mL PBS. Mix well, then add 0.1 g BSA, and mix.
3. Block slides in blocking buffer for 60 min (*see* **Note 10**).
4. Dilute primary antibody following the manufacturer's instructions in antibody dilution buffer.
5. Aspirate blocking solution, and apply diluted primary antibody.
6. Incubate overnight at 4 °C.
7. Rinse three times in PBS for 5 min each.
8. Incubate slides in fluorochrome-conjugated secondary antibody diluted in antibody dilution buffer for 1–2 h at room temperature in the dark.
9. Rinse three times in PBS for 5 min each.

10. Coverslip slides with Prolong Gold Antifade Reagent or Prolong Gold Antifade Reagent with DAPI.
11. For best results, cure overnight at room temperature. For long-term storage, store slides flat at 4 °C protected from light.

3.2 Methods of DAH

Pristane can induce lupus-associated DAH in C57BL/6 (B6) mice, manifested as macrophage, neutrophil, lymphocyte, and eosinophil infiltration; alveolar and perivascular inflammation; capillary inflammation; endothelial damage; and hemorrhage [6, 14, 39, 40]. Recruitment of macrophages and neutrophils started 3 days after pristane injection and peaked in 2 weeks [3]. Schedule of pristane-induced DAH can be seen in Fig. 2.

3.2.1 Administration of Viral Therapeutic Protein

- IP inject 100 μL, or appropriate volume, of putative therapeutic protein in saline solution into protein-treated group of female C57BL/6 (B6) mice right after pristane treatment described in Subheading 2.2. IP inject 100 μL of saline in saline-treated group (*see* **Note 2**).
- Repeat the injection every day for 14 days.
- The primary endpoint of the experiment is the 14th day.

3.2.2 Collection and Analysis of Bronchial Alveolar Lavage Fluid (BAL)

Pristane can lead to increased cytokines and inflammatory cells in mouse BAL. While the mouse is alive, the severity of DAH can be assessed by collecting BAL and analyzing the levels of cytokines and inflammatory cells in BAL. Cytokine detection method has been described in the previous paragraph.

Steps for BAL collection:

1. Anesthetize mice with avertin solution (2,2,2-tribromoethanol).
2. Expose the trachea and insert a cannula into the trachea.
3. Lavage the lungs three times with 1 mL sterile PBS (*see* **Note 11**).
4. Centrifuge BAL at 1000 rpm for 10 min at 4 °C to remove cells, and store the supernatants at −80 °C.
5. Resuspend the cell pellets in 1 mL PBS (*see* **Note 12**).

3.2.3 Lung Pathological Evaluation

DAH can be initially assessed by weight and gross observation of excised lungs. The severity of DAH can be classified into three degrees according to the percentage of hemorrhage in H&E staining: 1, no hemorrhage (0%); 2, partial hemorrhage (25–75%); and 3, complete hemorrhage (75–100%). IHC can be used to analyze infiltration of inflammatory cell with methods described above, including $CD45^+$, $CD64^+$, and $F4/80^+$ for macrophage; $CD11b^+$, $CXCR4^+$, $MHCII^-$, and $Ly\text{-}6G^+$ for neutrophil; and $B220^+$ for B cell. Detection of complement and immunoglobulins can also be achieved through IHC.

4 Notes

1. Take Serp-1 for example: Purified Serp-1 protein is expressed and secreted by a Chinese hamster ovary (CHO) cell line (Viron Therapeutics Inc., London, ON, CA). Good manufacturing product (GMP)-compliant purification was performed by sequential column chromatographic separation with greater than 95% purity, as determined by Coomassie-stained SDS-PAGE and reverse-phase HPLC. Serp-1 is endotoxin free by define LAL assay.
2. For example use Serp-1 at initial doses of 100 μg/kg/mouse.
3. Use the method described in Subheading 3.1, **step 2** to test the level of urine protein.
4. Reagents and samples should be fully equilibrated to room temperature (20–25 °C) before performing a test.
5. It is recommended that all standards and samples are prepared in duplicate.
6. The actual reaction time depends on the temperature. If temperature is low, reaction time should be extended. The ideal reaction temperature for optimum assay performance is 20–25 °C. Reliable standard curve is obtained when the OD450 absorbance value does not exceed 0.1 unit for the zero-standard concentration.
7. If nonspecific binding issues occur, additional blocking steps may be required (5% BSA-PBS, 30 min).
8. Flip the plate gently upside down on a paper towel to avoid washing solution left.
9. Pooled serum from the lupus-prone mouse (e.g., NZM2410) which has high anti-dsDNA titer can be used as standard.
10. All subsequent incubations should be carried out at room temperature unless otherwise noted in a humid lighttight box or covered dish/plate to prevent drying and fluorochrome fading.
11. 2.4–2.6 mL of the instilled lavage fluid can be routinely recovered.
12. Cells can be used for next counting with a standard hemocytometer or flow cytometry.

References

1. Almaani S, Meara A, Rovin BH (2017) Update on lupus nephritis. Clin J Am Soc Nephrol 12:825–835
2. Mittoo S, Fell CD (2014) Pulmonary manifestations of systemic lupus erythematosus. Semin Respir Crit Care Med 35:249–254
3. Barker TT, Lee PY, Kelly-Scumpia KM et al (2011) Pathogenic role of B cells in the development of diffuse alveolar hemorrhage induced by pristane. Lab Investig 91:1540–1550

4. Santos-Ocampo AS, Mandell BF, Fessler BJ (2000) Alveolar hemorrhage in systemic lupus erythematosus: presentation and management. Chest 118:1083–1090
5. Zamora MR, Warner ML, Tuder R et al (1997) Diffuse alveolar hemorrhage and systemic lupus erythematosus. Clinical presentation, histology, survival, and outcome. Medicine (Baltimore) 76:192–202
6. Shi Y, Tsuboi N, Furuhashi K et al (2014) Pristane-induced granulocyte recruitment promotes phenotypic conversion of macrophages and protects against diffuse pulmonary hemorrhage in Mac-1 deficiency. J Immunol 193:5129–5139
7. Jarrot PA, Tellier E, Plantureux L et al (2019) Neutrophil extracellular traps are associated with the pathogenesis of diffuse alveolar hemorrhage in murine lupus. J Autoimmun 100:120–130
8. Reeves WH, Lee PY, Weinstein JS et al (2009) Induction of autoimmunity by pristane and other naturally occurring hydrocarbons. Trends Immunol 30:455–464
9. Avigan J, Blumer M (1968) On the origin of pristane in marine organisms. J Lipid Res 9:350–352
10. Satoh M, Reeves WH (1994) Induction of lupus-associated autoantibodies in BALB/c mice by intraperitoneal injection of pristane. J Exp Med 180:2341–2346
11. Hamilton KJ, Satoh M, Swartz J et al (1998) Influence of microbial stimulation on hypergammaglobulinemia and autoantibody production in pristane-induced lupus. Clin Immunol Immunopathol 86:271–279
12. Satoh M, Kumar A, Kanwar YS et al (1995) Anti-nuclear antibody production and immune-complex glomerulonephritis in BALB/c mice treated with pristane. Proc Natl Acad Sci U S A 92:10934–10938
13. Wang ZL, Luo XF, Li MT et al (2014) Resveratrol possesses protective effects in a pristane-induced lupus mouse model. PLoS One 9: e114792
14. Chowdhary VR, Grande JP, Luthra HS et al (2007) Characterization of haemorrhagic pulmonary capillaritis: another manifestation of pristane-induced lupus. Rheumatology (Oxford) 46:1405–1410
15. Zhuang H, Han S, Lee PY et al (2017) Pathogenesis of diffuse alveolar hemorrhage in murine lupus. Arthritis Rheumatol 69:1280–1293
16. Buatois V, Fagete S, Magistrelli G et al (2010) Pan-CC chemokine neutralization restricts splenocyte egress and reduces inflammation in a model of arthritis. J Immunol 185:2544–2554
17. Bursill CA, McNeill E, Wang L et al (2009) Lentiviral gene transfer to reduce atherosclerosis progression by long-term CC-chemokine inhibition. Gene Ther 16:93–102
18. Chen S, Bacon KB, Li L et al (1998) In vivo inhibition of CC and CX3C chemokine-induced leukocyte infiltration and attenuation of glomerulonephritis in Wistar-Kyoto (WKY) rats by vMIP-II. J Exp Med 188:193–198
19. Bedard EL, Jiang J, Arp J et al (2006) Prevention of chronic renal allograft rejection by SERP-1 protein. Transplantation 81:908–914
20. Jiang J, Arp J, Kubelik D et al (2007) Induction of indefinite cardiac allograft survival correlates with toll-like receptor 2 and 4 downregulation after serine protease inhibitor-1 (Serp-1) treatment. Transplantation 84:1158–1167
21. Miller LW, Dai E, Nash P et al (2000) Inhibition of transplant vasculopathy in a rat aortic allograft model after infusion of anti-inflammatory viral serpin. Circulation 101:1598–1605
22. Tardif JC, L'Allier PL, Gregoire J et al (2010) A randomized controlled, phase 2 trial of the viral serpin Serp-1 in patients with acute coronary syndromes undergoing percutaneous coronary intervention. Circ Cardiovasc Interv 3:543–548
23. Satoh M, Richards HB, Shaheen VM et al (2000) Widespread susceptibility among inbred mouse strains to the induction of lupus autoantibodies by pristane. Clin Exp Immunol 121:399–405
24. Smith DL, Dong X, Du S et al (2007) A female preponderance for chemically induced lupus in SJL/J mice. Clin Immunol 122:101–107
25. Swirski FK, Libby P, Aikawa E et al (2007) Ly-6Chi monocytes dominate hypercholesterolemia-associated monocytosis and give rise to macrophages in atheromata. J Clin Invest 117:195–205
26. Means TK, Latz E, Hayashi F et al (2005) Human lupus autoantibody-DNA complexes activate DCs through cooperation of CD32 and TLR9. J Clin Invest 115:407–417
27. Savarese E, Chae OW, Trowitzsch S et al (2006) U1 small nuclear ribonucleoprotein immune complexes induce type I interferon in plasmacytoid dendritic cells through TLR7. Blood 107:3229–3234
28. Vollmer J, Tluk S, Schmitz C et al (2005) Immune stimulation mediated by autoantigen binding sites within small nuclear RNAs involves Toll-like receptors 7 and 8. J Exp Med 202:1575–1585

29. Herrmann M, Voll RE, Zoller OM et al (1998) Impaired phagocytosis of apoptotic cell material by monocyte-derived macrophages from patients with systemic lupus erythematosus. Arthritis Rheum 41:1241–1250
30. Zhuang H, Han S, Li Y et al (2016) A novel mechanism for generating the interferon signature in lupus: opsonization of dead cells by complement and IgM. Arthritis Rheumatol 68:2917–2928
31. Carlucci F, Ishaque A, Ling GS et al (2016) C1q Modulates the response to TLR7 stimulation by pristane-primed macrophages: implications for pristane-induced lupus. J Immunol 196:1488–1494
32. Nacionales DC, Kelly-Scumpia KM, Lee PY et al (2007) Deficiency of the type I interferon receptor protects mice from experimental lupus. Arthritis Rheum 56:3770–3783
33. Mitani Y, Takaoka A, Kim SH et al (2001) Cross talk of the interferon-alpha/beta signalling complex with gp130 for effective interleukin-6 signalling. Genes Cells 6:631–640
34. Richards HB, Satoh M, Jennette JC et al (2001) Interferon-gamma is required for lupus nephritis in mice treated with the hydrocarbon oil pristane. Kidney Int 60:2173–2180
35. Calvani N, Satoh M, Croker BP et al (2003) Nephritogenic autoantibodies but absence of nephritis in Il-12p35-deficient mice with pristane-induced lupus. Kidney Int 64:897–905
36. Lee PY, Weinstein JS, Nacionales DC et al (2008) A novel type I IFN-producing cell subset in murine lupus. J Immunol 180:5101–5108
37. Satoh M, Hamilton KJ, Ajmani AK et al (1996) Autoantibodies to ribosomal P antigens with immune complex glomerulonephritis in SJL mice treated with pristane. J Immunol 157:3200–3206
38. Li M, Shi X, Qian T et al (2015) A20 overexpression alleviates pristine-induced lupus nephritis by inhibiting the NF-kappaB and NLRP3 inflammasome activation in macrophages of mice. Int J Clin Exp Med 8:17430–17440
39. Myers JL, Katzenstein AA (1986) Microangiitis in lupus-induced pulmonary hemorrhage. Am J Clin Pathol 85:552–556
40. Satoh M, Weintraub JP, Yoshida H et al (2000) Fas and Fas ligand mutations inhibit autoantibody production in pristane-induced lupus. J Immunol 165:1036–1043

Chapter 15

Kidney Subcapsular Allograft Transplants as a Model to Test Virus-Derived Chemokine-Modulating Proteins as Therapeutics

Michelle Burgin, Jordan R. Yaron, Liqiang Zhang, Qiuyun Guo, Juliane Daggett, Jacquelyn Kilbourne, Kenneth M. Lowe, and Alexandra R. Lucas

Abstract

Solid tissue transplant is a growing medical need that is further complicated by a limited donor organ supply. Acute and chronic rejection occurs in nearly all transplants and reduces long-term graft survival, thus increasing the need for repeat transplantation. Viruses have evolved highly adapted responses designed to evade the host's immune defenses. Immunomodulatory proteins derived from viruses represent a novel class of potential therapeutics that are under investigation as biologics to attenuate immune-mediated rejection and damage. These immune-modulating proteins have the potential to reduce the need for traditional posttransplant immune suppressants and improve graft survival. The myxoma virus-derived protein M-T7 is a promising biologic that targets chemokine and glycosaminoglycan pathways central to kidney transplant rejection. Orthotopic transplantations in mice are prohibitively difficult and costly and require a highly trained microsurgeon to successfully perform the procedure. Here we describe a kidney-to-kidney subcapsular transplant model as a practical and simple method for studying transplant rejection, a model that requires fewer mice. One kidney can be used as a donor for transplants into six or more recipient mice. Using this model there is lower morbidity, pain, and mortality for the mice. Subcapsular kidney transplantation provides a first step approach to testing virus-derived proteins as new potential immune-modulating therapeutics to reduce transplant rejection and inflammation.

Key words Transplant, Rejection, Viral protein, Renal, Immunomodulatory, Therapeutics

1 Introduction

Organ transplantation is highly limited by a scarcity of matching donors from an already limited supply [1]. The importance of making every transplant count is ever-increasing. Acute and chronic rejection are two major causes of long-term graft loss, markedly increasing graft failure, and the need for repeat transplantation. Recurrent episodes of acute antibody-mediated immune rejection as well as persistent excess inflammation both lead to transplant

Alexandra R. Lucas (ed.), *Viruses as Therapeutics: Methods and Protocols*, Methods in Molecular Biology, vol. 2225, https://doi.org/10.1007/978-1-0716-1012-1_15,

allograft vasculopathy which causes late graft loss and scaring associated with chronic rejection [2–8]. Some immune responses, both innate or inflammatory and adaptive or antibody-mediated immune responses, are also induced by decreased blood flow (ischemia) and surgical trauma at the time of the transplantation, as well as the excess inflammation seen in brain-dead donors or with chronic infections. Traditionally, the approach to reduce transplant rejection is to flood the organ recipient with immunosuppressants. Many transplanted patients are reliant on immunosuppressants for the rest of their graft's functional life. However, much of the damage to the organ has already occurred and the upregulation of excess inflammatory and immune responses has begun early after transplant which can persist causing ongoing immune-mediated damage. Most therapies for transplants are postoperative and very few attempt treatment of the graft donor prior to transplantation. Treating the donor organ is a promising method for reducing the initial damage that leads to chronic rejection and thus graft loss [9, 10].

The utilization of viruses and virus-derived proteins as therapeutics and biologics has been under investigation throughout the past few decades. The capacity of viruses to evade host immune responses makes these virus-derived immune-modulating proteins a potential source of new drugs for use in treating a variety of immune-mediated diseases. From treating spinal muscular atrophy to targeting cancer cells, viruses have been shown to be capable of acting as therapeutics [11, 12]. Immunomodulatory proteins derived from viruses have evolved over millions of years to be highly active, providing a potential approach to controlling excess and potentially damaging immune responses that occur after organ transplantation. In prior studies, reducing endothelial heparan sulfation and blockade of chemokine-glycosaminoglycan (GAG) interaction with the myxoma virus-derived immune-modulating protein, M-T7, reduced aortic and renal graft vascular inflammation and rejection [9, 10, 13]. In European rabbits M-T7 acts as an interferon gamma receptor (IFNγR) homolog; however M-T7 only binds rabbit IFNγR; for example binding is rabbit species specific [13]. MT-7 is a broad-spectrum chemokine-binding protein that blocks the binding of C, CC, and CXC class chemokines to glycosaminoglycans. GAGs are a special type of polysaccharide with repeating disaccharide units and a modified amino sugar and negatively charged group [14]. GAGs make up a structure called the glycocalyx which is on the apical side of endothelial cells. They are involved in a variety of mechanisms including cell proliferation, inflammation, and cell adhesion. A critical function of the glycocalyx has been reported in graft rejection and specifically for the role of the glycocalyx in chemotaxis. Virus-derived proteins that block chemokine binding to GAGs and which prevent the recruitment of immune cells may provide new potential treatment approaches for acute antibody-mediated organ rejection [9, 10, 15].

Testing new therapeutics on kidney transplants can be a difficult procedure to perform on mice due to their small size. An alternative method for initial therapeutic target screening can be performed using subcapsular transplants. This is a simpler method than performing a full orthotopic transplant. Orthotopic transplants in mice require extensive training and skills, often requiring 6-month training for a microsurgeon, and also cost significantly more due to the need for extensive training, the prolonged surgical time, the need for larger numbers of mice, the higher surgical loss, and the increased need for surgical supplies. Subcapsular models have been used previously to test allograft rejection for example allograft transplants of pancreas, liver, kidney grafts, as well as many other cell types [16–22]. Transplantation of kidney tissues underneath the kidney capsule also provides a unique environment [23, 24]. Underneath the capsule there is access to circulating immune cells in the blood as well as the immune cells naturally present in the kidney. This environment provides an ideal target for transplantation of tissues or cells to test for transplant rejection. Here we describe a method to test virus-derived immune-modulating biologics as new therapeutic approaches for reducing rejection, using a subcapsular kidney transplant model.

2 Materials

2.1 Expression and Purification of M-T7

- Sterile deionized (DI) water.
- Sf21 (*Spodoptera frugiperda*) (unless otherwise specified we have used Invitrogen).
- High Five (*Trichoplusia ni*) cells (Invitrogen).
- pFastBacDual (Invitrogen) expression vector containing an eGFP (enhanced green fluorescent protein) reporter driven by the p10 promoter.
- DH10Bac (Invitrogen) cells.
- 100 mM Isopropyl-beta-D-thiogalactopyranoside (IPTG).
- 20 mg/mL X-gal (5-bromo-4-chloro-3-indolyl-b-D-galactopyranoside).
- Selection plates (Luria Broth (LB) agar plates with 50 mg/mL kanamycin, 10 mg/mL tetracycline, 7 mg/mL gentamicin).
- Standard DNA purification kits.
- Miniprep plasmid purification kit.
- High-fidelity polymerase chain reaction (PCR) reagents.
- Sf-900 II serum-free (SFM) media (Invitrogen).

- Six-well tissue culture plates.
- 1× Grace's Insect Media (Invitrogen).
- Cellfectin II reagent (Invitrogen).
- 30 mL of 1.3× Sf-900 media (Invitrogen).
- 10 kDa Molecular weight cutoff centrifugal filters (we use Amicon filters).
- Binding buffer: 50 mM NaH_2PO_4, 300 mM NaCl, 10 mM imidazole, pH 8.0.
- Wash buffer: 50 mM NaH_2PO_4, 300 mM NaCl, 20 mM imidazole, pH 8.0.
- Elution buffer: 50 mM NaH_2PO_4, 300 mM NaCl, 250 mM imidazole, pH 8.0.
- Storage buffer: 50 mM Tris, 150 mM NaCl, pH 8.
- Snakeskin 10 kDa cutoff dialysis tubing (we use Pierce tubing).
- Pre-equilibrated nickel nitrilotriacetic acid (Ni-NTA) slurry (equilibrated in 50 mM NaH_2PO_4, 300 mM NaCl, 10 mM imidazole, pH 8.0).
- HiLoad 16/60 Superdex 75 (we use GE Healthcare).
- Fast protein liquid chromatography (FPLC) column.
- Sodium dodecyl sulfate-polyacrylamide gel electrophoresis (SDS-PAGE) equipment.
- Coomassie stain (0.1–0.5%).
- 0.22 micron syringe filters and syringes.

2.2 Mice

All surgical procedures should be performed after institutional animal care and use committee (IACUC) approval and should conform to appropriate national, international, and university guidelines for animal care.

1. C57BL6/J and BALB/c mice are used and are commercially available.
   (a) Age of mice: 8–12 weeks.
   (b) Approximate weight is 20–25 g.
   (c) Equal numbers of male and female mice are ideal.

2.3 Surgery

All surgical instruments are autoclaved and sterile:

1. Insulin syringe (1 mL/CC, 28G 1/2).
2. Anesthetic: ketamine (100 mg/mL, 120 mg/kg) mixed with xylazine (20 mg/mL, 6 mg/kg).
3. Electric heating pads, maintained at 36–38 °C.
4. Electric hair clippers.

5. Sterile cotton swabs and pads.
6. Chlorhexidine gluconate 2% solution.
7. Sterile sodium chloride solution, 0.9% (w/v) (isotonic normal saline).
8. 70% Ethanol.
9. Veterinary ocular ointment.
10. Sterile drape with 15 mm centrally located fenestration.
11. Sterile iris forceps, serrated and 1 × 2 teeth, straight and curved.
12. Sterile dissecting scissors.
13. Surgical blades, size 10.
14. Scalpel handle, #3.
15. Sterile spring-loaded Vannas scissors.
16. Sterile Dumont forceps.
17. Oral gavage needle, 22 gauge.
18. Stereoscope.
19. Olsen-Hegar needle holder with suture scissors.
20. 4-0 Sterile absorbable suture with a 13 mm 1/2 c tapered needle.
21. Reflex clip applier.
22. Mosquito hemostatic forceps.
23. Wound reflex clips.
24. Buprenorphine 0.03 mg/mL, 0.05–1.0 mg/kg.

2.4 Materials for Histology

1. Formalin solution, neutral buffered, 10%.
2. Tissue cassettes.
3. Tissue processer (*see* **Note 1**).
4. Embedding station.
5. Paraffin cassette microtome.
6. Paraffin.
7. Oven kept at 55–65 °C.
8. Glass histology slides and coverslips.
9. Mixture of xylenes, histologic grade (also called xylene).
10. Ethanol (70%, 95%, 100%) diluted in DI water.
11. Gill No. 3 hematoxylin (6 g/L).
12. Eosin Y (5% in aqueous solution).
13. 1% Acid alcohol (1 mL hydrochloric acid and 100 mL 70% ethanol).

14. 0.2% Ammonia water (2 mL ammonium hydroxide and 1000 mL distilled water).
15. Microscope and imaging system.
16. Slide tray.
17. Tris-buffered saline, 0.1% Tween (TBST).
18. Protein block (we use 5% w/v bovine serum albumin diluted in TBST).
19. DAB (3,3-diaminobenzidine) substrate development kit (we use ImmPACT DAB from Vector Laboratories) (SK-4105).
20. Immunohistochemical markers:
    - Primary antibodies:
      - Rabbit polyclonal to C4d (Hycult Biotech, HP8033).
      - Rabbit polyclonal to F4/80 (Abcam, ab100790).
      - Rabbit polyclonal to CD3 (Abcam ab5690).
    - Secondary antibody:
      - Goat anti-rabbit, highly cross-adsorbed, HRP-conjugated (we use Jackson ImmunoResearch, #111-035-144).

3 Methods

3.1 Generation of M-T7 Baculovirus

1. Generate a pFastBacDual expression vector containing an eGFP reporter driven by the p10 promoter and a C-terminal His-tagged construct of M-T7 (NCBI Gene ID: 932170), using high-fidelity PCR and appropriate plasmid construction method (*e.g.*, Gibson assembly, Gateway reaction).
2. Transform DH10Bac cells with the pFastBacDual construct to generate a propagation library of bacmids (baculovirus shuttle vector).
3. Plate transformants onto LB agar selection plates.
4. Spread 100 μL of 100 mM IPTG and 40 μL of 20 mg/mL X-gal onto selection plates to screen the colonies by blue/white screening (blue colonies will contain the plasmids of interest). Similarly, PCR on 10–20 colonies can be used to identify positive transformants.
5. Purify bacmids using a standard DNA purification kit. Because of the size of the generated bacmids (~150 kb), it is recommended to elute with 100 μL elution buffer which has been pre-warmed to 50 °C.
6. If colony PCR was not used for colony screening, verify incorporation of M-T7 by PCR analysis.

7. One hour prior to transfection, seed Sf21 cells onto 6-well tissue culture plates with 1× Grace's insect media at 1 × 10^6 cells/mL in a total volume of 2 mL (*see* **Note 2**).
8. Preincubate 1–2 μg of purified bacmid DNA in 100 μL of 1× Grace's insect media in one tube.
9. Add 6 μL of Cellfectin II reagent to 100 μL of 1× Grace's insect media in a separate tube.
10. Incubate both tubes in the hood for 15 min.
11. Combine contents of the tubes, DNA, and Cellfectin II, together for 30 min at room temperature.
12. Bring up the transfection reaction mixture to a final volume of 1 mL with 1× Grace's insect media.
13. Carefully wash cells in 1× Grace's insect media and aspirate off media.
14. Add transfection solution and incubate cells at 27 °C for 5 h.
15. Replace the transfection reagent solution with Sf-900 II SFM.
16. Confirm the presence of transfectants 48 h after transfection by fluorescence microscopy. Foci of eGFP-expressing cells should be visible.
17. At 72-h post-transfection, collect the viral supernatant. Infect an additional population of Sf21 cells with the supernatant to amplify the virus stock.
18. Perform a viral plaque assay to obtain the viral titer.
19. Seed two milliliters of 5 × 10^5 Sf21 cells/mL onto 6-well plates and allow to adhere for 1 h on a flat surface for even cell distribution.
20. Create a serial dilution of the viral stock in Sf-900 II SFM containing a final plating volume of 1 mL per well.
21. Aspirate off media from the seeded 6-well plate.
22. Replace media with diluted virus and incubate for 1 h at room temperature.
23. Pour 10 mL of melted sterile and 4% plating agarose and add 30 mL of 1.3× Sf-900 media. Place in a 37 °C water bath.
24. Remove virus quickly from the seeded plate and add 2 mL of agarose mixture.
25. Allow agarose to harden for approximately 15 min.
26. Incubate plates in a container with wet paper towels in a 27 °C incubator for 4–10 days.
27. Observe formation of green florescent foci on a microscope daily.
28. When no new foci form for 48 h, calculate viral titer.

29. Infect suspension cultures of Sf21 cells at a MOI of 0.001–0.005 for 24 h.
30. Spin down cells and collect supernatant.
31. Store baculovirus at 4 °C for active use or −80 °C for long-term storage.

3.2 Expression and Purification of M-T7 from High Five Cells

1. Infect High Five cells at 1×10^6 cells/mL with baculovirus (MOI of 1) containing the M-T7 gene construct.
2. Incubate for 24–72 h and collect the protein-containing supernatant media.
3. Prior to purifying the protein from the media, M-T7 is concentrated with a 10 kDa molecular weight cutoff filter by centrifugation according to the manufacturer's procedure (e.g., Amicon) and then buffer-exchanged using 10 kDa Snakeskin dialysis tubing with a binding buffer solution (50 mM NaH2PO4, 300 mM NaCl, and 10 mM imidazole, pH 8.0).
4. Buffer-exchanged supernatant is cleared by centrifugation to remove particulates.
5. Add supernatant to a pre-equilibrated Ni-NTA slurry.
6. Incubate with rotation overnight at 4 °C.
7. Add supernatant to a small-volume gravity column or 1 mL pipette tip plugged with clean cotton and allow to flow through by gravity.
8. Wash samples three times with 1 mL exchanges of 50 mM NaH_2PO_4, 300 mM NaCl, and 20 mM imidazole, pH 8.0.
9. Elute with three fractions of 1 mL each with 50 mM NaH_2PO_4, 300 mM NaCl, and 250 mM imidazole, pH 8.0.
10. Concentrate fractions with a 10 kDa molecular weight cutoff centrifugal filter (*e.g.*, we use Amicon) and exchange buffer 3–5 times with storage buffer (50 mM Tris, 150 mM NaCl, pH 8) to remove imidazole and reduce salt content.
11. Purify further by size-exclusion chromatography with a HiLoad 16/60 Superdex 75 FPLC column and collect fractions (*see* **Note 3**).
12. Run SDS-PAGE gel and stain with Coomassie to confirm the purity of protein.
13. Filter sterilize final protein using a 0.22 micron filter and store at 4 °C for use within 1 week or store at −80 °C for long-term storage. Avoid freeze-thaw cycles.

3.3 Pretreatments with Virus-Derived Immune-Modulating Protein

1. Intraperitoneally inject C57BL6 mice for 7 days with 200 μL of chosen viral protein. For example, 200 μL M-T7 protein (1 ng/g body weight) diluted in sterile saline.

2. Include a control giving 200 μL of saline for 7 days or an inactive M-T7 mutant.
3. Monitor body weight by placing mice in a plastic cup on a scale to ensure no weight changes due to discomfort from drug.
4. On the day of surgery, euthanize donor mice via CO_2 asphyxiation at a rate of 1.8–2 L/min and follow up with a cervical dislocation.
5. Collect the donor organ from the pretreated C57BL6 mice about 1 h before transplant-engrafting surgery and place the donor kidney in oxygenated Dulbecco's modified Eagle medium (DMEM) (*see* **Note 4**).

3.4 Preparation of Mice for Surgery

1. Anesthetize mice by intraperitoneal (IP) injection of 100 μL of ketamine/xylazine mixture per 25 g mouse. Once anesthetized, place mouse on a heating pad to keep warm throughout the surgery (*see* **Note 5**).
2. Remove the fur using clippers to create a 2-in. area on the right side of the mouse over the area of the kidney. The kidney can be gently palpated before shaving to confirm proper placement.
3. Clean the surgical site by swabbing in a circular motion with chlorhexidine gluconate 2% solution and then with 70% ethanol. Repeat this step twice to ensure loose fur removal and a disinfected area of incision.
4. Apply veterinary ointment to the eyes of the mice to prevent drying.

3.5 Surgical Procedure

Surgeries are performed in a clean, and where possible, sterile field. General procedure is illustrated in Fig. 1.

1. With the mouse in left lateral recumbent position, place a sterile drape covering the mouse with the fenestration (hole) over the surgical site.
2. Using forceps and scissors make a vertical incision approximately 0.7–1.0 cm in length midway between the bottom of the rib cage and the iliac crest. Tent the skin prior to cutting to avoid cutting through the underlying abdominal wall.
3. Make a similar incision in the abdominal wall over the area of the kidney (*see* **Note 6**). It is best to keep the incision slightly smaller than the length of the kidney. Too large of an incision will cause the kidney to retract into the body easily during the procedure and can increase wound dehiscence later.
4. Using the thumb and forefinger apply pressure to dorsal and ventral sides of the abdominal cavity while applying gentle pressure to the abdominal wall incision with curved forceps to exteriorize the kidney. Applying direct pressure to the kidney may cause bruising or capsular rupture (*see* **Note 7**).

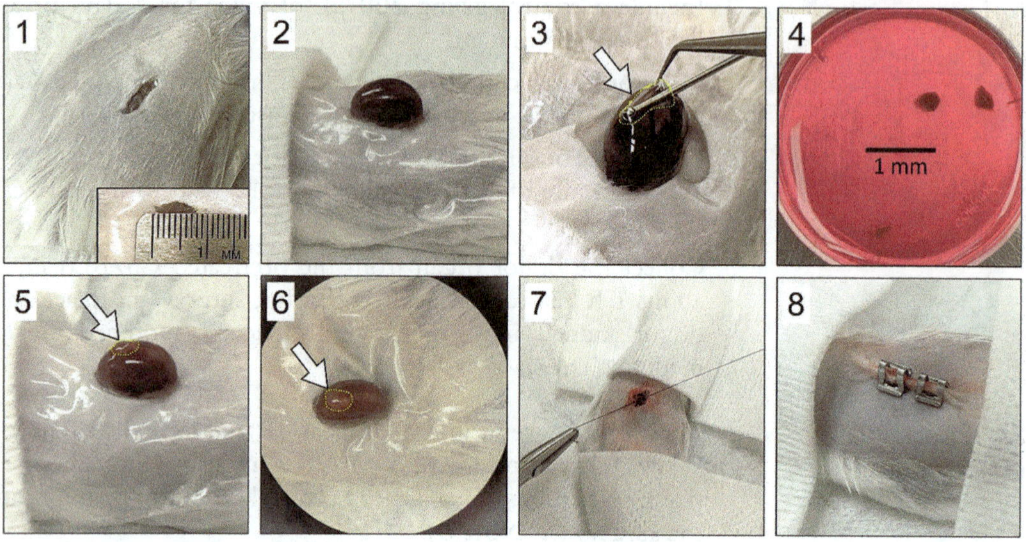

Fig. 1 General overview of surgical procedure. (*1*) Creation of vertical incision in skin and abdominal wall. (*2*) Exteriorized kidney. (*3*) Creation of pocket underneath the capsule for transplanted tissue. (*4*) Example size of donor tissue to be inserted underneath capsule. (*5*) Exteriorized kidney with donor tissue inserted beneath the capsule. (*6*) Kidney with donor tissue viewed under the stereoscope. (*7*) Closure of abdominal wall with sutures. (*8*) Closure of skin with wound clips

5. Once kidney is exteriorized, the area should be kept moist using sterile saline to prevent drying.
6. Using spring-loaded Vannas scissors make a very small incision about 2 mm in length on the capsule of the kidney. Size of incision can be changed depending on the size of the transplanted tissue (*see* **Note 8**). The incision should be performed under the stereoscope for a closer view and increased precision (Fig. 2).
7. Grasp the edge of the capsular incision with Dumont forceps and gently insert the bulb of the oral gavage needle between the capsule and the renal parenchyma to make a subcapsular pocket for the transplanted tissue. Care should be taken not to rip the capsule or damage the kidney underneath.
8. Push the donor tissue into the pocket using the oral gavage needle and Dumont forceps. Pressure from the capsule will ensure that the transplant is kept in place after the surgery (*see* **Note 9**). The capsule incision does not need to be closed if the capsule is not torn.
9. Return the kidney to the abdomen by grasping the edges of the abdominal incision and applying gentle pressure to the kidney. Take care not to press near the implanted tissue. Verify that the tissue is still under the kidney capsule before closing the abdominal wall.

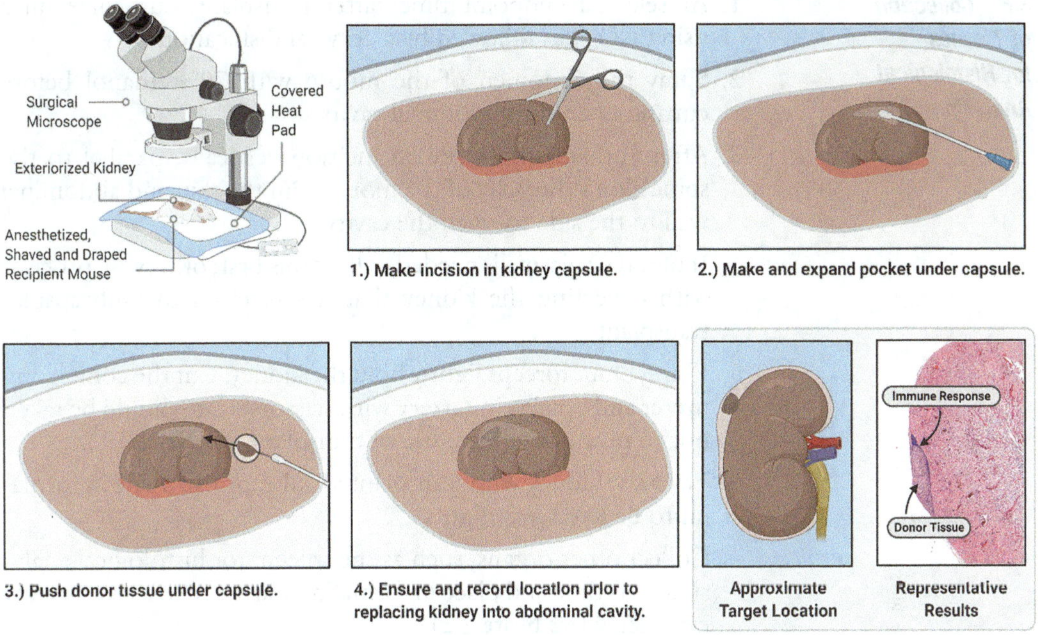

Fig. 2 Illustration of the procedure to insert donor tissue underneath the capsule of the recipient mouse and is performed underneath the stereoscope. After the kidney is exteriorized the mouse is placed underneath the scope for better lighting and closer view of the work being done. Images in the grayed area demonstrate approximate location of transplanted tissue with representative hematoxylin and eosin (H&E)-stained section of a subcapsular kidney transplant. Immune cells are stained dark purple and a boundary is clearly seen surrounding the transplanted tissue

10. Close the abdominal wall using 4-0 sterile absorbable suture with a 13 mm 1/2 c tapered needle. About 2–3 sutures is enough.
11. Close the skin using sterile wound clips. Typically, two clips are needed for a 20 g mouse.
12. To minimize pain, give an injection of buprenorphine about 0.1 mg per/kg mouse after the mouse has righting ability. Subcutaneous sterile saline (200–500 μL) can also be provided post-surgically to aid in recovery.
13. Mice should be kept on a heating pad until fully awake and then can be placed in a clean cage.
14. Check cages daily for signs of pain and discomfort (hunching, piloerection, weight loss, etc.). If signs of excess pain are seen, another dose of buprenorphine can be administered, or the mouse can be euthanized. Mice are monitored by the experienced veterinary staff as well as the experimental group. Daily weights are measured to identify any side effects or complications of surgery (*see* **Note 10**).

3.6 Collection of Tissues for Histological Analysis

1. At selected endpoint times after transplant, euthanize mice using CO_2 gas followed by a cervical dislocation.
2. Spray the abdomen of the mouse with 70% ethanol before cutting open the abdominal cavity.
3. After euthanizing, make an incision vertically, parallel to the spine along the belly of the mouse. Pin the skin and abdominal wall to the side to open the cavity.
4. If blood or plasma is needed collect this first; otherwise proceed with collecting the kidney that has received the subcapsular transplant.
5. Using blunt forceps lightly hold the kidney. Cut the connecting ureter and renal vein/artery with scissors. Care should be taken not to pinch or cut the site of transplant (*see* **Note 11**).
6. Place the kidney with transplanted subcapsular tissue in formalin to be fixed overnight.
7. Collect other organs, such as the spleen, for histological analysis at this time and place in formalin or arrange to assess by flow cytometry (*see* **Note 12**).
8. After the kidney with the subcapsular engraftment is fixed, place in a cassette and perfuse with paraffin wax to be sectioned for staining.

3.7 Histology and Immunohistology

1. Using a microtome cut embedded tissues into 5 μm thick sections.
2. Stain sections with hematoxylin and eosin (H&E):
   (a) Place sections in a 60 °C oven for 30 min to melt the paraffin. Transfer slides to xylene for 5 min, xylene for 5 min, 100% ethanol for 2 min, 100% ethanol for 2 min, 95% ethanol for 2 min, 70% ethanol for 2 min, and tap water for 2 min.
   (b) Place slides into hematoxylin for 5 min.
   (c) Move slides to water wash 1. Move slide rack up and down a few times and place slides into water wash 2 for 1 min. Move slides to water wash 3 for 5 min.
   (d) Dip slides into acid alcohol 12 times and place slides back into water wash 3 for 2 min.
   (e) Place slides into ammonium water for 30 s and move the slides back to water wash 3 for 2 min.
   (f) Place slides in eosin for 5 min.
   (g) Transition slides to 95% ethanol for 2 min, 95% ethanol for 2 min, 100% ethanol for 2 min, 100% ethanol for 2 min, xylene for 2 min, and xylene for 2 min.

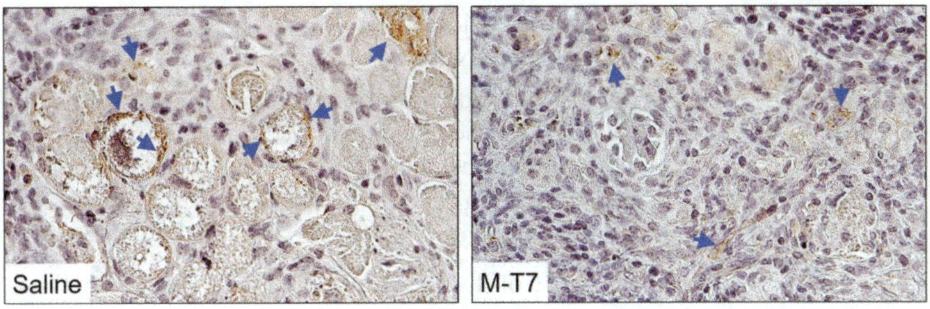

Fig. 3 IHC sections (40× objective) demonstrating C4d deposition in saline- and M-T7-treated subcapsular kidneys with a 15-day follow-up. C4d depositions indicate antibody-dependent rejection and are included in the Banff scoring criteria. Scoring of C4d is determined by the ratio of peritubular capillaries and vasa recta that has positive staining. Lesion scores range from 0 to 4 with no staining to more than 50% deposition (4). Blue arrows indicate C4d-positive staining. Saline section on the left shows greater amounts of positive staining around the peritubular capillaries. Approximately 50% of capillaries stain positive giving a Banff score of 2. In comparison, M-T7-treated section on the right shows minimal deposition of C4d and a maintained structure of a glomerulus in the center of the field. This section would receive a score of 1 according to the Banff lesion score. IHC staining using a C4d marker provides a molecular indication of antibody-mediated rejection and indicates poor graft survival

(h) Remove from xylene and allow slides to dry before mounting.

(i) Mount slides with glass coverslip.

3. Interstitial inflammation, scarring, and tubulitis can be observed on stained sections using a bright-field microscope. We use an Olympus BX51 microscope with 4–100× objectives, a Prior ProScan II stage, Olympus DP74 CMOS camera, and cellSens software analysis system. Comparisons to the Banff criteria can be made to score the level of rejection [4]. *It is recommended to have stained tissue sections read by a second reader, blinded to treatments used.*

4. Immunohistochemical (IHC) staining can be performed to detect specific immune markers (CD3 T cells, F4/80 macrophages) as well as markers for rejection specifically C4d [4, 9] (Fig. 3). Stained sections are measured by counting in ten random fields of staining the number of positive cells in a high-powered field (≥40× to 100×) or by measuring the area of positive staining compared to the area of transplant.

   (a) Place sections in a 60 °C oven for 30 min to melt the paraffin.

   (b) Place slides in xylene for 5 min, xylene for 5 min, 100% ethanol for 2 min, 100% ethanol for 2 min, 95% ethanol for 2 min, 70% ethanol for 2 min, and tap water for 2 min.

   (c) Transfer slides to Coplin jars containing epitope retrieval solution and place inside a steam chamber for 30 min. Remove and let cool for 20 min (*see* **Note 13**).

(d) Remove slides from Coplin jar and place them in DI water (*see* **Note 14**).

(e) Using a slide tray, place slides on the tray and add TBST to keep sections hydrated. Sections must be kept hydrated throughout the process.

(f) Tap off TBST and add hydrogen peroxide to each slide for 10 min.

(g) Rinse off hydrogen peroxide with TBST four times.

(h) Apply protein block for 1 h at room temperature.

(i) Add primary antibody according to the manufacturer's recommended concentration. Antibody can be diluted in protein block. 50 μL of primary antibody per tissue is enough to cover sections. Add paraffin coverslip to slides and place tray in a 4 °C fridge overnight.

(j) Rinse off primary antibody five times with TBST.

(k) Apply secondary antibody diluted in protein block for an hour at room temperature.

(l) Rinse off secondary antibody five times with TBST.

(m) Prepare DAB solution.

(n) Add DAB to slides 50 μL per tissue. Wait for visible color development and continue with the same timing for the rest of the slides.

(o) Rinse off DAB with TBST and place slides into DI water.

(p) Place slides into hematoxylin for 3 min.

(q) Move slides to water wash 1. Move slide rack up and down a few times and place slides into water wash 2 for 1 min. Move slides to water wash 3 for 5 min.

(r) Dip slides into acid alcohol 12 times and place slides back into water wash 3 for 2 min.

(s) Place slides into ammonium water for 30 s and move the slides back to water wash 3 for 2 min.

(t) Transition slides to 95% ethanol for 2 min, 95% ethanol for 2 min, 100% ethanol for 2 min, 100% ethanol for 2 min, xylene for 2 min, and xylene for 2 min.

(u) Remove from xylene and allow slides to dry before mounting.

(v) Mount slides with glass coverslip.

4 Notes

1. Specifics for equipment used in our lab: We use an embedding station (LEICA EG 1160), microtome (LEICA RM 2165), glass histology slides and coverslips (HARETA AHS90-WH), microscope and imaging system (Olympus BX51 microscope with 4×–100× objectives, a Prior ProScan II stage, Olympus DP74 CMOS camera, and cellSens software analysis system), and DAB kit (VECTOR SK-4105).
2. Generation of transfection complexes and execution of the transfection procedure should be performed in the absence of antibiotics. Once transfection has been performed, antibiotic-containing media can be exchanged.
3. This step is essential for ensuring that the only collected protein used for subsequent in vivo analysis is purified M-T7. In vitro studies may be performed without this step as the collected protein will still be 60–80% pure prior to FPLC purification.
4. Harvested organ from donor can be placed in a petri dish with the oxygenated DMEM. This allows for easy cutting of the tissue into appropriate-size pieces for transplantation. Cuts can be made using a sterile scalpel blade. The cut pieces should be about 1–1.5 mm in size with rounded edges. If the tissue edges are too sharp, then it may damage the capsule during implantation.
5. Level of anesthesia can be checked by firmly pinching the toes of the mouse. If the mouse moves more time should be taken until the mouse is completely under anesthetics and does not respond to toe pinch. Boosts of 20 μL of ketamine/xylazine can be given if mouse flinches or moves during surgery. This can be injected directly into the cavity if incision was already made.
6. Gently palpate the craniolateral aspect of the abdomen to locate the kidney.
7. If incision is too large the kidney will retract into the cavity. It is best to make the incision smaller and then expand it if needed. The kidney will remain outside the cavity from friction.
8. Size of transplant tissue should be kept to about 1–2 mm for an average 25 g mouse. Larger pieces will tear the capsule and the transplant will fall out inside cavity.
9. It is very helpful to note the location of each graft implant into the receiving kidney. Difficulty finding the site of transplant may arise later in the histology processing. Ear tagging before the surgery and noting each mouse's transplant location at the time of engrafting will make finding the site easier when sectioning.

10. Daily weights are used to assess complications along with observation for loss of appetite or pain with hunching and decreased activity. Urinary creatinine (Cr) to check for kidney function can be measured by collecting urine in a pan placed under the cage. Blood Cr levels can also be measured.
11. Attachments of the kidney to adjacent organs and abdominal wall are common. Extreme care should be used when collecting the organ. Often it is best to take part of the scar attachments with the kidney to avoid ripping the capsule and losing the transplanted tissue.
12. It is best to collect other organs into a 15 mL tube with formalin and/or RNALater in case the tissues are needed for further testing in the future. For flow cytometry spleen tissue samples are best isolated and assessed immediately, although frozen tissues have been used for some studies.
13. To prevent tissue from detaching from the slide during epitope retrieval a secondary uncharged slide can be placed on top of the tissue and paperclipped together. Remove the second slide before placing slides in DI water.
14. A pap pen can be used to create a hydrophobic barrier around the specimen. This reduces the amount of reagent needed to cover the tissue.

References

1. Abouna GM (2008) Organ shortage crisis: problems and possible solutions. Transplant Proc 40:34–38
2. Whalen HR, Glen JA, Harkins V et al (2017) High intrapatient tacrolimus variability Is associated with worse outcomes in renal transplantation using a low-dose tacrolimus immunosuppressive regime. Transplantation 101:430–436
3. Tanabe K, Takahashi K, Toma H (1996) Causes of long-term graft failure in renal transplantation. World J Urol 14:230–235
4. Roufosse C, Simmonds N, Clahsen-van Groningen M et al (2018) A 2018 Reference guide to the Banff classification of renal allograft pathology. Transplantation 102:1795–1814
5. VanBuskirk AM, Pidwell DJ, Adams PW et al (1997) Transplant Immunol 278:1993–1999
6. Shimizu A, Yamada K, Robson SC et al (2012) Pathologic characteristics of transplanted kidney xenografts. J Am Soc Nephrol 23:225–235
7. Filippone EJ, Farber JL (2015) Humoral immunity in renal transplantation: epitopes, Cw and DP, and complement-activating capability - an update. Clin Transpl 29:279–287
8. Guillén-Gómez E, Dasilva I, Silva I et al (2017) Early macrophage infiltration and sustained inflammation in kidneys from deceased donors are associated with long-term renal function. Am J Transplant 17:733–743
9. Chen H, Ambadapadi S, Wakefield D et al (2018) Selective deletion of heparan sulfotransferase enzyme, Ndst1, in donor endothelial and myeloid precursor cells significantly decreases acute allograft rejection. Sci Rep 8:13433
10. Yaron JR, Kwiecien JM, Zhang L et al (2019) Modifying the organ matrix pre-engraftment: a new transplant paradigm? Trends Mol Med 25:626–639
11. Pattali R, Mou Y, Li X-J (2019) AAV9 Vector: a novel modality in gene therapy for spinal muscular atrophy. Gene Ther 26:287–295
12. Chan WM, Rahman MM, McFadden G (2013) Oncolytic myxoma virus: the path to clinic. Vaccine 31:4252–4258
13. Mossman K, Nation P, Macen J et al (1996) Myxoma virus M-T7, a secreted homolog of the interferon-γ receptor, is a critical virulence factor for the development of Myxomatosis in European rabbits. Virology 215:17–30

14. Gandhi NS, Mancera RL (2008) The structure of glycosaminoglycans and their interactions with proteins. Chem Biol Drug Des 72:455–482
15. Epperson ML, Lee CA, Fremont DH (2012) Subversion of cytokine networks by virally encoded decoy receptors. Immunol Rev 250:199–215
16. Cunha GR, Baskin L (2016) Use of sub-renal capsule transplantation in developmental biology. Differentiation 91:4–9
17. Galun E, Burakova T, Ketzinel M et al (1995) Hepatitis C virus viremia in SCID → BNX mouse chimera. J Infect Dis 172:25–30
18. Hori J, Joyce N, Streilein JW (2000) Epithelium-deficient corneal allografts display immune privilege beneath the kidney capsule. Invest Ophthalmol Vis Sci 41:443–452
19. Mandel T, Koulmanda M, Kovarik J et al (1995) Transplantation of organ cultured fetal pig pancreas in non-obese diabetic (NOD) mice and primates (Macaca fascicularis). Xenotransplantation 2:128–132
20. Ricordi C, Flye MW, Lacy PE (1988) Renal subcapsular transplantation of clusters of hepatocytes in conjunction with pancreatic islets. Transplantation 45:1148–1150
21. Shultz LD, Goodwin N, Ishikawa F et al (2019) Subcapsular transplantation of tissue in the kidney. Cold Spring Harbor Protoc:737–741. https://doi.org/10.1101/pdb.prot078089
22. van den Berg CW, Ritsma L, Avramut MC et al (2018) Renal subcapsular transplantation of PSC-derived kidney organoids induces neo-vasculogenesis and significant glomerular and tubular maturation in vivo. Stem Cell Rep 10:751–765
23. Reece-Smith H, Du Toit DF, McShane P et al (1981) Prolonged survival of pancreatic islet allografts transplanted beneath the renal capsule. Transplantation 31:305–306
24. Foglia RP, DiPreta J, Statter MB et al (1986) Fetal allograft survival in immunocompetent recipients is age dependent and organ specific. Ann Surg 204:402–410

Chapter 16

A Mouse Model of Acute Liver Injury by Warm, Partial Ischemia-Reperfusion for Testing the Efficacy of Virus-Derived Therapeutics

Jordan R. Yaron, Liqiang Zhang, Qiuyun Guo, Hao Chen, and Alexandra R. Lucas

Abstract

Ischemia-reperfusion injury (IRI) drives early and long-term damage to organs as well as compounding damage from acute transplant rejection and surgical trauma. IRI initiates an aggressive and prolonged inflammation leading to tissue injury, organ failure, and death. However, there are few effective therapeutic interventions for IRI. The destructive inflammatory cell activity in IRI is part of an aberrant innate immune response that triggers multiple pathways. Hence, immune-modulating treatments to control pathways triggered by IRI hold great therapeutic potential. Viruses, especially large DNA viruses, have evolved highly effective immune-modulating proteins for the purpose of immune evasion and to protect the virus from the host immune defenses. A number of these immune-modulating proteins have proven therapeutically effective in preclinical models, many with function targeting pathways known to be involved in IRI. The use of virus-derived immune-modulating proteins thus represents a promising source for new treatments to target ischemia-reperfusion injury. Laboratory small animal models of IRI are well established and are able to reproduce many aspects of ischemia-reperfusion injury seen in humans. This chapter will discuss the methods used to perform the IRI procedure in mice, as well as clinically relevant diagnostic tests to evaluate liver injury and approaches for assessing histological damage while testing novel immune modulating protein treatments.

Key words Liver, Ischemia-reperfusion injury, Surgical injury, Serpin, Inflammation

1 Introduction

Ischemia-reperfusion injury (IRI) is a complex condition with broad clinical relevance. IRI is associated with a wide range of severe pathologies and can drive the devastating effects of both acute and chronic transplant rejection and post-surgical complications, leading causes of graft loss [1]. IRI is characterized by a two-step process. First, an initial, transient blockade of blood flow and oxygen delivery occurs, referred to as ischemia. During this period, sublethal injury to tissue occurs. In the second stage of IRI,

Alexandra R. Lucas (ed.), *Viruses as Therapeutics: Methods and Protocols*, Methods in Molecular Biology, vol. 2225, https://doi.org/10.1007/978-1-0716-1012-1_16, © Springer Science+Business Media, LLC, part of Springer Nature 2021

restoration of blood flow (reperfusion) occurs followed by a paradoxical acceleration of injury. The reperfusion stage of IRI initiates an aggressive, prolonged inflammation that can lead to tissue damage, organ failure and can even precipitate multiorgan dysfunction and death [2]. While several potential treatments are being tested in preclinical and Phase I/II trials, there are currently no effective treatments for IRI [1, 3]. One potential limitation in identifying effective treatments for IRI is the complex nature of the pathology. The destructive inflammatory cell activity in IRI is part of an aberrant innate immune response triggering multiple pathways, including apoptosis, necrosis, and the inflammasome, leading to pyroptosis [1, 4–7]. This ongoing damage can also amplify the humoral immune repsonse compounding episodes of acute and chronic antibody-mediated rejection [8]. Thus, single-pathway inhibition may be insufficient to limit the tissue destruction in IRI and a multi-pathway approach may be preferred.

Coevolution of viruses with their natural hosts invokes an adaptation of arms race, where a successful strategy for the virus relies on immune evasion, often targeting key pathways that drive immune activation [9]. Large DNA viruses, such as the leporipoxvirus Myxoma virus (MYXV) and other members of the poxvirus family, are adept at evading the innate immune system via a suite of virulence factors [10, 11]. Furthermore, because viruses are limited in their genomic space, it is common for immune-modulating proteins to exhibit multipotent functionality, targeting numerous pathways simultaneously. Translationally, these factors constitute a rich toolbox for developing immune modulators for treating disease.

The small animal model of hepatic ischemia-reperfusion is a popular approach for studying the physiology of sterile, surgery-associated acute liver injury. In contrast to laboratory transplantation models as well as static or machine perfusion-mediated organ preservation, the IRI models of perioperative and ischemia or surgery-associated injury have the advantage of greater technical ease, lower financial costs, and minimal logistical considerations (e.g., donor vs. recipient age, procedure synchronization). Furthermore, when compared to other laboratory models of IRI, the mouse model presents the key benefits of available genetic knockouts, exquisite control of diet and weight (important because obesity can affect acute outcomes [12]) and relative technical ease of the surgical procedure. The associated diagnostic tests for analyzing the effects of the surgical procedure are also relatively simple to perform and most equipment are available in the modern research setting.

We recently demonstrated the utility of the mouse model of IRI for assessing the efficacy of viral protein therapeutics by treatment with the Myxoma virus-derived serine protease inhibitor serpin, Serp-2. Serp-2 is a 34 kD intracellular (serpin clade B) cross-class serine and cysteine protease inhibitor which has inhibitory activity for caspase-1, -8, and -10 and granzyme B [13–15]. In other models, Serp-2 was found to suppress inflammation and improve outcomes in models of aortic transplant and carotid cuff compression injury [16–18]. Based on the overlap of the targeted protease pathways (inflammasome and apoptosis), Serp-2 was tested as a putative therapeutic to prevent liver injury and improve outcomes after 90 min of partial (70%) warm ischemia-reperfusion injury [19]. Using this model, we found that treatment with recombinant purified Serp-2 improves 10-day survival, ameliorates acute injury as measured by circulating liver markers, and preserves liver viability. Thus, the mouse model of IRI is a convenient and relevant model for assessing the efficacy of putative virus-derived therapeutics. Beyond recombinant protein treatments, this model may be adapted to testing putative therapeutics delivered by a wide variety of methods such as adenovirus (AdV), adeno-associated virus (AAV), and nonviral plasmid DNA delivery. The relatively low complexity and short timeline of this model make modifications simple to perform (treatment timing, end-point timing, etc.).

This chapter describes the methods for (i) performing the mouse model of ischemia-reperfusion injury, (ii) assessing acute injury by circulating diagnostic markers, and (iii) evaluating the degree of liver injury by histological analysis. The methods described here for evaluating the efficacy of putative therapeutics in this model are to be considered the minimum for experimental follow-up, and each laboratory may find that supplemental analysis by other methods is necessary to completely answer experimental questions (e.g., Western blotting, ELISA).

2 Materials

2.1 Ischemia-Reperfusion Surgical Procedure

- Experimental treatment (*see* **Note 1**).
- 8–12-Week-old mouse (any strain appropriate for the experimental question).
- Heating pad.
- Sterile syringe and needle.
- Isoflurane or cocktail of ketamine and xylazine as anesthetic (dose determined by local animal care and use committee guidelines).
- Hair clipper.
- Chlorhexidine wash.

- 70% Ethanol.
- Cotton swabs.
- Sterile small animal surgical tools (e.g., 11.5 cm Metzenbaum scissors, 8.5 cm blunt-tip dissecting scissors, 12 cm Adson forceps, 10 cm Iris forceps, 14 cm Olsen-Hegar needle holder).
- Sterile atraumatic clamp (e.g., 8 mm Schwartz vessel clip).
- Sterile cotton gauze.
- Sterile 0.9% normal saline solution.
- Sterile absorbent suture (e.g., 4–0 coated VICRYL Polyglactin 910 sutures).
- Sterile nylon sutures.
- Buprenorphine analgesic.

2.2 Pre-histology Tissue Processing

- CO_2 asphyxiation chamber or injectable euthanasia solution.
- Dissection instruments.
- 50 mL Tubes or appropriate container for collecting tissues.
- 10% Neutral-buffered formalin.
- Tissue-processing cassettes.
- Automated tissue processor or the following solutions for manual tissue processing in 20 mL scintillation vials:
  - 1× 10% Neutral-buffered formalin
  - 1× 100% Deionized water
  - 1× 30% Ethanol in deionized water
  - 1× 50% Ethanol in deionized water
  - 1× 70% Ethanol in deionized water
  - 1× 90% Ethanol in deionized water
  - 2× Absolute ethanol
  - 1× 1:1 Mixture of ethanol and xylene
  - 2× 100% Xylene
  - 1× 1:1 Mixture xylene and melted paraffin wax
  - 1× Melted paraffin wax.
- Tissue-embedding station.
- Microtome.
- Charged glass slides.
- Heated float bath.

2.3 Hematoxylin and Eosin (H&E) Staining

- 60 °C Oven or air convection incubator.
- Paper towels.

- Chemical fume hood.
- Staining jars with slide racks containing the following solutions in order:
  - 100% Xylene
  - 100% Xylene
  - Absolute ethanol.
  - Absolute ethanol.
  - 95% Ethanol
  - 70% Ethanol
  - Tap water.
  - Hematoxylin (Gill's formula #3 or preferred).
  - 3x Tap water
  - Acid alcohol (0.5% hydrochloric acid in 70% ethanol).
  - Tap water.
  - Ammonium water (2% ammonium hydroxide in deionized water).
  - Tap water.
  - Acidic, aqueous eosin Y solution.
  - 95% Ethanol
  - Absolute ethanol.
  - Absolute ethanol.
  - 100% Xylene
  - 100% Xylene
- Xylene-based mounting media (e.g., CytoSeal XYL).
- #1.5 Cover glass.

2.4 Periodic Acid-Schiff (PAS) Staining

- 60 °C oven or air convection incubator.
- Paper towels.
- Chemical fume hood.
- Staining jars with slide racks containing the following solutions in order:
  - 100% Xylene
  - 100% Xylene
  - Absolute ethanol.
  - Absolute ethanol.
  - 95% Ethanol
  - 70% Ethanol
  - Tap water.

  - 0.5% Periodic acid solution
  - 2× Deionized water
  - Schiff reagent.
  - 3× Lukewarm tap water
  - Hematoxylin (Gill's formula #3 or preferred).
  - 3× tap water
  - Acid alcohol (0.5% hydrochloric acid in 70% ethanol).
  - Tap water.
  - Ammonium water (2% ammonium hydroxide in deionized water).
  - Tap water.
  - 95% Ethanol
  - Absolute ethanol.
  - Absolute ethanol.
  - 100% Xylene
  - 100% Xylene.
- Xylene-based mounting media (e.g., CytoSeal XYL).
- #1.5 Cover glass.

2.5 Suzuki Scoring

- Microscope with low- and high-power objectives.
- Microscope-mounted color documentation camera.

2.6 Liver Enzyme Assays

- Serum or plasma collected from the experimental mouse after euthanasia.
- L-Aspartate (for AST assay only).
- L-Alanine (for ALT assay only).
- α-Ketoglutarate (also called 2-oxoglutarate).
- NADH.
- Lactate dehydrogenase.
- Malate dehydrogenase (for AST assay only).
- Sodium phosphate (100 mM) or Tris (20 mM) buffer, pH 7.4–7.8.
- Plate reader capable of time-course measurements of absorbance at 340 nm and heated incubation.
- 96-Well UV-transparent plates.
- Pipettes and tips.
- Tubes, bottles, or flasks for buffer preparation.
- pH meter.

3 Methods

3.1 Ischemia-Reperfusion Surgical Procedure

All animal procedures should be approved by appropriate institutional animal care and use committees and conform to local, national, and international guidelines on the ethical use of animals in laboratory experiments.

1. Anesthetize a mouse with isoflurane (1–3% to affect) or a ketamine/xylazine mixture (120 mg/kg ketamine, 5 mg/kg xylazine at a dose of 100 μL per 25 g mouse or as indicated by institutional guidelines). Place mouse on controlled heating pad.
   - Administration of experimental treatment is appropriate at this step (*see* **Note 2**).
2. Shave and wash the abdominal area with a three-stage chlorhexidine soap/alcohol/chlorhexidine topical wash.
3. Make an incision using sterile technique from the xiphoid process to the symphysis pubis and then expose the portal vein.
4. Carefully apply an atraumatic clamp to interrupt the artery/portal venous blood supply to the left and middle (medial) liver lobes (i.e., only the left and middle hepatic artery and portal vein are occluded by the clip to achieve 70% occlusion, while the right branch of portal vein and hepatic artery are patent providing normal blood flow). No intestinal ischemia should be seen with this model because the right branch of hepatic blood flow remains open (Fig. 1a).
5. Cover the incision with wet gauze (wet with 0.9% normal saline) during ischemia-reperfusion injury. Blanching of the left and middle lobes may be observed as confirmation of ischemia (Fig. 1b, c).
6. Remove the clamp after 30–90 min of ischemia (length of ischemia is dependent on experimental question) to allow reperfusion. Confirm reperfusion by return of blood flow and return of color.
7. Inject 0.9% normal saline (200–500 μL) subcutaneously as a rehydration bolus at a site remote from the surgical incision on the dorsum of the mouse.
8. Close the inner muscle and connective tissue with absorbent suture (e.g., 4–0 coated VICRYL Polyglactin 910 Absorbable Suture).
9. Close the dermal layers with sterile nylon suture.
10. Give buprenorphine (0.05–0.1 mg/kg or as indicated by institutional guidelines) subcutaneously (SC) postoperatively as an analgesic.

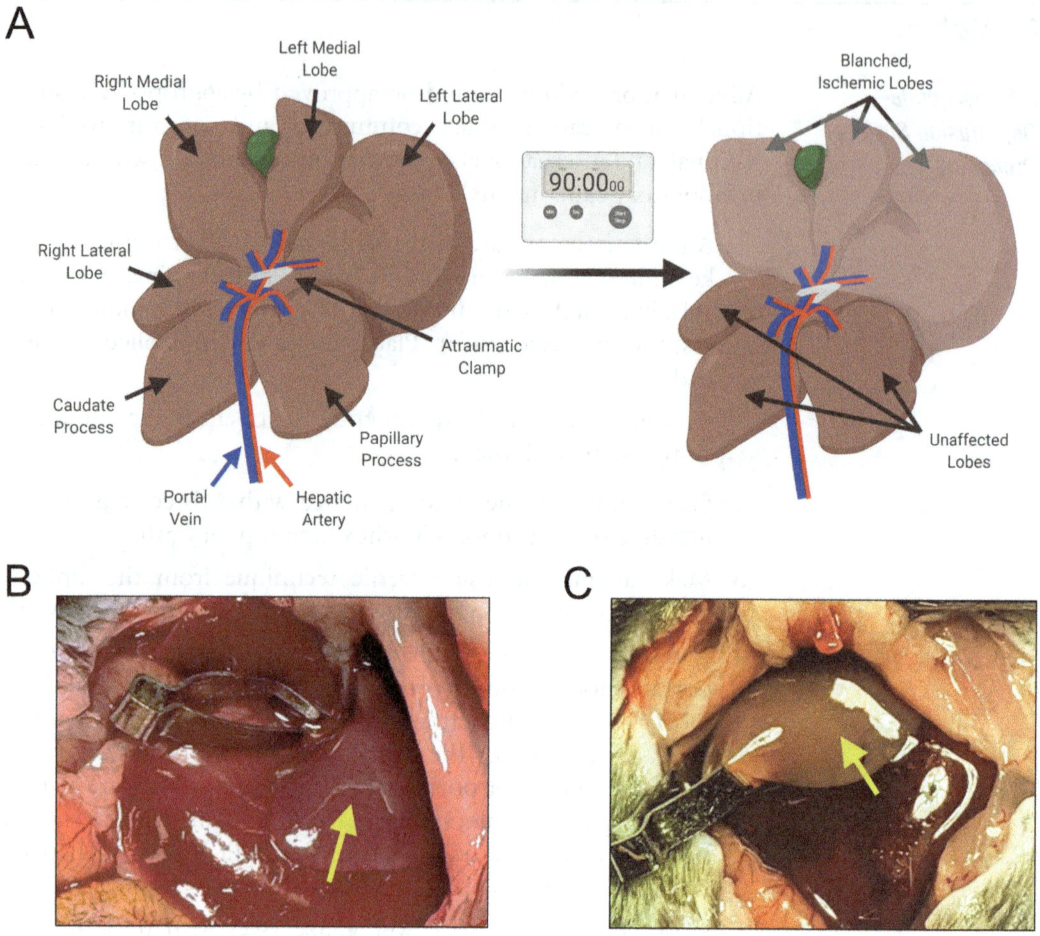

Fig. 1 Anatomical guide for liver ischemia-reperfusion procedure. (**a**) Cartoon representation of mouse liver with left lateral and medial lobes lifted, exposing the target region of the portal vein and hepatic artery. The atraumatic clamp is carefully applied to the left anatomical branch of the paired vessels, causing an occlusion and ischemia to the lifted lobes that can be maintained for up to 90 min. (**b**) Early ischemia is clearly seen in the left lateral lobe (yellow arrow) within approximately 5 min of application of the atraumatic clamp. (**c**) Well-established ischemia is apparent beyond 15–20 min with ischemic lobes appearing extremely blanched (yellow arrow)

- Administration of experimental treatment is appropriate at this step.

11. Follow up for time points relevant to the experimental question. For example, 8, 12, and 24 h are common time points for determining circulating liver enzymes as a marker of acute damage. Ten-day follow-up is appropriate for survival analysis.
    - During longer term follow-ups, remove sutures at 7–10 days post-surgery under isoflurane anesthetic.

3.2 Pre-histology Tissue Processing

1. Euthanize mouse at selected time point by asphyxiation in a CO_2 chamber at a fill rate of 1.8–2 L/min, or as indicated by institutional guidelines. Alternatively, injectable euthanasia solution may be used. Confirm by secondary euthanasia with cervical dislocation.
2. If mice are to be assessed for circulating liver enzymes as an indication of acute liver injury (Subheading 3.6 in this chapter), collect blood at this stage by cardiac puncture (*see* **Note 3**).
3. Dissect the mouse and harvest the liver. Care should be taken while handling the liver to avoid puncturing the tissue with forceps and producing handling artifacts.
4. Place liver tissue in 4–5 mL of 10% buffered formalin in an appropriate container (such as 15 or 50 mL centrifuge tube). Ischemic lobes (left lateral, left medial, and right medial) may be isolated in a separate tube if desired. Comparisons between ischemic lobes and nonischemic lobes (right lateral, caudate process, papillary process) may be of experimental interest as internal controls.
5. Allow tissue to fix at room temperature for 3–5 days.
6. Place tissues into a histology cassette and autoprocess into a paraffin-perfused state or manually process according to the following conditions in an appropriate container, such as 50 mL tube or a 20 mL scintillation vial (Table 1).
7. Drain and out-gas processed tissues in a fume hood for 30–60 min.
8. If available, embed tissues in a mold using a tissue-embedding station for preparation to mount on microtome. If no embedding station is available, embed tissues manually with melted paraffin wax onto a tissue cassette for application to a microtome and trim excess paraffin to create a square-to-rectangular shape amenable to histological ribbon cutting.

3.3 Hematoxylin and Eosin (H&E) Staining

H&E staining should be considered the minimum histological analysis technique for liver ischemia-reperfusion injury analysis. In the liver, hematoxylin stains nuclei blue/purple while eosin stains collagen pale pink, intact hepatocyte cytoplasm strong pink/purple, necrotic cytoplasm faint pink, and red blood cells a vivid bright red (Fig. 2, left panel).

1. Cut 4–6 μm tissue sections on a microtome, capture on charged glass slides using a float bath, and dry overnight. It may be necessary to make a series of cuts at different depths of the tissue by shaving 50–100 μm slides between sets of sections (*see* **Note 4**).
2. Heat slides at 55–60 °C for 20–30 min in an air convection incubator or appropriate alternative (e.g., a scientific oven).

Table 1
Manual tissue processing routine

Pos.	Solution	Time	Temp.	Notes
1	10% Neutral-buffered formalin	30–60 min	Room temp.	Agitate frequently
2	100% Deionized water	60 min	Room temp.	Agitate frequently
3	30% Ethanol in deionized water	60 min	Room temp.	Agitate frequently
4	50% Ethanol in deionized water	60 min	Room temp.	Agitate frequently
5	70% Ethanol in deionized water	60 min	Room temp.	Agitate frequently
6	90% Ethanol in deionized water	60 min	Room temp.	Agitate frequently
7	Absolute ethanol	60 min	Room temp.	Agitate frequently
8	Absolute ethanol	60 min	Room temp.	Agitate frequently **Appropriate to stop overnight**
9	1:1 Mixture of ethanol and xylene	60 min	Room temp.	Agitate frequently
10	100% Xylene	60 min	Room temp.	Agitate frequently
11	100% Xylene	30–60 min	Room temp.	Agitate frequently
12	1:1 Mixture xylene:melted paraffin wax	60 min	55–60 °C	Agitate frequently
13	Melted paraffin wax	60 min	55–60 °C	Agitate frequently

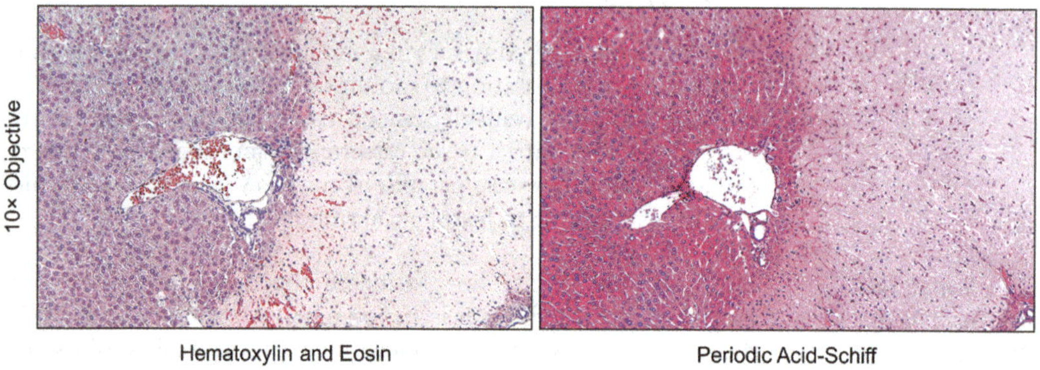

Fig. 2 Comparison of H&E and PAS staining for ischemic liver tissue analysis. Serial sections of mouse liver subjected to 90 min of ischemia followed by reperfusion injury over 24 h. (Left panel) Hematoxylin and eosin staining (H&E) clearly indicates the boundary of healthy tissue on the left, versus necrotic tissue on the right. Red blood cells are intensely stained and clearly indicate congestion. Inflammatory reaction around the portal triad (portal vein, bile duct, portal artery) is visible by darkly stained immune cells. (Right panel) Periodic Acid-Schiff staining (PAS) is also capable of clearly demarcating healthy and necrotic tissue. Inflammatory reaction around the portal triad and congestion are less apparent. Variability in hepatocyte storage of glycogen is visible with PAS stain, but not H&E staining, and can be an early indication of hepatocyte response to ischemia-reperfusion injury

Table 2
H&E staining

Pos.	Solution	Time	Notes
1	100% Xylene	2–3 min	Transfer immediately from heating incubator
2	100% Xylene	2–3 min	
3	Absolute ethanol	2–3 min	
4	Absolute ethanol	2–3 min	
5	95% Ethanol	2–3 min	
6	70% Ethanol	2–3 min	
7	Tap water	2–3 min	
8	Hematoxylin (Gill's formula #3 or preferred)	5 min	Adjust time according to the formula of hematoxylin
9	Tap water	3 dips	Quickly dip
10	Tap water	1 min	
11	Tap water	2–3 min	
12	Acid alcohol (0.5% hydrochloric acid in 70% ethanol)	10–15 dips	Amount of dipping will depend on how fresh the acid alcohol is
13	Tap water	2–3 min	
14	Ammonium water (2% ammonium hydroxide in deionized water)	30 s	Amount of time will depend on how fresh the ammonium water is
15	Tap water	2–3 min	
16	Acidic, aqueous eosin Y solution	3–5 min	Amount of time will depend on how fresh the eosin Y is
17	95% Ethanol	2–3 min	
18	Absolute ethanol	2–3 min	
19	Absolute ethanol	2–3 min	
20	100% Xylene	2–3 min	
21	100% Xylene	2–3 min	

3. Rehydrate tissues and stain with hematoxylin and eosin, tapping rack of slides on a stack of paper towels in between each solution, according to the following order (Table 2).
4. Allow slides to completely dry.
5. Mount with xylene-based mounting medium (e.g., CytoSeal XYL), taking care to avoid bubbles on the tissue.
6. Allow to dry for at least 20–30 min before manipulating to avoid the cover glass sliding.

3.4 Periodic Acid-Schiff (PAS) Staining

PAS staining without amylase differentiation can be a useful indicator of hepatocyte viability by staining glycogen stores in the parenchyma of the liver. Ischemia-reperfusion injury can reduce glycogen stores in hepatocytes early, prior to necrotic death, and can be a useful indicator of hepatocyte viability prior to death [20]. PAS staining should not be considered a replacement for H&E staining as other indicators of liver viability such as red blood cell congestion and inflammatory infiltration are not as clearly observed (Fig. 2, right panel).

1. Cut 4–6 μm tissue sections on a microtome, capture on charged glass slides using a float bath, and dry overnight (*see* **Note 4**).
2. Heat slides at 55–60 °C for 20–30 min in an air convection incubator or appropriate alternative (e.g., a scientific oven).
3. Rehydrate tissues and stain in a Periodic acid-Schiff reaction, tapping rack of slides on a stack of paper towels in between each solution, according to the following order (Table 3).
4. Allow slides to completely dry.
5. Mount with xylene-based mounting medium (e.g., CytoSeal XYL), taking care to avoid bubbles on the tissue.
6. Allow to dry for at least 20–30 min before manipulating to avoid the cover glass sliding.

3.5 Suzuki Scoring

1. Stain tissues with H&E according to Subheading 3.3 of this chapter (Fig. 3).
2. In a blinded manner, assess slides at 20–40× magnification with a microscope equipped with a color camera for documentation. Micrographs of sections may be collected and read blinded by numerous lab members or, preferably, a blinded pathologist. Consensus counts should be obtained by the means of numerous fields per mouse, average across numerous readers. Scoring should be performed based on the method developed by Suzuki et al. [21] (Table 4).
3. Statistical analysis and reporting of scores should be performed separately for congestion, vacuolization, and necrosis as well as a composite score of each category summed.

3.6 Liver Enzyme Assays

Hepatocytes make up 70% of the cell number and 80% of the volume of the liver [22]. Injury to the liver can be conveniently measured by circulating levels of two enzymes, aspartate aminotransferase (AST) and alanine aminotransferase (ALT). However, AST is also released during injury to heart, skeletal muscle, kidney, pancreas, spleen, and lungs. Similarly, ALT is released during injury to the kidney, skeletal muscle, and heart. In an experimental setting, such as the liver ischemia-reperfusion model described in this

Table 3
PAS staining

Pos.	Solution	Time	Notes
1	100% Xylene	2–3 min	Transfer immediately from heating incubator
2	100% Xylene	2–3 min	
3	Absolute ethanol	2–3 min	
4	Absolute ethanol	2–3 min	
5	95% Ethanol	2–3 min	
6	70% Ethanol	2–3 min	
7	Tap water	2–3 min	
8	0.5% Periodic acid solution	5 min	
9	Deionized water	1 min	
10	Deionized water	2–3 min	
11	Schiff reagent	15 min	Schiff reagent should be stored at 4 °C (*see* **Note 5**)
12	Lukewarm tap water	2–3 min	
13	Lukewarm tap water	2–3 min	
14	Lukewarm tap water	2–3 min	
15	Hematoxylin (Gill's formula #3 or preferred)	5 min	Adjust time according to the formula of hematoxylin
16	Tap water	3 dips	Quickly dip
17	Tap water	1 min	
18	Tap water	2–3 min	
19	Acid alcohol (0.5% hydrochloric acid in 70% ethanol)	10–15 dips	Amount of dipping will depend on how fresh the acid alcohol is
20	Tap water	2–3 min	
21	Ammonium water (2% ammonium hydroxide in deionized water)	30 s	Amount of time will depend on how fresh the ammonium water is
22	Tap water	2–3 min	
23	95% Ethanol	2–3 min	
24	Absolute ethanol	2–3 min	
25	Absolute ethanol	2–3 min	
26	100% Xylene	2–3 min	
27	100% Xylene	2–3 min	

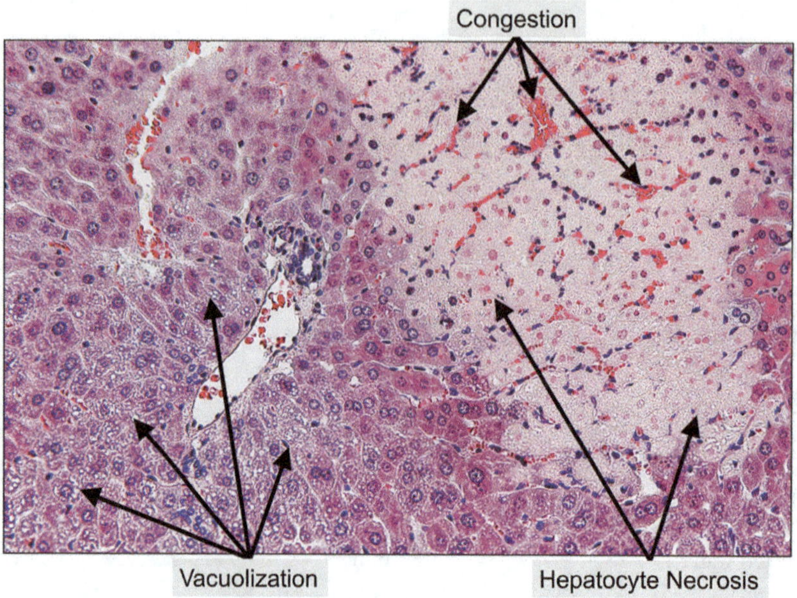

Fig. 3 Example of major characteristics of Suzuki staining. Mouse liver subjected to 90 min of ischemia followed by reperfusion injury over 24 h and stained by H&E. Vacuolization (small, white holes/bubbles in hepatocytes), hepatocyte necrosis (faint nuclei, faint pink cytoplasm, disrupted membrane boundaries), and congestion (densely packed stretches of red blood cells) are indicated by arrows

Table 4
Suzuki scoring criteria for assessing liver damage after ischemia-reperfusion injury by histology

Numerical assessment	Congestion	Vacuolization	Necrosis
0	None	None	None
1	Minimal	Minimal	Single-cell necrosis
2	Mild	Mild	≤30%
3	Moderate	Moderate	>30% but ≤60%
4	Severe	Severe	>60%

chapter, it is unlikely other types of tissue injury are occurring and thus AST and ALT may be considered reliable for liver injury. AST and ALT are conveniently measured in a coupled, two-step kinetic assays wherein the aminotransferase activity (in the assay reaction) produces a substrate for readout of dehydrogenase activity (indicator reaction) (Fig. 4).

1. Prepare reaction buffer according to the following recipes (*see* **Note 6**):
   - **AST assay buffer.**

     L-Aspartic acid 250 mM,

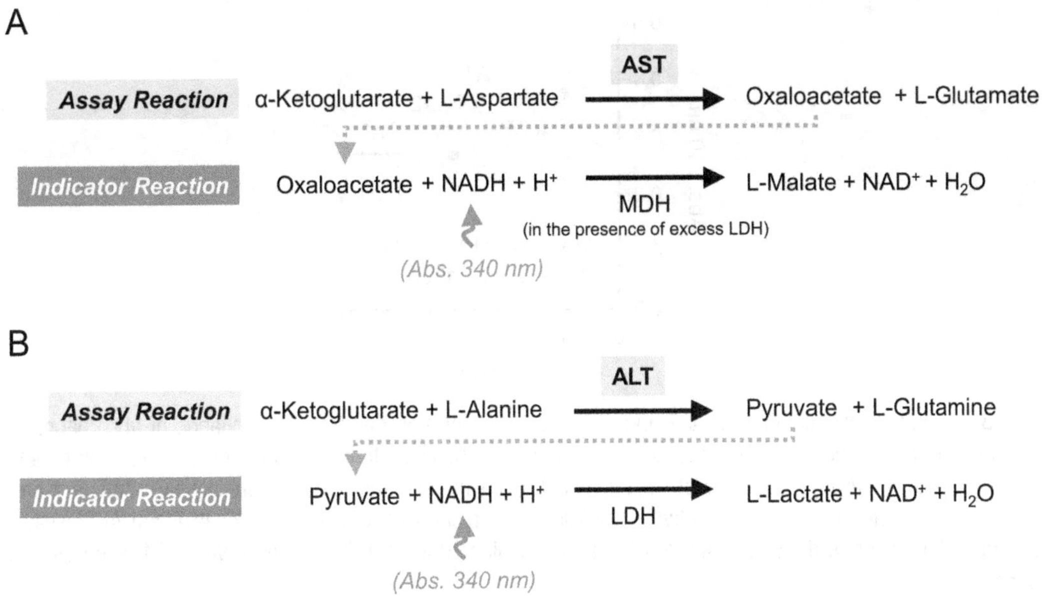

Fig. 4 Reaction schema for AST- and ALT-coupled reactions. The reactions for (**a**) AST and (**b**) ALT proceed as a coupled reaction wherein AST or ALT mediates an aminotransferase reaction of either L-aspartate to L-glutamine (AST) or L-alanine to L-glutamine (ALT). By-products of the reaction, oxaloacetate for the AST reaction and pyruvate for the ALT reaction, serve as substrates for dehydrogenases in the presence of NADH. The conversion of NADH in the dehydrogenase reaction is monitored by absorbance at 340 nm, serving as an indicator of the rate of production of the dehydrogenase substrate, which in turn is produced at a rate equivalent to the concentration of aminotransferase present

Lactate dehydrogenase 1000 U/L or greater.

Malate dehydrogenase 800 U/L or greater.

NADH 0.2 mM.

Sodium phosphate (100 mM) or Tris (20 mM) buffer, pH 7.4–7.8.

- **ALT Assay Buffer.**

  L-Alanine 500 mM,

  Lactate dehydrogenase 2000 U/L or greater.

  NADH 0.2 mM.

  Sodium phosphate (100 mM) or Tris (20 mM) buffer, pH 7.4–7.8.

- **Start Solution.**

  α-Ketoglutarate 150 mM,

  Sodium phosphate (100 mM) or Tris (20 mM) buffer, pH 7.4–7.8.

2. Dilute serum or plasma sample (*see* **Note 3**) at a ratio of 1:10 or 1:20 in 100 μL of AST or ALT assay buffer in a 96-well plate.

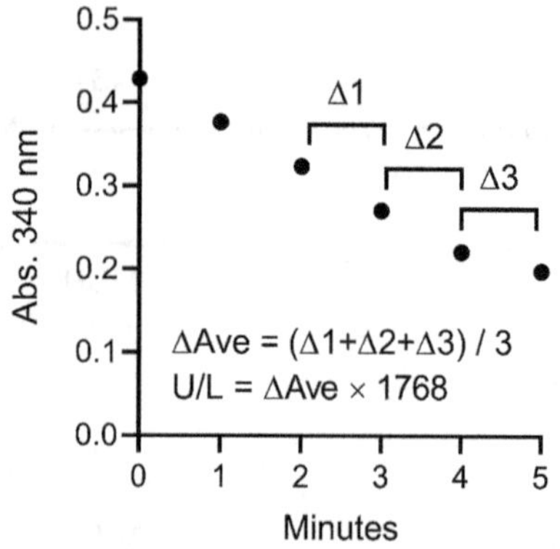

Fig. 5 Example of calculations for AST and ALT assays. Determine the amount of change of absorbance at 340 nm for each of the last 4 measurements (the difference between time points 2 and 3, 3 and 4, and 4 and 5). Determine an average change of absorbance at 340 nm per minute and multiply by 1768, a factor determined by the millimolar absorptivity of NADH, the indicator of reaction rate for AST and ALT assays. Example data are real data from serum of a mouse subjected to 90-min ischemia with a 24-h reperfusion period

3. Incubate diluted sample at 37 °C for 10 min. Pre-warm start solution to 37 °C at the same time.
4. Add 10 μL of start solution (containing α-ketoglutarate) to the pre-warmed, diluted sample. Gently tap the reaction plate, make an initial measurement of absorbance at 340 nm, and incubate further at 37 °C.
5. Record the absorbance at 340 nm every minute for 5 min (six total measurements).
6. Calculate the average ΔAbs/minute for the last 3 measurements and multiply by 1768 (based on the millimolar absorptivity of NADH as 6.22) to quantify the amount of AST or ALT in the measured sample in U/L (Fig. 5).

4 Notes

1. We previously tested this model with purified recombinant Serp-2 protein [19]. To produce the recombinant protein, the coding sequence for Serp-2 was tagged with an N-terminal 10×-histidine affinity tag and transformed into HeLa cells with a vaccinia/T7 vector [17, 23]. Expressed protein was purified over a His-Bind resin and confirmed for purity >90% by SDS-12% PAGE followed by silver staining and immunoblotting. The method described in this chapter is amenable to testing a wide variety of putative therapeutic treatments including recombinant proteins, adenovirus (AdV)- and adeno-associated virus (AAV)-delivered therapeutic gene

therapies, and nonviral plasmid DNA-delivered therapeutic genes. Adaptation of the described methods should be made per lab preference and appropriateness to the treatment being tested. For example, AAV-delivered therapies may require treatment 1–2 weeks prior to model induction to allow sufficient expression of experimental treatment.

2. Amount, frequency, and route of administration of experimental therapeuticproteins should be determined empirically for each lab. For Serp-2, we gave protein at a dose of 100 ng/g bodyweight by intraperitoneal injection in 100 μL saline with a 30-gauge needle 30 min prior to the LIRI procedure and then on alternating days for a total of five doses across 10 days.
3. We routinely collect approximately 300 μL of blood for serum/plasma analysis. For serum, allow blood to clot at room temperature for 15–30 min. Remove the clot by centrifugation at 1000–2000 × *g* for 10 min at 4 °C. For plasma, collect blood with syringes and tubes pre-coated with EDTA, citrate, or endotoxin-free heparin. Centrifuge collected blood at 1000–2000 × *g* for 10 min at 4 °C to remove cells (or at 2000 × *g* for 15 min at 4 °C to additionally remove platelets). After centrifugation of serum or plasma, transfer supernatant to a new tube, carefully avoiding collecting any of the cell pellet. Aliquot and store at −80 °C until use.
4. Ensure that labels are made with xylene-resistant marker or graphite pencil prior to rehydration.
5. Schiff reagent degrades over time. Storage at 4 °C prolongs the degradation of the reagent. Activity prior to staining can be performed by depositing a few drops onto a watch dish containing 3–5 mL 37% formaldehyde. Active reagent will turn a red/purple color. A delayed reaction and a blue/purple color indicate degraded reagent.
6. Numerous companies offer pre-made assay components for performing AST and ALT assays in colorimetric and fluorometric assay types. We have had success using assays from Pointe Scientific, but other companies have similar reactions.

References

1. Eltzschig HK, Eckle T (2011) Ischemia and reperfusion-from mechanism to translation. Nat Med 17:1391–1401
2. Park SW, Kim M, Brown KM et al (2011) Paneth cell-derived interleukin-17A causes multiorgan dysfunction after hepatic ischemia and reperfusion injury. Hepatology 53:1662–1675
3. Hausenloy DJ, Yellon DM (2013) Myocardial ischemia-reperfusion injury : a neglected therapeutic target. J Clin Invest 123:92–100
4. Gujral JS, Bucci TJ, Farhood A et al (2001) Mechanism of cell death during warm hepatic ischemia-reperfusion in rats: apoptosis or necrosis? Hepatology 33:397–405
5. Clavien PA, Rüdiger HA, Selzner M (2001) Mechanism of hepatocyte death after ischemia:

apoptosis versus necrosis. Hepatology 33:1555–1557
6. Kim HY, Kim SJ, Lee SM (2015) Activation of NLRP3 and AIM2 inflammasomes in Kupffer cells in hepatic ischemia/reperfusion. FEBS J 282:259–270
7. Mastrocola R, Penna C, Tullio F et al (2016) Pharmacological inhibition of NLRP3 inflammasome attenuates myocardial ischemia/reperfusion injury by activation of RISK and mitochondrial pathways. Oxid Med Cell Longev 2016:1-11
8. Fuquay R, Renner B, Kulik L, McCullough JW, Amura C, Strassheim D, Pelanda R, Torres R, Thurman JM, (2013) Renal Ischemia-Reperfusion Injury Amplifies the Humoral Immune Response. J Am Soc Nephrol 24 (7):1063–1072.
9. Best SM, Kerr PJ (2000) Coevolution of host and virus: the pathogenesis of virulent and attenuated strains of myxoma virus in resistant and susceptible European rabbits. Virology 267:36–48
10. Kerr P, McFadden G (2002) Immune responses to myxoma virus. Viral Immunol 15:229–246
11. Spiesschaert B, McFadden G, Hermans K et al (2011) The current status and future directions of myxoma virus, a master in immune evasion. Vet Res 42:76
12. Liss KHH, McCommis KS, Chambers KT et al (2018) The impact of diet-induced hepatic steatosis in a murine model of hepatic ischemia/reperfusion injury. Liver Transplant 24:908–921
13. Petit F, Bertagnoli S, Gelfi J et al (1996) Characterization of a myxoma virus-encoded serpin-like protein with activity against interleukin-1 beta-converting enzyme. J Virol 70:5860–5866
14. MacNeill AL, Turner PC, Moyer RW (2006) Mutation of the Myxoma virus SERP2 P1-site to prevent proteinase inhibition causes apoptosis in cultured RK-13 cells and attenuates disease in rabbits, but mutation to alter specificity causes apoptosis without reducing virulence. Virology 356:12–22
15. Turner PC, Sancho MC, Thoennes SR et al (1999) Myxoma virus Serp2 is a weak inhibitor of granzyme B and interleukin-1beta-converting enzyme in vitro and unlike CrmA cannot block apoptosis in cowpox virus-infected cells. J Virol 73:6394–6404
16. Davids JA, Dai E, Chen H et al (2014) Viral anti-inflammatory proteins target diverging immune pathways with converging effects on arterial dilatation, plaque and apoptosis. Eur J Inflamm 12:131–145
17. Viswanathan K, Bot I, Liu L et al (2012) Viral cross-class serpin inhibits vascular inflammation and T lymphocyte fratricide: a study in rodent models in vivo and human cell lines in vitro. PLoS One 7:e44694
18. Bot I, Von der Thüsen JH, Donners MMPC et al (2003) Serine protease inhibitor Serp-1 strongly impairs atherosclerotic lesion formation and induces a stable plaque phenotype in ApoE−/− mice. Circ Res 93:464–471
19. Yaron JR, Chen H, Ambadapadi S et al (2019) Serp-2, a virus-derived apoptosis and inflammasome inhibitor, attenuates liver ischemia-reperfusion injury in mice. J Inflamm 16:12
20. Quintana AB, Guibert EE, Rodríguez JV (2005) Effect of cold preservation/reperfusion on glycogen content of liver. Concise review. Ann Hepatol 4:25–31
21. Suzuki S, Toledo-Pereyra LH, Rodriguez FJ et al (1993) Neutrophil infiltration as an important factor in liver ischemia and reperfusion injury. Modulating effects of FK506 and cyclosporine. Transplantation 55:1265–1272
22. Gao B, Jeong W-I, Tian Z (2008) Liver: an organ with predominant innate immunity. Hepatology 47:729–736
23. Turner PC, Sancho MC, Thoennes SR et al (1999) Myxoma virus Serp2 is a weak inhibitor of Granzyme B and Interleukin-1β-converting enzyme in vitro and unlike CrmA cannot block apoptosis in cowpox virus-infected cells. J Virol 73(8):6394–6404

Index

Alexandra R. Lucas (ed.), *Viruses as Therapeutics: Methods and Protocols*, Methods in Molecular Biology, vol. 2225, https://doi.org/10.1007/978-1-0716-1012-1, © Springer Science+Business Media, LLC, part of Springer Nature 2021